THE CELL CYCLE

Gene-Enzyme Interactions

CELL BIOLOGY: A Series of Monographs

EDITORS

D. E. Buetow
*Department of Physiology
and Biophysics
University of Illinois
Urbana, Illinois*

I. L. Cameron
*Department of Anatomy
University of Texas
Medical School at San Antonio
San Antonio, Texas*

G. M. Padilla
*Department of Physiology and Pharmacology
Duke University Medical Center
Durham, North Carolina*

G. M. Padilla, G. L. Whitson, and I. L. Cameron. THE CELL CYCLE: *Gene-Enzyme Interactions,* 1969

Other volumes in preparation

THE CELL CYCLE

Gene-Enzyme Interactions

Edited by   **G. M. PADILLA**   **G. L. WHITSON**

DEPARTMENT OF PHYSIOLOGY
AND PHARMACOLOGY
DUKE UNIVERSITY MEDICAL
CENTER
DURHAM, NORTH CAROLINA

DEPARTMENT OF ZOOLOGY AND
ENTOMOLOGY
THE UNIVERSITY OF TENNESSEE
KNOXVILLE, TENNESSEE

I. L. CAMERON

DEPARTMENT OF ANATOMY
UNIVERSITY OF TEXAS
MEDICAL SCHOOL AT SAN ANTONIO
SAN ANTONIO, TEXAS

ACADEMIC PRESS   New York and London   1969

ACADEMIC PRESS, INC.
111 Fifth Avenue, New York, New York 10003

United Kingdom Edition published by
ACADEMIC PRESS, INC. (LONDON) LTD.
Berkeley Square House, London W.1

LIBRARY OF CONGRESS CATALOG CARD NUMBER: 68-26646

PRINTED IN THE UNITED STATES OF AMERICA

The editors dedicate this volume to the memory of
DR. OTTO H. SCHERBAUM
one of the pioneers in synchronization studies.
His contributions will stand as models for us to follow.

List of Contributors

Numbers in parentheses refer to the pages on which the authors' contributions begin.

E. C. ANDERSON, Biomedical Research Group, Los Alamos Scientific Laboratory, University of California, Los Alamos, New Mexico (341)

NORMAN I. BISHOP, Department of Botany and Plant Pathology, Oregon State University, Corvallis, Oregon (179)

J. S. CLEGG, Laboratory for Quantitative Biology, University of Miami, Coral Gables, Florida (249)

GERALDINE H. COHEN, Laboratory for Quantitative Biology, Department of Biology, University of Miami, Coral Gables, Florida (299)

JOSEPH E. CUMMINS,[1] Department of Zoology, University of Washington, Seattle, Washington (141)

W. D. DONACHIE,[2] Medical Research Council, Microbial Genetics Research Unit, Hammersmith Hospital, London, England (37)

SIGRID ELSAESSER, Biochemisches Institut, University of Freiburg, Freiburg, Germany (279)

DAVID EPEL, Hopkins Marine Station of Stanford University, Pacific Grove, California (279)

RAY EPSTEIN, Laboratory of Molecular Biology, University of Wisconsin, Madison, Wisconsin (101)

F. J. FINAMORE, Biology Division, Oak Ridge National Laboratory, Oak Ridge, Tennessee (249)

---

[1] *Present address:* Department of Histology, The Karolinska Institute, Stockholm, Sweden (*until May, 1969*).

[2] *Present address:* Medical Research Council, Microbial Genetics Research Unit, Department of Molecular Biology, University of Edinburgh, Scotland.

HARLYN O. HALVORSON, Laboratory of Molecular Biology, University of Wisconsin, Madison, Wisconsin (101)

PATRICIA HARRIS, Department of Zoology, Oregon State University, Corvallis, Oregon (315)

CHARLES E. HELMSTETTER, Roswell Park Memorial Institute, Buffalo, New York (15)

T. W. JAMES, Department of Zoology, University of California, Los Angeles, California (1)

R. M. IVERSON, Laboratory for Quantitative Biology, Department of Biology, University of Miami, Coral Gables, Florida (299)

JOHN R. KENNEDY, JR., Department of Anatomy, Bowman Gray School of Medicine of Wake Forest University, Winston-Salem, North Carolina (227)

MILLICENT MASTERS,[3] Medical Research Council, Microbial Genetics Research Unit, Hammersmith Hospital, London, England (37)

J. M. MITCHISON, Department of Zoology, University of Edinburgh, Edinburgh, Scotland (361)

D. F. PETERSEN, Biomedical Research Group, Los Alamos Scientific Laboratory, University of California, Los Alamos, New Mexico (341)

BERTON C. PRESSMAN, Johnson Research Foundation Medical School, University of Pennsylvania, Philadelphia, Pennsylvania (279)

ROBERT R. SCHMIDT, Department of Biochemistry and Nutrition, Virginia Polytechnic Institute, Blacksburg, Virginia (159)

ECKHART SCHWEIZER, Laboratory of Molecular Biology, University of Wisconsin, Madison, Wisconsin (101)

HORST SENGER, Botanisches Institut der Universität, Marburg/Lahn, Germany (179)

RALPH A. SLEPECKY, Biological Research Laboratories, Department of Bacteriology and Botany, Syracuse University, Syracuse, New York (77)

G. J. STINE,[4] Biology Division, Oak Ridge National Laboratory, Oak Ridge, Tennessee (119)

PATRIC TAURO,[5] Laboratory of Molecular Biology, Department of Bacteriology, University of Wisconsin, Madison, Wisconsin (101)

[3] *Present address:* Medical Research Council, Microbial Genetics Research Unit, Edinburgh University, Edinburgh, Scotland.

[4] *Present address:* Department of Botany, University of Tennessee, Knoxville, Tennessee.

[5] *Present address:* Donner Laboratory, University of California, Berkeley, California.

R. A. TOBEY, Biomedical Research Group, Los Alamos Scientific Laboratory, University of California, Los Alamos, New Mexico (341)

ANTHONY M. WEAVER, Hopkins Marine Station of Stanford University, Pacific Grove, California (279)

ARTHUR M. ZIMMERMAN, Department of Zoology, University of Toronto, Toronto, Canada (203)

Preface

The primary level of regulation of growth and differentiation most likely occurs at the interaction of the genetic and enzymatic complements of a cell. New insights into the mechanisms which govern such interactions have evolved since Jacob and Monod proposed the operon model for the control of genetic expression in bacterial cells. This model has not only served as a guide for a clearer understanding of the regulatory mechanisms of gene function but, largely through the use of synchronous cell populations, it has also been extended to the analysis of the sequential processes which comprise the cell cycle. This volume compiles a number of investigations under such a unifying theme. These contributions represent the culmination of intensive research by investigators who use cytological as well as biochemical criteria to analyze the primary regulatory mechanisms of the cell cycle. A wide variety of procaryotic and eucaryotic cell types have been studied. Theoretical and methodological discussions have been included.

This work will not only complement an earlier volume "Cell Synchrony—Studies in Biosynthetic Regulation" [I. L. Cameron and G. M. Padilla, eds. (1966), Academic Press, New York] but is the first in the Cell Biology Monograph Series (D. E. Buetow, I. L. Cameron, and G. M. Padilla, eds., Academic Press, New York). We hope that this work will provide a stimulus for students and investigators to continue to consider the cell cycle as the meeting ground for the various disciplines of modern biology.

We wish to express our gratitude to Dr. R. F. Kimball, Director of the Biology Division, Oak Ridge National Laboratory, and to Dr. W. G. Anlyan, Dean of the School of Medicine, Duke University, for aiding in the organization of the 2nd International Conference on Cell Synchrony. The Conference was held at Oak Ridge, Tennessee, in April, 1967, under the joint sponsorship of these two institutions. It was at this conference that the plans to produce this volume were initially formulated. We also thank Dr. Alexander Hollaender for his inspiration enthusiasm, and continuing interest in the pursuit of excellence in basic biological research to which we are all dedicated.

January, 1969

G. M. PADILLA
G. L. WHITSON
I. L. CAMERON

Contents

4. Synchrony and the Formation and Germination of Bacterial Spores

Ralph A. Slepecky

5. Synthesis of Macromolecules during the Cell Cycle in Yeast

Patric Tauro, Eckhart Schweizer, Ray Epstein, and Harlyn O. Halvorson

6. Investigations during Phases of Synchronous Development and Differentiation in *Neurospora crassa*

G. J. Stine

7. Nuclear DNA Replication and Transcription during the Cell Cycle of *Physarum*

Joseph E. Cummins

8. Control of Enzyme Synthesis during the Cell Cycle of *Chlorella*

Robert R. Schmidt

9. Light-Dependent Formation of Nucleic Acids and Its Relation to the Induction of Synchronous Cell Division in *Chlorella*

Horst Senger and Norman I. Bishop

10. Effects of High Pressure on Macromolecular Synthesis in Synchronized *Tetrahymena*

Arthur M. Zimmerman

11. The Role of Microtubules in the Cell Cycle

John R. Kennedy, Jr.

12. Biochemical Aspects of Morphogenesis in the Brine Shrimp, Artemia salina

F. J. Finamore and J. S. Clegg

13. The Program of Structural and Metabolic Changes following Fertilization of Sea Urchin Eggs

David Epel, Berton C. Pressman, Sigrid Elsaesser, and Anthony M. Weaver

THE CELL CYCLE

Gene-Enzyme Interactions

CHAPTER 1

Thoughts on Cell Evolution and Thermodynamics*

T. W. James

I. INTRODUCTION

This chapter is an attempt to stimulate thinking about cell evolution and thermodynamics. The degree to which it fulfills its purpose can best be judged by the individual who upon reading it is prompted to generate new schemes and to design experiments to test them. One may not defend the chapter against the criticism that it is speculative, but one must insist that it contains several important fundamentals. These include the duplication-mutation-retention hypothesis and the outline of a role for biochemical homology (Margoliash and Smith, 1965) in an evolutionary mechanism. "Sequence homology," as I will call it, is meaningful in that it makes possible the operation of a directing mechanism for deriving enzymatic series and their corresponding metabolic

* Supported by National Science Foundation Grant 4275 awarded to T. W. James.

1

pathways (Horowitz, 1945, 1965), as well as allosteric control mechanisms (Monod et al., 1963). Furthermore, a study of these homology-related mechanisms is a legitimate experimental area. The relationships between cellular control mechanisms, steady state minimum entropy production (Prigogine, 1955), and the evolutionary selection of control mechanisms are emphasized, and these mechanisms are considered also as expressions of thermodynamic selection. Hopefully, these ideas may suggest a profitable direction for future thought and investigation.

The evolution of the cell is discussed in two sections—one bearing on the expansion of the cell genome and its consequences; and the other on the role of thermodynamics in selection, specifically in the development of control mechanisms.

II. GENOME EVOLUTION

It is generally accepted that at least three mechanisms could have been instrumental in the creation of more complicated cell genomes. One of these might be the fusion of different procaryotic cell types leading to a corresponding increase through the summation of the genomes. Another possible mechanism involves a symbiotic union of procaryotes with eucaryotes, which is often considered the mechanism by which mitochondria, chloroplasts, and other DNA-carrying particulates appeared in the eucaryotes (Gibor and Granick, 1964). The most primitive mechanism for the expansion of a cell's genome most likely depended on phenomena that are part of ordinary cellular processes. To emphasize this mechanism, we will call it the duplication-mutation-retention sequence and discuss it in some detail. Our emphasis should not be interpreted as a rejection of the likelihood that the other mechanisms may be equally relevant.

Consider the question of the expansion of a cell's genome in the evolutionary sense. Starting with a genome rather than a gene, the mechanism of genome expansion can be illustrated and the mechanisms for selection and maintenance can be outlined. Given an operating procaryotic genome, what might have increased its size, and what were the consequences of these increases as observed in the evolving cell system? First, from cytological evidence we know that a cell can double its genes by *duplication* without an ensuing cell division. Second, to be meaningful in terms of new functions, this event would have had to be followed by *mutation* in one of the gene pairs to yield at least one cistron that differed from the one from which it was copied. Third, the *retention* of this mutation would thus provide the cell with a new protein. Depending on the nature of the mutation, the new protein would have either an altered function, a completely new function, or an absence of function. Which of these occurs will govern the retention of the cistron.

Now consider the details of this process. If one of the duplicated pairs of cistrons undergoes a mutation that consists of a change in only one codon and it is retained and integrated into the genome, it will display a codon sequence that is only slightly different from the cistron from which it was derived. It will, therefore, display a high degree of *sequence homology*. The term "homology" here is used in its evolutionary sense implying a similarity that is ancestrally derived. Cistron sequence homology will, of course, lead to amino acid sequence homologies, in the corresponding polypeptides or proteins, and we will therefore apply the terms interchangeably. Work on amino acid sequences in proteins from a variety of sources shows the extensive existence of amino acid sequence homologies as displayed in the book edited by Bryson and Vogel (1965). The degree to which such relationships exist in the different enzymes or proteins of the same cell is, however, still to be examined in depth.

A. THE ANCESTRY OF METABOLIC PATHWAY ENZYMES

In our illustration we will speak only of one gene even though our scheme states that an entire genome was duplicated. Perhaps as an expedient we can depend on the usual dogma of evolution that states that useless excess genes will be eliminated by selection, whereas useful new genes will be maintained. In any event, the mechanism by which such new genes would be perpetuated in the cell's genome will be discussed later, but it should be emphasized that integration may not have been accomplished by physical linkage of the new gene to its precursor gene. If duplication were the first step in this process, it would not be expected that the new cistron be part of the same linkage group. Association of genes into clusters might occur if duplication were followed first by a cistron shift in the genome and then by a crossover between the new and old linkage group in which the mutation occurred. This latter mechanism might account for the cluster phenomenon proposed by Horowitz (1945, 1965) and found by Demerec and Hartman (1959). The tandem alignment of cistrons in a linkage group, each of which codes for an enzyme of a metabolic pathway, may not be the result of early events in this process. End-to-end cistron association is contradictory to the parallel mode of duplication. However, it can be argued that clusters of genes that govern associated processes will contribute to controlled gene function, and therefore clustering should be preserved on the grounds that control mechanisms are selected for. This will be discussed later.

B. Metabolic Pathways

If one analyzes the mechanism by which the ancestry of an homologous sequence might contribute to the development of an enzymatic series, it can be seen that their evolution will be more positively biased than one might suspect. The process we suggest can be outlined as follows. Starting with a cistron α_1 that is coded to produce an enzyme A_1, which is capable of transforming substrate a_1 to a product b_1 (refer to a diagram below), let cistron α_1 be duplicated and one of the pair undergo a mutation to α_2. After this, the cistron is retained by the cell and integrated into the genome. The cistron α_2 will, of course, produce a new protein A_2. A_2 will, by virtue of the mutation, show an altered or lost activity toward substrate a_1. If the mutation has involved a codon that codes for an amino acid in the region of the active site, the activity of the enzyme toward a_1 may well be lost. But such an altered enzyme will most likely retain many of its properties, i.e., conformational structure, interchain bridges, perhaps even its capacity to bind a_1 or b_1 just by virtue of the sequence similarities that have been retained. The possibility that it may bind a_1 but not catalyze its breakdown to b_1 is one of several probable consequences of the mutation. Conversely, it may have retained a sufficient degree of sequence homology to bind b_1, particularly if b_1 has, by reason of its conversion from a_1, retained some of the groupings that were responsible in its original binding to the enzyme A_1.

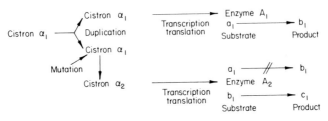

More concisely, there is a rather high probability that sooner or later a mutation will occur which will change A_1 to an A_2 of such a form as to give rise to a new enzyme that will bind and react with b_1 to yield a product c_1. Thus an enzyme series appears and the gradual evolution of a metabolic pathway can be visualized. Its evolution is dependent on residuals of sequences that are part of the first enzyme in the series. The scheme by which gene complement expansion leads to an enzyme series may thus depend on this probabilistic model. That is, mutations that appear in a retained sister cistron will sooner or later give rise to an enzyme which, by virtue of its ancestral similarity, will catalyze the conversion of the product of the sibling enzyme.

It is obvious that this mechanism could have worked most favorably in cases where strong structural similarities exist between substrate and product. For this to have happened, feasible structural conversions of the product b_1 to a new product c_1 would have had to be possible. There is nothing essential about restricting the change to one mutation for the derived enzyme to show activity for the conversion of product b_1 to c_1. It may have been necessary for its cistron to undergo more than one mutation before it was endowed with an active site that was reactive in an otherwise unreactive binding site for product b_1. This scheme might then account for the production of an enzyme series of a metabolic pathway. Showing to what degree DNA sequence homologies have been retained in an enzyme series might add credence to this proposal.

The nature of the selection pressure for the appearance and integration of new steps in an enzyme complex may be of many kinds. The value of the product in promoting the synthesis of its own enzyme is an obvious first consideration since it would provide a positive feedback for the evolution of a complex cellular system. If the enzymatic step is one of a series that yields energy or molecular precursors that can either be stored or used immediately by the cell, the competitive advantage thus acquired leads to success. Alternately, it might allow the cell a new niche in its otherwise fixed nutrient environment. Furthermore, it is possible that enzymatic metabolic pathways could be added on from either end. At the low entropy substrate end, the addition of an enzyme might channel in a new source of nutrients, while at the opposite end it may provide a more complete breakdown and utilization of a given substrate.

C. METABOLIC CYCLES

This mechanism of evolution of enzymatic series may apply to the appearance of metabolic cycles. In fact, the existence of metabolic cycles which are maintained by the condensation of one substrate to another as the first step in the cycle may depend upon the retention of similarities of both the substrate and the amino acid sequence in the enzymes at each step of the pathway. The analysis of such enzymes for amino acid sequence similarities and differences would be an alternate test of this hypothesis. Enzymes responsible for side reactions would also arise as a result of this scheme. One recognizes that wherever a branching pathway occurs there might be some degree of sequence and structural homology in the enzymes.

Sequence homology between adjacent enzymes in a metabolic pathway may also be selected against if the mechanism for improving their effectiveness depends on the gradual introduction of other amino acids in the sequences that improve the enzyme in its function. Presumably this might gradually destroy similarities between such enzymes, but, alternately, since it occurs at the gene level, it would provide for new cistrons as raw material in which the alterations would give a wider range of function than if sequence homologies were rigidly maintained.

D. ALLOSTERIC CONTROLLED PATHWAYS

Another possibility that should be considered as a ramification of this scheme is the role it might play in the appearance of allosterically controlled metabolic pathways (Monod et al., 1963). Here too, the mechanism depends on sequence similarities that may be retained in pathway enzymes. How did end product control mechanisms arise? One possibility is that an ancestral region that once bound a substrate molecule remains competent to bind an end product derived from the first substrate because of remnants of structural similarities between the two. In the course of evolutionary time, this site that now binds the end product may be shifted in the control enzyme relative to its active site for the first substrate that enters the pathway. The result of this would be the gradual appearance of the capability of the end product interacting with the control point enzyme in a manner that would alter the effectiveness of the active center to the entering substrate. Alteration of the activity would be controlled through the end product's capacity to interfere stereochemically or to induce conformational changes in the enzyme which would affect its activity.

Additionally, allosteric phenomena that are a result of peptide-peptide associations may also be a result of sequence homologies common to the associated units. Such similarities may play a role in a crystalliztion-like formation of associations and therefore affect enzymatic behavior toward substrates. That some such aggregate may alter its activity by changing its degree of association and thus affect its activity to the substrate or other controlling molecules is again consistent with the concept of a common ancestral cistron.

III. SELECTION AND THERMODYNAMICS

To some biologists the most interesting developments in recent years are the concepts of open-system thermodynamics (Denbigh, 1951; Prigogine, 1955; de Groot, 1959; Katchalsky and Curran, 1965). The gradual acceptance of these idealized systems, which have been formalized mathematically, may provide the route by which many of the properties of living systems, i.e., cells and organisms, can be analyzed. Of first importance to our present discussion is a theorem of Prigogine (1955). It states that in an open system the steady state is a condition of minimum entropy production (least dissipation of free energy). This theorem was discovered by employing some of the fundamentals derived by Onsager (Prigogine, 1955) in his statistical mechanical analysis of the so-called phenomenological equations. The application of these principles to biology must, of course, be limited to systems that fulfill conditions under which the principles were derived. A brief excursion into some of these relationships will illustrate the possibilities of this application.

The phenomenological equations are generalizations of relationships that are often expressed as laws. They take on the form $J = LX$, i.e., the flux J is produced by a force X and these are related through an appropriate coefficient L. Many well-known laws, i.e., Ohm's, Fick's, and Fourier's, can be generalized to fit this form, namely, equating a conjugate flux to its force through a linear coefficient. Chemical reaction rates as affected by chemical potentials can also be expressed in a similar form provided they are close to equilibrium. Establishing this relationship for chemical reactions requires a rather involved thermodynamic argument (Prigogine, 1955). Onsager extended these generalizations to include the effect of one conjugate set of fluxes and forces on another such that the force of one set would affect the flux of the other. In other words, the two sets were coupled. To express this in a two-component system, two such sets can be interrelated by the following set of equations:

$$J_1 = L_{11}X_1 + L_{12}X_2$$
$$J_2 = L_{21}X_1 + L_{22}X_2$$

In both cases the fluxes are the result of two forces that are related through a set of coefficients. Onsager discovered that this coupling was symmetrical and proved that the coefficient L_{12} equaled L_{21}. Thus relationships between phenomenological equations could be dealt with in a considerably simpler form. It is important to note that this applies to open systems in which such coupling occurs. In a multicomponent system, phenomenological equations relating all such interacting systems can be set down in a matrix form. These might include a wide variety of phenomena that in our discussion are essential to the operation of a cell, i.e., they could be physiological, biophysical, and biochemical.

A fundamental thermodynamic idea is related to this notation, namely, that the rate of entropy production can be expressed as the product of the flux and the force from the phenomenological equations:

$$dS/dt = JX$$

The equality of the coupling coefficients, the Onsager reciprocity relationship, can be made to hold if they are selected so as to satisfy this condition of entropy production. The total entropy production for the two-component system above is

$$dS/dt = J_1 X_1 + J_2 X_2$$

Prigogine discovered that by substituting the values for the fluxes from the two-component set of simultaneous equations into this entropy production equation and applying the Onsager reciprocity relationship $L_{12} = L_{21}$ a quadratic expression resulted. When this expression was solved for a minimum with one of the forces held constant and its conjugate flux going to zero, a positive definite but minimum entropy production was obtained. This concept is beautifully illustrated in the monograph by Katchalsky and Curran (1965). These authors presented the equations in conjunction with a geometric illustration of this fundamental theorem.

Various broad suggestions have been made that this minimum entropy production statement has evolutionary as well as ontogenetic significance (Prigogine and Wiame, 1946). The great difficulty in such generalizations, as Katchalsky and Curran have pointed out, is that cells and organisms cannot be shown to meet the conditions in which their associated phenomena and the laws governing them obey the linearity requirements, or, for that matter, the Onsager reciprocity condition or the independence of the coefficients from the forces. The basis for these difficulties is the great complexity of a living system in which the nature of coupled phenomena is not as yet understood or appreciated. It is, however, our thesis that the minimum entropy production relationship might best be applied closer to the molecular level, namely, to the selection of genes and their products, particularly enzymes by virtue of their participation in metabolic pathways. But before continuing, we will examine some properties of mutation and selection.

A. GENE SELECTION

The foundation of cellular evolution is, as in higher organisms, the mutation and selection of genes. These primary unitary changes operate by alteration of a single codon of a cistron and in turn this change affects the properties of the protein by an alteration in an amino acid. Since proteins are composed of a rather large number of amino acids, the number of single mutations that will alter a protein can be very large. A protein, and in particular an enzyme, may thus express the effect of a mutation over a wide range of variations in its properties. These may range from minute changes in its properties due to a mutation in a region unrelated to its function to drastic changes expressed in a region crucial to its activity. Thus variations resulting from a single mutation provide a continuum of properties upon which evolutionary selection pressures may have been brought to bear.

But the question that is most obvious is, "What specific property will be selected for?" Given two cells identical except for a single mutation in one cistron, what will determine the direction the selection will take? Consider the fact that a mutation which occurs in a chemostat will predominate if it reduces the generation time of the cell relative to that of its wild type (Novick and Szilard, 1950). In this instance, however, the chemostat is designed to select for short generation time cells, and therefore gene mutations that are in some way related to the effective use of the limiting nutrient will be selected for. The chemostat thus serves to reacquaint us with the reality of selection, namely, it must occur in the context of a specific environment. Which of the various agents involved in selection takes precedence? Can one predict the relative survival of genes? Consider the chemostat example, i.e., the selection of a specific mutation in the chemostat is achieved by employing a single growth-limiting nutrient that controls the growth rate. If a mutation occurs, selection for it will not occur if the nutrient to which it is related is not the factor that controls the growth rate, namely, if that nutrient is present in excess. Consequently, if the mutation occurs which would decrease the generation time (increase the growth rate) by more effective use of a nutrient but if the nutrient is present in excess, the mutation would not be seen. In other words, the mutation must be put into a competitive situation relative to the wild type in order for it to display itself. In this case the competition is for rapid growth rate, and since the chemostat is a growth rate selecting device, the shortest generation time cells will come to predominate. From this example one might be lured into defining evolutionary success as depending on the rate of increase of number since this appears to operate in this situation, but it would be a trivial definition. The importance of growth rate can be challenged on the grounds that the larger the population, the more rapid the exhaustion of nutrients with the eventual decline or demise of the mutant type and its precursor. This dilemma forces one to appreciate that evolutionary success is not as tied to multiplication rate as it is to the effective use of energy sources that exist in the environment. The capacity of a species to pace its expansion at such a rate that a low average rate of energy expenditure can be drawn for a maximum period of time is a characteristic of considerable evolutionary advantage. This characteristic is linked directly to the existence of control mechanisms, and such control mechanisms are genetic characters that provide an organism a maximum evolutionary advantage. They give the appearance of having the capacity to anticipate.

Genes are duplicated, mutated, transcribed, and translated into enzymes and other proteins and are selected for or eliminated over an extended period of time and in the context of other gene-enzyme systems. The apparent capacity of a control mechanism to anticipate is largely a result of integrating the effects of these factors that occur in time into the design of the cell. It can be argued that evolutionary selection pressures, therefore, operate in favor of systems that procure and channel matter and energy from some source into their duplication, and select for systems that regulate the utilization of the matter and energy pool. We will argue that the selection for control mechanisms operates at several levels in the cell. The two most important are the control of metabolism and the control of gene activity. We will concern ourselves only with the selection pressures for metabolic controls, although a similar argument could be developed for the selection of gene control systems.

B. Selection for Control Mechanisms

Given the fact that an evolving metabolic pathway may arise by a scheme similar to the one outlined, i.e., using sequence homology, it is obvious that the flux of metabolism which is conjugated to the primary source of chemical potential cannot flow to its lowest free energy state devoid of coupling to other processes and still have any role in the decrease in entropy that accompanies the growth of structure and form in the cell. Coupling thus implies an interaction of processes. The coupling will, in general, reduce or interfere with the flux of the primary system. The degree to which such interference constitutes a control mechanism can be evaluated only against the criterion of how well a system resists distortion. The state from which a distortion can best be measured is the steady state. The steady state can best be defined as the state of a system through which there is a flux of matter and energy but which is invariant with respect to its composition in matter and energy as a function of time. Defined this way, it is obviously an ideal state. Many may argue that the steady state condition is not the rule and rarely obtains in biological systems, but it is this same steady state that displays a minimum entropy production, and we propose it as a working ideal. Our argument is as follows. First, the state of minimum entropy production is intrinsically connected to the degree to which an open system has evolved control mechanisms. Second, it is the condition of minimum entropy production or of least dissipation of free energy that will preserve the environmental energy pools for the longest time and still permit the system to operate. Third, gene mutations that participate in the development of control devices will be selected for and preserved. It must be granted that the steady state is an ideal end point rarely achieved, but it can be maintained that controls are essential to the approach as well as the achievement of a steady state condition. Since entropy production is positive, definite, and goes to a minimum, systems that are evolving control mechanisms but in which steady-state behavior is not completely achieved will presumably have a selective advantage over those that lack that degree of control.

The evolution of enzymes, metabolic pathways, metabolic cycles, and allosteric control systems are, it would appear, a consequence of these thermodynamic relationships. In the case of enzymes and metabolic pathways, there are several considerations. First, it is the existence of enzymes as such that provides a considerable measure of control. The upper limit of the rate of substrate conversion is controlled by the number of enzyme active sites available. This means that at a fixed level of enzyme concentration, rates are controlled within a broad set point, i.e., from zero to the specific turnover number per molecule. Second, metabolic pathways are also control mechanisms that aid in the achievement of steady-state behavior, for it readily can be seen that the extension of a pathway not only gives rise to a greater energetic yield in a given environment, but also adds specificity to the ways in which the energy is utilized. When one speaks of entropy production the informational definition of entropy (Shannon and Weaver, 1949) should not be overlooked. Entropy decreases that occur at the synthetic end of the scale must also be added into the overall entropy production rate. Configurational and conformational changes that represent increases in order are related to the broad general definition of specificity, and therefore pathways that contribute to that end are also contributing to minimum entropy production. These latter points require considerations that are far too intricate to be covered in this discussion and, in fact, may need development beyond the present realm of knowledge.

It should be pointed out that phenomenological equations for enzyme catalyzed biochemical reaction rates will, perhaps, be quite different from the equations governing the rate of ordinary chemical reactions. The restrictions on the applicability of these equations will, therefore, be quite different. Insofar as enzymes and pathways contribute to the controlled use of energies and play a role in the direction of achieving steady-state behavior, they will contribute to minimum entropy production and will, therefore, be preserved.

Metabolic cycles can also be shown to contribute to steady-state operations and would thus be favored evolutionally. If one looks at such cycles devoid of side reactions, it is clearly obvious that they are rate limited by the availability of end products for condensation and recycling. Cycles looked at in this way are analogous to the simple mechanical escapement used to maintain the time constant in clocks. Furthermore, metabolic cycles, if considered in the presence of side reaction pathways, are subject to the controlling influence of additions or removals of pathway intermediates. It need not be emphasized again that steady-state behavior is the end product that may be achieved by these types of controls.

Finally, the evolutionary invention of allosteric control mechanisms such as end product inhibitions can be attributed to the same basic selection factors, namely, the selective advantage obtained through minimum entropy production which is in turn achieved by selection of control mechanisms that assist the cell or organism in approaching steady state operation. The process of selection for allosteric mechanisms can be considered as an extension or supplement to the controls that are inherent in a simple enzymatic step and metabolic pathways.

The overall thesis of this latter section of the discussion may apply equally well to organisms above the molecular control levels in the cell, although it is obvious that the phenomenological equations could not be applied to homeostatic mechanisms since they are several levels of organization removed from molecular phenomena. It should be emphasized though that this process for the selection of control mechanisms rationalizes one of the observations common to all forms of organic evolution. This is that the process of organic evolution is always accompanied by an increased complexity in the evolved form. For example, evolution of each of the many organ systems both invertebrates and vertebrates regulate the cellular environment to a relatively high degree of constancy. All these systems that are associated with water balance, respiratory gases, homeothermy, etc., are control mechanisms that to operate must of necessity increase complexity.

There should be little wonder then why one feels justified in applying the minimum entropy production idea to the evolution of control mechanisms and to gene-related mechanisms. It should, perhaps, be recalled that the proof of many laws of nature rests chiefly on the statement, "It is so because they work." In other words, violations have thus far not been found to the general statement of the law. The interdependence of the operation of evolutionary mechanisms and the laws of thermodynamics may well fall to a similar nonproof.

REFERENCES

Bryson, V., and Vogel, H. J. (eds.). (1965). "Evolving Genes and Proteins." Academic Press, New York.
de Groot, S. R. (1959). "Thermodynamics of Irreversible Processes." North-Holland Publ., Amsterdam.

Demerec, M., and Hartman, P. E. (1959). *Ann. Rev. Microbiol.* **13**, 377.
Denbigh, K. G. (1951). "The Thermodynamics of the Steady State." Methuen, London.
Gibor, A., and Granick, S. (1964). *Science* **145**, 890.
Horowitz, M. H. (1945). *Proc. Natl. Acad. Sci. U.S.* **31**, 153.
Horowitz, M. H. (1965). *In* "Evolving Genes and Proteins" (V. Bryson and H. J. Vogel, eds.), pp. 15–23. Academic Press, New York.
Katchalsky, A., and Curran, P. F. (1965). "Nonequilibrium Thermodynamics in Biophysics." Harvard Univ. Press, Cambridge, Massachusetts.
Margoliash, E., and Smith, E. L. (1965). *In* "Evolving Genes and Proteins" (V. Bryson and H. J. Vogel, eds.), pp. 221–242. Academic Press, New York.
Monod, J., Changeux, J. P., and Jacob, F. (1963). *J. Mol. Biol.* **6**, 306.
Novick, A., and Szilard, L. (1950). *Proc. Natl. Acad. Sci. U.S.* **36**, 708.
Prigogine, I. (1955). "Introduction to Thermodynamics of Irreversible Processes." Thomas, Springfield, Illinois.
Prigogine, I., and Wiame, J. M. (1946). *Experientia* **2**, 451.
Shannon, C. E., and Weaver, W. (1949). "The Mathematical Theory of Communication." Univ. Illinois Press, Urbana, Illinois.

CHAPTER 2

Regulation of Chromosome Replication and Cell Division in *Escherichia coli*

Charles E. Helmstetter

I. INTRODUCTION

Genetic information necessary for duplication of the bacterial cell is contained within continuous, DNA-containing structures that have been termed "bacterial chromosomes." The number of chromosomes in a cell, and the extent of their replication, depend on the age and physiological state of the cell. To establish and maintain a particular pattern of chromosome replication during the division cycle, there must be a fixed relationship between chromosome replication and cell division. With regard to this relationship, Maaløe and Kjeldgaard (1966) have suggested that the rate of cell division is determined by the frequency of initiation of chromosome replication.

15

This report is concerned with the regulation of initiation of chromosome replication, and the coordination between this event and cell division, in *Escherichia coli*. A feasible control mechanism is presented, and
it is used to predict the response of cells to a variety of treatments.

II. COORDINATION BETWEEN CHROMOSOME REPLICATION AND CELL DIVISION

A. DNA Synthesis during the Division Cycle

The first step in our study of the linkage between chromosome replication and cell division involved locating the stages in the division cycle
at which rounds of chromosome replication began and ended in cultures
of *E. coli* B/r growing at various rates. The experiments were based
on the idea that the start and finish of a round of replication would
appear as an abrupt increase and decrease, respectively, in the rate
of DNA synthesis during the division cycle (Clark and Maaløe, 1967;
Helmstetter, 1967; Helmstetter and Cooper, 1968). Exponentially growing cultures were pulse-labeled with radioactive thymidine, and the
amount of label incorporated into cells of different ages was found by
measuring the radioactivity in their progeny. The progeny of the pulse-
labeled cells were withdrawn from the culture after it was bound to
the surface of a membrane filter as described in detail elsewhere (Helmstetter, 1968a).

Figure 1 shows the results of measurements of the time for a round
of chromosome replication (C), the time between the end of a round
of replication and the following cell division (D), and the sum $(C + D)$.
The values for these parameters were fairly constant and equal to about
41, 22, and 63 minutes, respectively, in cells growing with generation
times between 22 and 53 minutes. Above a generation time (τ) of 63
minutes, the broken lines are drawn such that $(C + D)$ equals τ (start
of rounds coincident with division), C equals $\frac{2}{3}\tau$, and D equals $\frac{1}{3}\tau$.
The experimental values are in the vicinity of the lines except that
the values for $(C + D)$ generally fall above the line at the lower growth
rates, indicating that rounds of replication began slightly before division
in these cells.

B. Origin and Sequence of Chromosome Replication

The second step in this study involved locating the chromosomal site
at which each new round of replication was initiated. There is considerable evidence that replication of the bacterial chromosome begins at
a fixed point on the genome (Meselson and Stahl, 1958; Cairns, 1963;
Bonhoeffer and Gierer, 1963; Lark *et al.*, 1963; Nagata, 1963; Yoshikawa

and Sueoka, 1963a,b; Donachie and Masters, 1966; Cutler and Evans, 1967; Berg and Caro, 1967; Abe and Tomizawa, 1967; Wolf *et al.*, 1968; Helmstetter, 1968b; Donachie and Masters, Chapter 3 of this volume). The location of the replication origin and the sequence of repli-

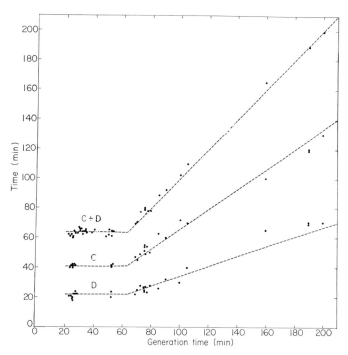

FIG. 1. Kinetics of chromosome replication in *Escherichia coli* B/r growing at various rates. C = time for a replication point to traverse the chromosome (a round of replication); D = time between the end of a round of replication and the subsequent cell division; $(C + D)$ = time between the start of a round of replication and the cell division following completion of that round. Cultures (100 ml) growing exponentially with 1×10^8 cells/ml in minimal salts medium containing various carbon sources were exposed to 0.02 μc of thymidine-^{14}C per milliliter, filtered onto 15-cm diameter, grade GS Millipore membrane filters, washed with 100 ml of minimal medium, and eluted with minimal medium. The midpoints of abrupt decreases and increases in the radioactivity per effluent cell were considered to correspond to the start and end of rounds of chromosome replication, respectively.

cation in *E. coli* B/r was investigated by measuring the sequence in which the capacities for induced synthesis of different enzymes increased during a round of chromosome replication (Helmstetter, 1968b). It was assumed that a sudden increase in the initial rate of induced synthesis of an enzyme was a consequence of the duplication of its structural gene (Masters *et al.*, 1964; Kuempel *et al.*, 1965; Donachie and Masters, 1966; Donachie and Masters, Chapter 3 of this volume). The results indicated an origin at approximately 8 o'clock (60 minutes) on the genetic map (Taylor and Thoman, 1964) and a clockwise direction of replication.

From the results summarized in this section, a simple picture of the coordination between chromosome replication and cell division in *E. coli* B/r can be deduced based on the parameters C and D. Chromosome replication is initiated at a fixed point on the genome, and a cell division follows $(C + D)$ minutes later. The arrival of a replication point at the terminus of the chromosome appears to trigger a sequence of events leading to cell division (Helmstetter and Pierucci, 1968). By means of this linkage, the rate of initiation of replication governs the rate of cell division. Thus the control of initiation must be understood, at least phenomenologically, before a reasonable description of the regulation of chromosome replication and cell division can be constructed.

III. REGULATION OF CHROMOSOME REPLICATION AND CELL DIVISION

A. INITIATION OF CHROMOSOME REPLICATION

Chromosome replication is initiated when an unusual situation arises in the cell. Among the ingredients which might be involved in this process are: localized conversion of DNA to a denatured or primer form (Bollum, 1963; Lark, 1963), opening of a point on the chromosome (Jacob and Monod, 1962; Taylor, 1963), duplication of the protein portion of the chromosome (Wolff cited in Kuempel and Pardee, 1963), fluctuation in enzyme levels (Johnson and Schmidt, 1966), formation of an active DNA polymerase (Lee-Huang and Cavalieri, 1965), relocation of the polymerase and formation of a DNA–enzyme complex (Billen, 1962; Littlefield *et al.*, 1963), production or disappearance of activators or repressors (Jacob and Monod, 1962), achievement of a critical cell mass or related parameter (Hanawalt *et al.*, 1961; Maaløe, 1963; Maaløe and Kjeldgaard, 1966), or formation of a covalent linkage between the terminus of the parental strand and the starting nucleotide for the new strand (Yoshikawa, 1967).

A major development in the search for the "signal" for initiation of chromosome replication was the finding of a requirement for protein synthesis to achieve each initiation event. It was suggested by Maaløe and Hanawalt (1961), and experimentally verified by Lark *et al.*, (1963), that protein synthesis is required to initiate chromosome replication but not to maintain it. Another major step was the development of the "replicon" model by Jacob and Brenner (1963). According to their model, the chromosome contains specific loci that control the initiation of its own replication. A structural gene controls the synthesis of a cytoplasmic initiator which acts on a segment of the chromosome to start replication. They also proposed an association between the chromosome and membrane structures, and suggested that daughter chromosomes could be distributed to daughter cells through growth of membrane between the points of attachment. Lark and co-workers (Lark and Lark, 1964, 1965, 1966; Pritchard and Lark, 1964; K. G. Lark, 1966; Lark *et al.*, 1967) have also suggested that the initiation of chromosome replication requires the participation of a cytoplasmic initiator protein and a structural element to which the chromosome is attached during replication. Considerable evidence for a DNA-membrane linkage has now been reported (Jacob *et al.*, 1963; Ganesan and Lederberg, 1965; Ryter and Jacob, 1966; Jacob *et al.*, 1966; Cuzin and Jacob, 1967; Smith and Hanawalt, 1967).

Thus, the synthesis of at least two species of protein appears to be required for initiation of chromosome replication. In addition, there must be a relationship between the cellular growth rate and the ability of cells to initiate replication. K. G. Lark (1966) has suggested that chromosome replication is governed, in part, by the rate of synthesis of the initiator and the structural element. In any event, the production of chromosomes and the rate of cell division are determined by the frequency of initiation; this, in turn, appears to be regulated by the rate of protein synthesis (Maaløe and Kjeldgaard, 1966; Maaløe, 1966).

B. REGULATION OF INITIATION

Present evidence suggests that initiation is *not* influenced by the presence, absence, position, or rate of movement of replication points on the chromosome. For example, the initiation of a new round of replication is not triggered by the completion of the previous round since new rounds can begin long before the previous round has ended in rapidly growing cells (Yoshikawa *et al.*, 1964; Helmstetter and Cooper, 1968). The processes leading to initiation appear to take place whether or not the chromosome is replicating, and the rate of DNA synthesis per replication point can be changed without affecting the rate of production of the special state necessary for initiation (Maaløe and Rasmussen, 1963).

An important clue to the regulation of initiation was the finding by Hanawalt *et al.* (1961) of a close relationship between cell mass and initiation in *E. coli* 15T⁻. When amino acids were restored to a starved culture, the following behavior was observed: (a) the rate of DNA synthesis increased gradually until the definitive rate was reached after about one generation of growth; (b) the period of increasing rate of DNA synthesis corresponded to an increase in the fraction of cells synthesizing DNA; and (c) if DNA synthesis was inhibited for about one generation and then restored, DNA synthesis began immediately at the definitive rate. These studies suggested that the capacity for initiation of chromosome replication can be achieved in the absence of DNA synthesis, and that DNA synthesis is not initiated until a critical cellular mass/DNA ratio is reached.

The initiation of chromosome replication in cells with a generation time of I minutes may thus be described as follows: (i) the cells initiate chromosome replication every I minutes; (ii) the capacity for initiation is achieved as a result of a total of I minutes of unrestricted protein synthesis; and (iii) if the I minutes of protein synthesis are interrupted by periods of inhibition, initiation is delayed for a time equal to the sum of the interruptions. The period of required protein synthesis probably consists of a complex sequence of events (e.g., synthesis of a complex structure organized in the membrane or a set of individual molecules) which result in the achievement of a critical cell mass. The important point is that initiation behaves as if it were the end result of an *accumulation* of some hypothetical protein(s) (see discussions by Nakada, 1960; Maaløe, 1963; Pritchard and Lark, 1964; Lark and Lark, 1964; Hayes, 1965; Pritchard, 1966; Maaløe and Kjeldgaard, 1966; Ehret and Trucco, 1967; Donachie and Masters, Chapter 3 in this volume).

In summary, regulation of chromosome replication and cell division can be described as a three-step process: $I + C + D$, where I is the time for synthesis (accumulation) of an initiator complex of *constant size per initiation event* at a given temperature. In an exponentially growing culture, the initiator is synthesized continuously, initiation takes place when a complete complex has accumulated, and the cell divides $(C + D)$ minutes later. Consequently, the generation time of cells growing exponentially in a particular medium is equal to, and determined by, the time required for the cell to synthesize the initiator in that medium (i.e., $\iota = I$). During balanced growth, the generation time is independent of C and D.

C. Chromosome Replication and Cell Division

1. Construction of a Chromosome Replication Pattern

The chromosome replication pattern during the division cycle of cells in any physiological state can be determined by applying the $(I + C + D)$ rule to a hypothetical cell containing a single chromosome. An example of such a construction is shown in Fig. 2 for $I = 20$, $C = 40$, and $D = 20$ minutes. The construction begins with a single chromosome at the moment of initiation of replication. The first replication point reaches the terminus of the chromosome at 40 minutes and a division follows in 20 minutes. Between 0 and 20 minutes, a complete initiator complex accumulates per origin resulting in the second initiation event at 20 minutes and another division at 80 minutes. The steady state pattern of chromosome replication during the division cycle in this physiological state is given between divisions.

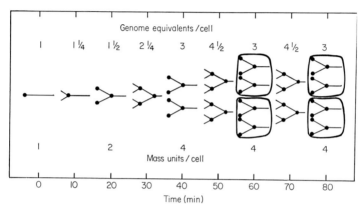

FIG. 2. Construction of a chromosome replication pattern during the bacterial division cycle. The construction is based on the $(I + C + D)$ rule with $I = 20$, $C = 40$, and $D = 20$ minutes. Filled circles indicate replication points and the configurations in rectangles indicate a division. For simplicity, chromosomes have been shown as linear structures with the origin at the left and the terminus at the right. Initiations of chromosome replication are indicated by the appearance of replication points at the origins, and the time between each of these events is equal to I. See text for details.

Since rounds of replication are assumed to begin when a fixed amount of initiator has accumulated, independent of the rate of accumulation, the quantity of initiator in a cell might be reflected by a constant cell mass per initiation event at all growth rates (Hanawalt *et al.*, 1961; Donachie and Masters, Chapter 3 in this volume). The numbers below the chromosome configurations in Fig. 2 indicate mass units per cell where one unit is the mass of a cell containing a single chromosome at the moment of initiation. If the cell mass per initiation is constant, there would be considerable variation in the average mass/DNA ratio in exponential phase cultures growing at different rates. Assuming that cell mass increases exponentially during the division cycle, the predicted mass/DNA ratio can be calculated using Eq. (5) of Cooper and Helmstetter (1968) for the average genome equivalents of DNA per cell. For cultures growing at rates of 100, 40, and 20 minutes per doubling, the cell mass at initiation would be 1, 2, and 4 mass units so that the average mass/DNA ratio in mass units per genome equivalent would be about 0.9, 1.0, and 1.3, respectively. The variation in these ratios is virtually the same as the reported values for *Salmonella typhimurium* (see Maaløe and Kjeldgaard, 1966). However, if the initiator is entirely protein, the accumulation concept might be reflected in total protein rather than total mass and the above considerations should be replaced by protein units per cell and protein/DNA ratios.

Finally, twice as many rounds are initiated at 40 minutes as at 20 minutes in Fig. 2. If twice as much initiator is made during the second 20-minute interval, and if the rate of synthesis depends on the dosage of initiator genes, then this implies that these genes are located at, or very near the origin.

2. *Chromosome Replication during the Division Cycle*

The pattern of chromosome replication during the division cycle of cells growing at various rates can be constructed using the values for C and D in Fig. 1 and applying the $(I + C + D)$ rule. Several examples are shown in Fig. 3. The rate of DNA synthesis per replication point is considered constant (Clark and Maaløe, 1967; Helmstetter, 1967). The pattern of DNA synthesis and the predicted number of genomes per cell are consistent with the experimental results (Cooper and Helmstetter, 1968).

Cells with generation times shorter than about 40 minutes contain three replication points per chromosome during a portion of their division cycle, as has been found experimentally in *Bacillus subtilis* (Yoshikawa *et al.*, 1964) and *E. coli* B/r (Helmstetter and Cooper, 1968; Helmstetter, 1968b).

Cells with generation times longer than about 40 minutes contain a period devoid of DNA synthesis at the end of their division cycle (Helmstetter, 1967). However, according to the model described here, periods devoid of DNA synthesis could exist at both the beginning and end of the division cycle if $(C + D)$ was less than the generation time. In this case, the division cycle could be divided into G_1, S, and G_2 periods as is commonly found in higher organisms (Howard and Pelc, 1953). Thus, it is predicted that cells of *E. coli* B/r could contain a G_2 without a G_1, but a G_1 could never exist without a G_2. C. Lark (1966) and Kubitschek *et al.* (1967) have reported G_1 periods in slowly growing *E. coli* 15T⁻, and the latter authors also observed a brief G_2 period. However, Maaløe (1966) and Clark and Maaløe (1967) have

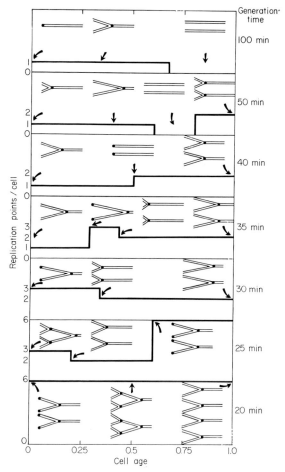

FIG. 3. Schematic illustrations of the pattern of chromosome replication during the division cycle of *Escherichia coli* B/r. The illustrations were constructed as in Fig. 2, using approximate values of C and D for *E. coli* B/r given in Fig. 1 (i.e., $C = 40$ minutes and $D = 20$ minutes except when $\tau = 100$ minutes, in which case $C = \frac{2}{3}\tau$ and $D = \frac{1}{3}\tau$). The rate of DNA synthesis is shown in units of replication points per cell.

suggested the possibility of a relationship between the start of rounds and the next cell division such that division cannot take place until the condition for initiation has been reached. Since we have found that rounds of replication begin near division in slowly growing cells, this possibility cannot be ruled out.

A major advantage of the $(I + C + D)$ model is that it can be used to predict the response of cells to changes in their environment. The remainder of this chapter deals with the predicted response of cells to three commonly employed treatments: (a) nutritional shifts, (b) inhibition of DNA synthesis, and (c) inhibition of protein and RNA synthesis.

IV. CELLULAR RESPONSE TO NUTRITIONAL ALTERATIONS

A. NUTRITIONAL SHIFT-UP

Kjeldgaard et al. (1958) and Kjeldgaard (1961) have employed nutritional shifts to study the regulation of macromolecular synthesis and cell division in bacteria. One of the most striking observations was that when cells were transferred from a poorer to a richer medium (a "shift-up"), the preshift rate of cell division was maintained for 60–70 minutes before the new rate was achieved (Kjeldgaard et al., 1958; Kjeldgaard, 1961; Schleif, 1967; S. Cooper, personal communication). The rate of mass increase assumed the new value shortly after the shift (about 5 minutes), but the definitive postshift rate of DNA synthesis was not reached for 20–25 minutes (Kjeldgaard et al., 1958; Kjeldgaard, 1961).

Maaløe (1966) has proposed a model to explain the behavior of cells following a shift-up. He suggested that the increase in rate of DNA synthesis was due to an increase in the frequency of initiation, beginning with the introduction of extra replication points after approximately one postshift exponential-phase doubling time. The characteristic pattern of cell division was explained by assuming that the time between initiation of chromosome replication and cell division was constant (20 minutes) and that division was not triggered unless two separate DNA units were present at the time of initiation.

According to the $(I + C + D)$ model, the increase in rate of DNA synthesis after a shift-up is also due to an increase in the frequency of initiation, but the regulation of this process (and cell division) is explained somewhat differently. The rate of cell division would be expected to remain unchanged for $(C + D)$ minutes after a shift-up if C and D were constant over the range of the shift. Once a round of chromosome replication has been initiated, a division would take place $(C + D)$ minutes later, independent of the shift-up.

After $(C + D)$ minutes in the new medium, the rate of cell division would increase as a consequence of the increased rate of initiator synthesis. For example, if a cell were X minutes prior to initiation in the preshift medium, a division would normally occur in $(X + C + D)$ minutes, but if this cell were shifted up at that time, the rate of initiator synthesis would eventually increase and less than X minutes would elapse before initiation. Thus, a division would occur in less than $(X + C + D)$ minutes.

The manner in which a culture would respond to a shift-up is shown schematically in Fig. 4. Four representative cells from a culture growing with a 40-minute doubling time ($I = 40$, $C = 40$, and $D = 20$ minutes) are shifted at 0 minutes to a nutritional state in which $I = 20$, $C = 40$, and $D = 20$ minutes. It is assumed that the new rate of initiator synthesis is achieved immediately. At the time of the shift-up, the cell in the top row contains a chromosome which began replicating 20 minutes earlier in a medium in which $I = 40$ minutes. Therefore, it contains one-half of a complete initiator complement per origin, and it initiates

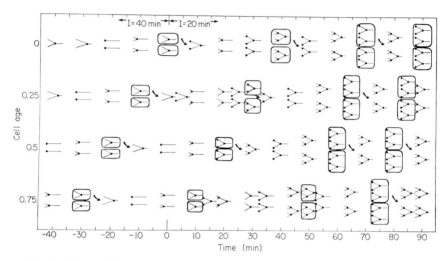

FIG. 4. Theoretical response of individual cells to a nutritional shift-up. Four representative cells from an exponentially growing culture with $I = 40$, $C = 40$, and $D = 20$ minutes are shifted after 40 minutes to $I = 20$, $C = 40$, and $D = 20$ minutes. The ordinate shows the ages of the cells containing the chromosome configurations shown in the left-hand column. After a division, only one of the two cells is followed as indicated by the arrows.

when another one-half is synthesized; this requires 10 minutes since $I = 20$ minutes after the shift. (To simplify description, I am assuming that initiator increases linearly.) After this time, initiation occurs every 20 minutes and a division follows in $(C + D)$ minutes.

The cell in the second row at 0 minutes has been accumulating initiator for 30 minutes, and therefore it contains three-quarters of an initiator complex per origin. After the shift-up, initiation takes place when the remaining one-quarter is synthesized, which requires 5 minutes in the new medium. Thus, a division occurs at $(5 + C + D)$ minutes instead of $(10 + C + D)$ minutes as would have been the case in the absence of the shift.

The cumulative effect of the shift-up on the entire population in terms
of cell number, total DNA, and total mass is shown in Fig. 5. The
rate of cell division increases abruptly at 60 minutes $(C + D)$ and the
rate of DNA synthesis increases gradually due to a gradual increase

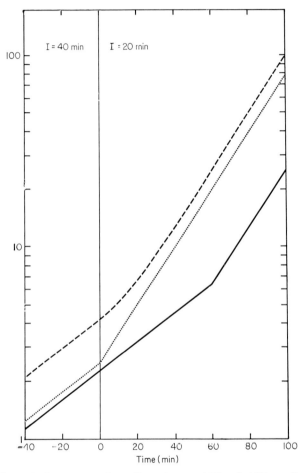

FIG. 5. Theoretical response of a culture to a nutritional shift-up. Summation of
the effect of the shift-up shown in Fig. 4 ($I = 40$ to $I = 20$ minutes) over an
exponentially growing culture in terms of cell number, ———; total mass, · · · · ;
and total DNA, – – –.

in number of replication points. The theoretical curves are quite similar
to the experimental results (Kjeldgaard *et al.*, 1958; Kjeldgaard, 1961).

It has been assumed that the definitive postshift rate of initiator synthesis is achieved immediately. Experimentally, the rate of protein syn-

thesis increases gradually, and therefore, the definitive rate of cell division may not be achieved until after $(C + D)$ minutes. In addition, if C and/or D were not constant over the range of the shift, the cell number curve would be modified accordingly (Marr, personal communication; Cooper, personal communication).

B. INHIBITION OF DNA SYNTHESIS

If initiation is independent of the rate of movement of replication points, the frequency of initiation should not change if the rate of DNA

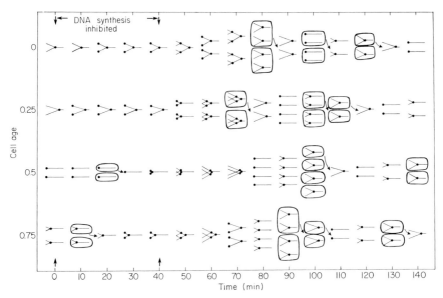

FIG. 6. Theoretical response of individual cells to inhibition and restoration of DNA synthesis. DNA synthesis is inhibited at 0 minutes and restored at 40 minutes with $I = 40$, $C = 40$, and $D = 20$ minutes.

synthesis per replication point is altered. The cellular response to a change in the rate of DNA synthesis will be discussed in terms of the limiting case: complete inhibition of DNA synthesis.

Figure 6 illustrates the effects of inhibiting DNA synthesis for **40** minutes in a population of cells in which $I = 40$, $C = 40$, and $D = 20$ minutes. The cell of age 0 at 0 minutes contains a half-replicated chromosome and one-half of an initiator complex per origin since it has progressed 20 minutes since its last initiation. During inhibition of DNA synthesis, the preexisting replication point does not move. At **20** minutes a full complement of initiator has accumulated in this cell, resulting in the appearance of replication points. At 40 minutes, DNA synthesis is restored and the three replication points begin moving along the chromosome. At this time, the cell contains one-half of an initiator complex, and therefore, initiation occurs again at 60 minutes. The other cells have been treated in a similar way.

Cells continue to divide for D minutes following inhibition of DNA synthesis (Helmstetter and Pierucci, 1968). When DNA synthesis is restored at 40 minutes all cells contain one full initiator complement per origin plus the amount which was present at 0 minutes. Thus, new rounds of replication are initiated at the origins of all chromosomes in the culture, and every cell divides $(C + D)$ minutes later.

The effects of selective inhibition of DNA synthesis summed over the entire culture are shown in Fig. 7. When the capacity for DNA synthesis is restored, the amount of DNA increases at three times the normal rate because of the presence of three replication points on each chromosome (Fig. 6). Cells begin dividing again D minutes after synthesis is restored. At 100 minutes, there is a sudden doubling in cell number as a result of the simultaneous initiations which took place on all chromosomes $(C + D)$ minutes earlier. This is consistent with the stepwise increase in cell number observed by Barner and Cohen (1956) following thymine starvation in $E.$ $coli$ 15T⁻.

The effects illustrated in Figs. 6 and 7 are generally consistent with a number of previous experiments. Enhanced synthesis of DNA has been observed following thymine starvation (Barner and Cohen, 1956; Nakada, 1960; Pritchard and Lark, 1964; Hardy and Binkley, 1967), after exposure to ultraviolet light (Swenson and Setlow, 1966), and after exposure to nalidixic acid (Boyle et al., 1967). Pritchard and Lark (1964) reported that the high level of DNA synthesis after thymine starvation was due to excessive initiation of chromosome replication, and the experiments of Boyle et al. (1967) with nalidixic acid were in agreement with this concept. Hewitt and Billen (1964, 1965) found that DNA replication proceeded from a site other than the original active site after ultraviolet irradiation. Each of these treatments (thymine starvation, ultraviolet irradiation, and nalidixic acid) are relatively specific inhibitors of DNA synthesis, and their effects are in accord with the continued accumulation of initiator in the absence of DNA synthesis.

It has also been reported that growth of cells in 5-bromouracil induces initiation at the origin (Abe and Tomizawa, 1967; Wolf et al., 1968). It has been suggested that their results could be a consequence of the continued synthesis of initiator at a relatively normal rate while the presence of 5-bromouracil reduced the rate of DNA synthesis per replication point (O. Pierucci, personal communication). Consequently, new rounds would be initiated before the previous replication point had traveled the normal distance on the chromosome, and chromosomes with multiple replication forks would appear in cells in which they did not previously exist.

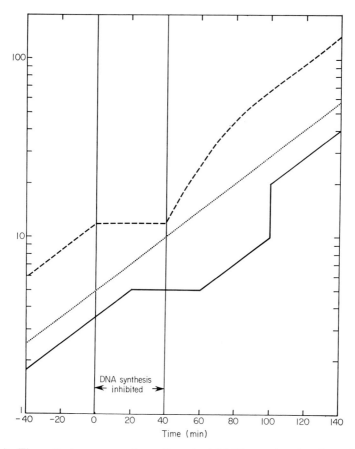

FIG. 7. Theoretical response of a culture to inhibition and restoration of DNA synthesis. Summation of the effects of the inhibition shown in Fig. 6 over an exponentially growing culture in terms of cell number, ———; total mass, · · · · ; and total DNA, – – –.

Pritchard and Lark (1964) and Pritchard (1966) originally considered that the effect of thymine starvation could be due to the accumulation of initiator during selective inhibition of DNA synthesis, but they expressed reservations concerning this possibility since other treatments which inhibited DNA synthesis (i.e., exposure to deoxyadenosine or cytosine arabinoside, and deoxyriboside starvation) did not cause the same effect. They suggested that the enhanced initiation following thymine starvation could be a direct consequence of the absence of thymine because of the production of an initiator which is normally repressed by a thymine-containing compound. However, as the number of experiments suggesting accumulation of an initiator during inhibition of DNA synthesis has increased, the former interpretation has become more appealing. Maaløe and Kjeldgaard (1966) have also speculated that the cellular reaction to thymine starvation is primarily a response to an increase in mass during starvation.

C. Inhibition of RNA and Protein Synthesis

The effects of inhibiting protein and RNA synthesis (initiator synthesis) in a culture with $I = 40$, $C = 40$, and $D = 20$ minutes are shown

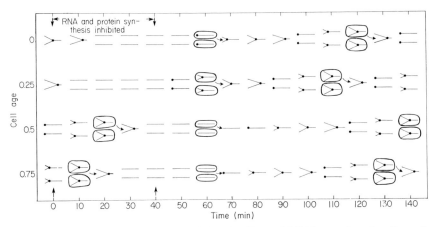

Fig. 8. Theoretical response of individual cells to inhibition and restoration of protein and RNA synthesis. Protein and RNA synthesis are inhibited at 0 minutes and restored at 40 minutes with $I = 40$, $C = 40$, and $D = 20$ minutes.

in Fig. 8. It has been assumed that completion of a round of chromosome replication "triggers" division, that the "trigger" requires protein and/or RNA synthesis, and that cells in the D period of the division cycle divide in the absence of RNA and protein synthesis (Lark et al., 1963; C. Lark, 1966; Billen and Hewitt, 1966).

During the inhibition, rounds of chromosome replication are completed (Maaløe and Hanawalt, 1961; Lark et al., 1963); as a result, all cells divide D minutes after protein and RNA synthesis are restored. DNA synthesis resumes in individual cells when synthesis of the initiator complement is completed. [The experimentally observed lag before resumption of DNA synthesis has been neglected (Lark *et al.*, 1963; Billen and Hewitt, 1966).] Cells that contained a nearly complete initiator complex before the inhibition begin to synthesize DNA shortly after protein synthesis is restored, and cells which had just started a round of replication do not initiate until 40 minutes later. For instance, the cell of age 0 at 0 minutes contains one-half of the initiator complex, and therefore, it requires 20 minutes to complete initiator accumulation after protein and RNA synthesis are restored. Thus, DNA synthesis is initiated at different times in cells of different ages during the first 40 minutes after resumption of protein and RNA synthesis. This interpretation is identical to that made by Hanawalt *et al.* (1961), who originated the idea of a critical mass for initiation.

The cumulative effect of this treatment is shown in Fig. 9. The pattern of DNA synthesis is similar to that described in previous reports Maaløe and Hanawalt, 1961; Lark *et al.*, 1963; Billen and Hewitt, 1966). The important point in this figure is the prediction of a synchronous doubling in cell number. Some recent reports of synchronous growth following starvation could be explained in this way, although not more than one stepwise increase in cell number would have been expected (Matney and Suit, 1966; Stonehill and Hutchison, 1966; Cutler and Evans, 1966).

The period of increasing rate of DNA synthesis in Fig. 9 corresponds to an increase in the fraction of cells synthesizing DNA. In contrast, Lark *et al.* (1963) and K. G. Lark (1966) have suggested that when DNA synthesis resumes after an extended period of amino acid starvation, a large fraction of the cells are synthesizing DNA but at different rates. In this case, the period of increasing rate of DNA synthesis would correspond to increases in the rate of synthesis at individual replication points. According to K. G. Lark (1966) initiation is regulated by two protein components. One of these (the replicator) forms the attachment of one strand of the chromosome to a membrane structure. When a round of chromosome replication is initiated, the other strand is opened and becomes attached by another protein (the initiator) to a site (proreplicator) on the membrane. At completion of a round of replication, the proreplicator is converted to a replicator and the initiator is released. During amino acid starvation, replication continues but proreplicators cannot be converted to replicators. When amino acids are restored, this conversion takes place liberating initiator which can start a new round. Thus, he suggests that chromosome replication is regulated by two mechanisms: (a) the rate of synthesis of initiator and proreplicator sites, and (b) the completion of a round of chromosome replication which converts the proreplicator into a replicator.

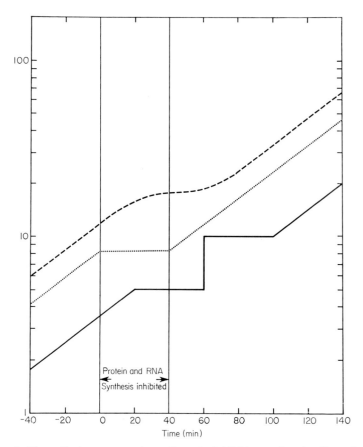

FIG. 9. Theoretical response of a culture to inhibition and restoration of protein and RNA synthesis. Summation of the effect of the inhibition shown in Fig. 8 over an exponentially growing culture in terms of cell number, ———; total mass, · · · · ; and total DNA, – – –.

The relationship between growth rate and the rate of formation of initiator proposed by Lark is essentially the same as the model described here, but the suggested involvement of the completion of one round of replication in the initiation of a subsequent round is at variance. In this report it is suggested that initiation is governed solely by the rate of formation of an initiator complex which probably includes both a cytoplasmic "initiator" and a structural component.

V. CONCLUSIONS

Koch and Schaechter (Koch and Schaechter, 1962; Koch, 1966) have proposed a deterministic model of bacterial growth which postulates that cells must attain a critical size before division can occur. The control mechanism discussed in this report is very similar to their proposal since it is based on the accumulation of a specific quantity of material leading to division after a fixed sequence of events $(C + D)$. The precise linkage between these events (particularly the D period) is unknown, but this problem has been discussed in detail by Kuempel and Pardee (1963) and Walker and Pardee (1968) with emphasis on the relationship between chromosome synthesis and septation. (See also Slepecky, this volume, Chapter 4.)

The $(I + C + D)$ rule is intended solely as a phenomenological description of the control of chromosome replication and cell division in *E. coli* B/r. However, strain-dependent differences in cell size or chemical composition do not imply different regulatory mechanisms since these properties may be determined by C and D. At a given doubling time, a larger value for $(C + D)$ would result in greater mass and DNA content. For instance, the DNA content and chromosome replication pattern of glucose-grown *E. coli* 15T⁻ reported by Lark and Lark (1965) are consistent with $C = 40$ and $D = 40$ minutes, i.e., two chromosomes which begin replication at division and replicate simultaneously throughout the division cycle.

Since the regulatory mechanism described for strain B/r may be applicable to other strains of *E. coli*, it seems worth considering the means for determining C and D in these strains. D can be found by measuring the period of continued cell division following inhibition of DNA synthesis in an exponentially growing culture. Once D is known, C can be determined either by measuring the DNA content and using Eq. (5) of Cooper and Helmstetter (1968), or by performing a nutritional shift-up and measuring the time between the shift and the abrupt increase in the rate of cell division which is $(C + D)$.

REFERENCES

Abe, M., and Tomizawa, J. (1967). *Proc. Natl. Acad. Sci. U.S.* **58,** 1911.
Barner, H. D., and Cohen, S. S. (1956). *J. Bacteriol.* **72,** 115.
Berg, C. M., and Caro, L. G. (1967). *J. Mol. Biol.* **29,** 419.
Billen, D. (1962). *Biochem. Biophys. Res. Commun.* **7,** 179.
Billen, D., and Hewitt, R. (1966). *J. Bacteriol.* **92,** 609.
Bollum, F. J. (1963). *J. Cell. Comp. Physiol.* **62** (Suppl. 1), 61.
Bonhoeffer, F., and Gierer, A. (1963). *J. Mol. Biol.* **7,** 534.
Boyle, J. V., Goss, W. A., and Cook, T. M. (1967). *J. Bacteriol.* **94,** 1664.
Cairns, J. (1963). *Cold Spring Harbor Symp. Quant. Biol.* **28,** 43.
Clark, D. J., and Maaløe, O. (1967). *J. Mol. Biol.* **23,** 99.
Cooper, S., and Helmstetter, C. E. (1968). *J. Mol. Biol.* **31,** 519.
Cutler, R. G., and Evans, J. E. (1966). *J. Bacteriol.* **91,** 469.
Cutler, R. G., and Evans, J. E. (1967). *J. Mol. Biol.* **26,** 91.

34   CHARLES E. HELMSTETTER

Cuzin, F., and Jacob, F. (1967). *Ann. Inst. Pasteur* **112**, 529.
Donachie, W. D., and Masters, M. (1966). *Genet Res.* (Cambridge) **8**, 119.
Ehret, C. F., and Trucco, E. (1967). *J. Theoret. Biol.* **15**, 240.
Ganesan, A. T., and Lederberg, J. (1965). *Biochem. Biophys. Res. Commun.* **18**, 824.
Hanawalt, P. C., Maaløe, O., Cummings, D. J., and Schaechter, M. (1961). *J. Mol. Biol.* **3**, 156.
Hardy, C., and Binkley, S. B. (1967). *Biochemistry* **6**, 1892.
Hayes, W. (1965). *Symp. Soc. Gen. Microbiol.* **15**, 294.
Helmstetter, C. E. (1967). *J. Mol. Biol.* **24**, 417.
Helmstetter, C. E. (1968a). *In* "Methods in Microbiology" (D. W. Ribbons and J. R. Norris, eds.). Academic Press, New York, in press.
Helmstetter, C. E. (1968b). *J. Bacteriol.* **95**, 1634.
Helmstetter, C. E., and Cooper, S. (1968). *J. Mol. Biol.* **31**, 507.
Helmstetter, C. E., and Pierucci, O. (1968). *J. Bacteriol.* **95**, 1627.
Hewitt, R., and Billen, D. (1964). *Biochem. Biophys. Res. Commun.* **15**, 588.
Hewitt, R., and Billen, D. (1965). *J. Mol. Biol.* **13**, 40.
Howard, A., and Pelc, S. R. (1953). *Heredity Suppl.* **6**, 261.
Jacob, F., and Brenner, S. (1963). *Compt. Rend. Acad. Sci.* **256**, 298.
Jacob, F., and Monod, J. (1962). *In* "Biological Organization at the Cellular and Supercellular Level" (R. J. C. Harris. ed.), *Unesco Symp.,* pp. 1–24. Academic Press, New York.
Jacob, F., Brenner, S., and Cuzin, F. (1963). *Cold Spring Harbor Symp. Quant. Biol.* **28**, 329.
Jacob, F., Ryter, A., and Cuzin, F. (1966). *Proc. Roy. Soc.* **B164**, 267.
Johnson, R. A., and Schmidt, R. R. (1966). *Biochim. Biophys. Acta* **129**, 140.
Kjeldgaard, N. O. (1961). *Biochim. Biophys. Acta* **49**, 64.
Kjeldgaard, N. O., Maaløe, O., and Schaechter, M. (1958). *J. Gen. Microbiol.* **19**, 607.
Koch, A. L. (1966). *J. Gen. Microbiol.* **43**, 1.
Koch, A. L., and Schaechter, M. (1962). *J. Gen. Microbiol.* **29**, 435.
Kubitschek, H. E., Bendigkeit, H. E., and Loken, M. R. (1967). *Proc. Natl. Acad. Sci. U.S.* **57**, 1611.
Kuempel, P. L., and Pardee, A. B. (1963). *J. Cell. Comp. Physiol.* **62** (Suppl. 1), 15.
Kuempel, P. L., Masters, M., and Pardee, A. B. (1965). *Biochem. Biophys. Res. Commun.* **18**, 858.
Lark, C. (1966). *Biochim. Biophys. Acta* **119**, 517.
Lark, C., and Lark, K. G. (1964). *J. Mol. Biol.* **10**, 120.
Lark, K. G. (1963). *In* "Molecular Genetics" (J. H. Taylor, ed.), Part I. pp. 153–206. Academic Press, New York.
Lark, K. G. (1966). *Bacteriol. Rev.* **30**, 3.
Lark, K. G., and Lark, C. (1965). *J. Mol. Biol.* **13**, 105.
Lark, K. G., and Lark, C. (1966). *J. Mol. Biol.* **20**, 9.
Lark, K. G., Repko, T., and Hoffman, E. J. (1963). *Biochim. Biophys. Acta* **76**, 9.
Lark, K. G., Eberle, H., Consigli, R. A., Minocha, H. C., Chai, N., and Lark, C. (1967). *In* "Organizational Biosynthesis" (H. J. Vogel, J. O. Lampen, and V. Bryson, eds.), pp. 63–89. Academic Press, New York.
Lee-Huang, S., and Cavalieri, L. F. (1965). *Science* **148**, 1474.
Littlefield, J. W., McGovern, A. P., and Margeson, K. B. (1963). *Proc. Natl. Acad. Sci. U.S.* **49**, 102.

Maaløe, O. (1963). *J. Cell. Comp. Physiol.* **62** (Suppl. 1), 31.

Maaløe, O. (1966). *In* "Phage and the Origins of Molecular Biology" (J. Cairns, G. S. Stent, and J. D. Watson, eds.), pp. 265–272. Cold Spring Harbor Laboratory of Quantitive Biology, Cold Spring Harbor, New York.

Maaløe, O., and Hanawalt, P. C. (1961). *J. Mol. Biol.* **3**, 144.

Maaløe, O., and Kjeldgaard, N. O. (1966). "Control of Macromolecular Synthesis." Benjamin, New York.

Maaløe, O., and Rasmussen, K. V. (1963). *Colloq. Intern. Centre Natl. Rech. Sci.* **124**, 165.

Matney, T. S., and Suit, J. C. (1966). *J. Bacteriol.* **92**, 960.

Masters, M., Kuempel, P. L., and Pardee, A. B. (1964). *Biochem. Biophys. Res. Commun.* **15**, 38.

Meselson, M., and Stahl, F. W. (1958). *Proc. Natl. Acad. Sci. U.S.* **44**, 671.

Nagata, T. (1963). *Proc. Natl. Acad. Sci. U.S.* **49**, 551.

Nakada, D. (1960). *Biochim. Biophys. Acta* **44**, 241.

Pritchard, R. H. (1966). *Proc. Roy. Soc.* **B164**, 258.

Pritchard, R. H., and Lark, K. G. (1964). *J. Mol. Biol.* **9**, 288.

Ryter, A., and Jacob, F. (1966). *Ann. Inst. Pasteur* **110**, 801.

Schleif, R. (1967). *J. Mol. Biol.* **27**, 41.

Smith, D. W., and Hanawalt, P. C. (1967). *Biochim. Biophys. Acta* **149**, 519.

Stonehill, E. H., and Hutchison, D. J. (1966). *J. Bacteriol.* **92**, 136.

Swenson, P. A., and Setlow, R. B. (1966). *J. Mol. Biol.* **15**, 201.

Taylor, A. L., and Thoman, M. S. (1964). *Genetics* **50**, 659.

Taylor, J. H. (1963). *In* "Molecular Genetics" (J. H. Taylor, ed.), Part I, pp. 65–111. Academic Press, New York.

Walker, J. R., and Pardee, A. B. (1968). *J. Bacteriol.* **95**, 123.

Wolf, B., Newman, A., and Glaser, D. A. (1968). *J. Mol. Biol.* **32**, 611.

Yoshikawa, H. (1967). *Proc. Natl. Acad. Sci. U.S.* **58**, 312.

Yoshikawa, H., and Sueoka, N. (1963a) *Proc. Natl. Acad. Sci. U.S.* **49**, 559.

Yoshikawa, H., and Sueoka, N. (1963b) *Proc. Natl. Acad. Sci. U.S.* **49**, 806.

Yoshikawa, H., O'Sullivan, A., and Sueoka, N. (1964). *Proc. Natl. Acad. Sci. U.S.* **52**, 973.

CHAPTER 3

Temporal Control of Gene Expression in Bacteria

W. D. Donachie and Millicent Masters

I. INTRODUCTION

A knowledge of the overall composition of a cell and an outline of the regulation of its individual components does not by itself provide a solution to the problem of what makes a cell a growing and self-replicating unit. This problem is one of understanding the mutual control and integration of the syntheses of large numbers of different components, so that the cell remains a stable unit under diverse environmental conditions.

In order to understand the integration of subcellular processes, it is necessary to have some knowledge of the temporal sequence of events during a single cell cycle. Techniques that make possible the production of synchronously dividing populations of bacteria in a metabolically undisturbed state have made it possible to begin to produce this information. In this chapter we will review observations and experiments on the timing of events in the cell cycles of some bacteria and, wherever possible, attempt to provide explanations in terms of known control systems.

II. RATE OF INDUCED ENZYME SYNTHESIS IN SYNCHRONOUS POPULATIONS

The rate of synthesis of most enzymes is greatly affected by any change in the overall rate of growth of the cells. Consequently observations on the rate of synthesis of enzymes must be made in cells that are growing at a constant rate in a constant environment. Otherwise periodic changes in rates of enzyme production might be swamped by metabolic adjustments of the cells to a changing environment, or even be a part of this adjustment process.

In the past, procedures used to produce synchronously dividing populations of cells have often relied on pretreatments that alternately slowed down and then speeded up the growth of the cells until a synchronous pattern of division had been imposed on the whole population. In such populations, synchronized by temperature cycling or alternate periods of feeding and starvation, the effects of the initial synchronizing procedure might be expected to profoundly affect the synthesis of enzymes during at least the initial synchronous division cycles. Since in most such systems synchrony does not itself persist for more than a few divisions, such "synchronized" populations (Abbo and Pardee, 1960) can generally not be used to study events that occur periodically during enzyme synthesis in unperturbed cells.

Three techniques that produce well-synchronized bacterial populations and that reduce metabolic disturbance to a greater or lesser degree have been employed in the last two or three years. In the first and simplest of these, asynchronous populations of cells are allowed to grow in a volume of medium until stationary phase is reached. The cells are then diluted into fresh medium after a particular time in stationary phase. This procedure may be repeated a second time to improve the synchrony obtained, but often a single cycle is enough (Yanagita and Kaneko, 1961; Masters et al., 1964; Masters, 1965; Cutler and Evans, 1967; Tevethia and Mandel, 1967). The mechanism by which this method produces a synchronously dividing population is not known but may involve the termination of chromosome replication by all cells at the same chromosomal locus when they enter stationary phase (see Yoshikawa and Sueoka, 1963). It can be easily seen that this method results in a disturbed pattern of enzyme synthesis for the first few division cycles, but in *Bacillus subtilis* W23 the division synchrony may persist for up to seven successive cycles and periodic patterns of enzyme synthesis can be observed in later cycles (Masters, 1965; Masters and Pardee, 1965; Donachie, 1965; Masters and Donachie, 1966). An extremely elegant method for obtaining synchronous populations of metabolically undisturbed cells is the selective elution of newly formed daughter cells from a membrane on which the parent asynchronous population is adsorbed (Helmstetter and Cummings, 1963, 1964). The adsorbed cells are exposed to a constantly changing supply of medium and their rate of division is the same as in normal culture. Unfortunately this technique has as yet been successful with only a single strain of *E. coli* (B/r) and, unless elaborate apparatus is used (Cummings, 1965), produces only very small numbers of synchronously dividing cells. A rapid and simple alternative is provided by the selection of a homogeneous fraction of cells by sucrose gradient centrifugation (Mitchison and Vincent, 1965; 1966). The exposure of *E. coli* to centrifugation and to low concentrations of metabolically inert sucrose results in very little disturbance to their pattern of growth. If all operations are carried out at constant temperature and the sucrose gradients are made up in growth medium, disturbance of the cells is minimized and has usually disappeared by the first synchronous division. The advantages of this method are that it can be used to prepare large populations of cells at an identical stage in the cell cycle and that it works equally well with many different strains of *E. coli* (as well as other cell types). We have therefore used this method for *E. coli*.

Such synchronously dividing populations of metabolically undisturbed cells can be used to study the ability of cells to produce enzymes at different stages in their division cycle. Samples can be taken at intervals from the culture and cells induced for the synthesis of a particular enzyme. The rate at which this enzyme is synthesized in the sample can be measured and taken as a measure of the "inducibility" of the enzyme in cells at that particular stage in the cell cycle (Kuempel *et al.*, 1965).

Experiments of this kind have shown clearly that cell populations can be induced to synthesize any of a large number of enzymes at all stages of growth and division (Kuempel *et al.*, 1965; Masters and Pardee, 1965; Masters and Donachie, 1966; Cummings, 1965; Ferretti and Gray, 1967, 1968; Nishi and Horiuchi, 1966). This behavior at the population level could reflect the behavior of the individual cells, or alternatively might mean no more than asynchrony among these cells for a periodic and limited ability to produce the enzyme. However, this second possibility has been effectively eliminated by an analysis of a change in rate of enzyme production during synchronous growth and division. Thus the rate of enzyme synthesis immediately following the addition of inducer, changes discontinuously and periodically. This rate of induced synthesis (inducibility) is generally constant for a period of time equal in duration to the cell division cycle and then rises abruptly to a new value. Thus the inducibility of synchronous bacterial populations rises discontinuously in a series of steps.

A. *Bacillus subtilis*

The change in rate of induced synthesis of the enzyme sucrase has been measured during synchronous growth and division of *B. subtilis* W23 (Masters, 1965; Masters and Pardee, 1965). The stepwise increase of inducibility obtained is shown in Fig. 1.

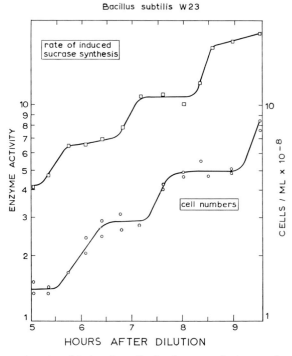

FIG. 1. Change in rate of induced synthesis of sucrase during synchronous growth of *Bacillus subtilis* W23. Aliquots were removed at intervals from the culture and induced with sucrose. The initial rates of production of enzyme in these samples are shown (Masters, 1965).

B. *Escherichia coli*

The kinetics of inducibility during synchronous growth have now been studied in several strains of *E. coli* and, with a single exception, the same general pattern has been seen in all of them. Kuempel *et al.* (1965) found the rate of derepressed synthesis of alkaline phosphatase and aspartate transcarbamylase to increase discontinuously once per cell cycle

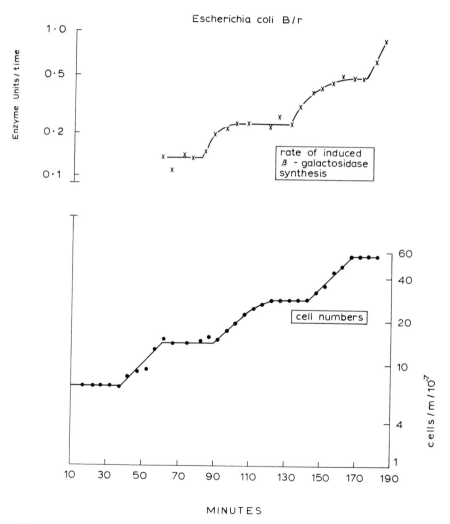

Fig. 2. Change in rate of induced synthesis of β-galactosidase during synchronous growth of *Escherichia coli* B/r. The procedure was as in Fig. 1. Thiomethyl galactoside was used as an inducer.

TABLE I

The Induction or Derepression of Enzymes in Synchronous Populations

Reference	Enzyme	Bacterial strain		Change in inducibility
1. Kuempel et al. (1965), Kuempel (1965)	β-Galactosidase	E. coli K12	HfrCS101	Stepwise
	Aspartate transcarbamylase	E. coli K12	HfrCS101	Stepwise
	Aspartate transcarbamylase	E. coli K12	HfrH	Stepwise
	Aspartate transcarbamylase	E. coli K12	HfrAB312	Stepwise
	Alkaline phosphatase	E. coli K12	HfrCS101	Stepwise
	Tryptophanase	E. coli K12	HfrCS101	Stepwise
2. Donachie and Masters (1966 and in preparation)	β-Galactosidase	E. coli K12	F⁻	Stepwise
	β-Galactosidase	E. coli K12	F'lac⁺/lac⁺	Stepwise (double)
	β-Galactosidase	E. coli B/r		Stepwise
	β-Galactosidase	E. coli 15		Stepwise
	Tryptophanase	E. coli K12	F⁻	Stepwise
	Tryptophanase	E. coli B/r		Stepwise
	d-Serine deaminase	E. coli B/r		Stepwise
	d-Serine deaminase	E. coli 15		Stepwise
	Aspartate transcarbamylase	E. coli B/r		Stepwise
	Dihydroorotase	E. coli B/r		Stepwise
	OMP pyrophosphorylase	E. coli B/r		Stepwise
	Dihydroorotic dehydrogenase	E. coli B/r		Stepwise
3. Nishi and Horiuchi (1966)	β-Galactosidase	E. coli K12	HfrH	Stepwise
	β-Galactosidase	E. coli K12	F⁻(E64)	Exponential
	β-Galactosidase	E. coli K12	(E64)F'lac⁺/del. lac	Stepwise
	d-Serine deaminase	E. coli K12	HfrH	Stepwise
	d-Serine deaminase	E. coli K12	F⁻(E64)	Exponential
	d-Serine deaminase	E. coli K12	(E64)F'lac⁺/del. lac	Exponential
4. Masters and Pardee (1965), Masters (1965)	Sucrase	B. subtilis	W23	Stepwise

in the K12 strain Hfr CS101. Donachie and Masters (1966) found the same to be true of the induced synthesis of β-galactosidase, tryptophanase, and D-serine deaminase in F⁻ strains of *E. coli* K12 (58-161), B/r, and 15. Later work (Donachie *et al.* in preparation) has extended these observations to other enzymes (aspartate transcarbamylase, orotidine monophosphate pyrophosphorylase, dihydroorotase and dihydroorotic dehydrogenase) in *E. coli* B/r (F⁻). The change in rate of induced synthesis of β-galactosidase during synchronous growth of *E. coli* B/r is shown in Fig. 2.

An exception to this general rule, by which the rate of induced synthesis of an enzyme rises suddenly once in each cell cycle, has been reported by Nishi and Horiuchi (1966). In an F⁻ strain of *E. coli* K12 ("E64") the inducibility of both β-galactosidase and D-serine deaminase rose continuously and exponentially during synchronous division of the cells. In contrast, the same workers found the inducibility of both enzymes to increase discontinuously once per cell cycle in *E. coli* K12 HfrH. This exception will be referred to again in the following section, where the interpretation of patterns of enzyme inducibility will be discussed.

A list of the enzymes and bacterial strains which have been studied in this way is given in Table I. In every case (except that of E64, above) the same pattern of increase in rate from one constant value to the next has been found.

III. CHANGE IN RATE OF INDUCED ENZYME SYNTHESIS AND SEQUENTIAL GENE REPLICATION

A. CHROMOSOME REPLICATION

The simplest interpretation of the observed pattern of enzyme inducibility is that the rate of induced (or derepressed) synthesis of an enzyme is proportional to the number of copies of its corresponding structural gene, and that the number of these genes doubles at a particular time in the cell cycle. This would explain the constancy in enzyme inducibility during a whole cycle of cell growth and division (since this rate of synthesis would be unaffected by cell mass) and its periodic abrupt increase at the time of synchronous gene replication. We will now describe a few experiments that have been made to test this hypothesis.

In synchronous cultures of *B. subtilis* W23 the rate of induced synthesis of sucrase and the number of sucrase structural genes (as measured by transformation) increase periodically in parallel with one another (Masters and Pardee, 1965). Thus both events are periodic and in phase.

If DNA replication is specifically inhibited, cells will continue to grow in length and mass for more than a generation time. Thus one can distinguish between the effects of general growth and synthesis (of RNA, protein and other cellular machinery) from those which are directly dependent on DNA replication. According to the gene-dosage hypothesis, which is proposed to explain the periodic changes in inducibility of enzymes, inhibition of DNA replication should prevent any increase in rate of induced synthesis of enzymes. Experiments in which DNA synthesis has been specifically inhibited by thymine deprivation have shown that the inducibility of at least two enzymes (β-galactosidase and D-serine deaminase) remains constant during growth without DNA replication (Donachie and Masters, 1966 and unpublished). One such experiment is shown in Fig. 3. Evidently the regular stepwise increase in inducibility

Escherichia coli 15T −

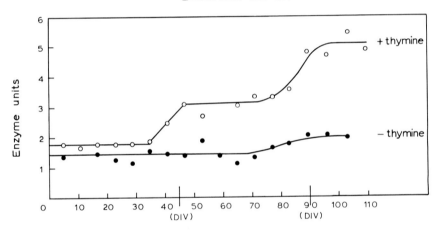

Minutes after washing

Fig. 3. Change in rate of induced synthesis of D-serine deaminase during synchronous growth of *Escherichia coli* 15T⁻. At 0 minutes the cells were washed free of thymine and resuspended in medium with or without thymine. Aliquots were taken from both cultures at intervals, and the ability to form D-serine deaminase was measured by induction with D-serine.

of enzymes is completely dependent on continuing DNA replication.

Such evidence as is at present available therefore indicates that the gene-dosage hypothesis for inducibility of enzymes is correct. If all genes replicated simultaneously, then the inducibility of all enzymes should increase at the same time. If, however, the replication of the genome is sequential, starting at the same locus and proceeding in the same direction in every cell, the times of the steps in inducibility should be different for different enzymes. Moreover the sequence of enzyme steps should reflect the sequence of replication of the genes. The work of Sueoka and his collaborators has now amply demonstrated the sequential replication of the *B. subtilis* genome from a fixed origin, but there is still debate about the mechanism of replication of the *E. coli* chromosome (see Berg and Caro, 1967, 1968; Abe and Tomizawa, in press). Cairns has very beautifully shown that replication of individual *coli* chromosomes is sequential (1963), but not whether there is a fixed origin in all cells or whether the point of origin changes from time to time in successive rounds of replication. It can be inferred from the density-label experiments of Meselson and Stahl (1958) and Lark *et al.* (1963) that there is, at the least, a strong tendency for the same piece of DNA to be replicated first in successive rounds of replication. However, although it is strongly suggested by these experiments, it cannot be formally concluded that there is a unique origin of replication at the same locus in every cell. Still less is it possible to locate any such origin on the linkage map.

Since the inducibility of an enzyme increases at the same time in all members of a synchronously dividing population, this implies that the replication of the corresponding gene is also synchronous. In cells of *E. coli* B/r growing under the conditions used for testing inducibility, the synthesis of DNA takes place continuously. Figure 4 shows that the *rate* of this synthesis is constant for a period equal to one division cycle, after which it suddenly doubles. Since it is known that chromosome replication is sequential at a constant rate in this strain (Cairns, 1963) these kinetics are consistent with successive rounds of sequential replication of the DNA with all cells beginning their rounds at the same stage in the cell cycle (see Clark and Maaløe, 1967; Helmstetter, 1967; Cooper and Helmstetter, 1968). Such a process could give rise to the observed synchronous increases in inducibility only if the replication of the genome began at the same chromosomal locus in every cell and proceeded at the same rate in the same direction around the chromosome. Thus the stepwise increases in inducibility of enzymes are themselves good evidence that all the cells of a particular bacterial strain begin replication at a fixed chromosomal location. Thus both Hfr (Kuempel *et al.*, 1965) and F⁻ strains (Donachie and Masters, 1966) appear to have fixed origins of replication.

The one exceptional report (Nishi and Horiuchi, 1966) referred to above, in which synchronously dividing F⁻ cells did not show a stepwise increase in inducibility of enzymes, was interpreted by its authors as indicating that there was no fixed chromosomal origin of replication in this strain. Since other F⁻ strains do not behave in this way, alternative explanations should also be looked for. One such explanation arises from the observation (unpublished experiments in this laboratory) that synchrony of cell division is no guarantee of synchrony of DNA replication. Thus in many cases synchronously dividing populations of K12 strains have been obtained in which the rate of DNA synthesis (as measured by rate of incorporation of labeled thymidine) increased exponentially. In such cases the procedure used to select the synchronously dividing population seems to have upset the normal coupling between

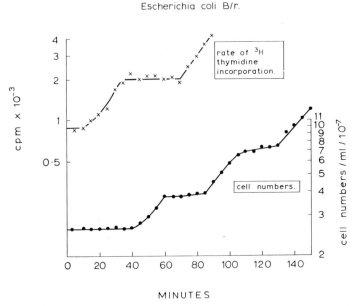

FIG. 4. Rate of incorporation of thymidine-³H into DNA during synchronous growth of *Escherichia coli* B/r. The rate of incorporation was measured by pulsing label for 4 minutes in samples taken from the culture at intervals, and measuring counts in the TCA-precipitable fraction.

the cell cycle and the DNA replication cycle. Against this explanation must be set Nishi and Horiuchi's observation of synchrony of episome replication (see below) in the same cells which showed asynchrony of chromosome replication. Either the timing of episome replication is less easily disturbed, or E64 is a genuine case of a bacterial strain without a fixed replicative origin. A decision must wait on the simultaneous demonstration of synchrony of rounds of DNA replication in this strain together with a measure of the increase in inducibility.

In theory a detailed analysis of the sequence of changes in the inducibility of several enzymes during a single cell division cycle could give information about both the order of chromosome replication and the location of the origin. An attempt has been made to do this for *E. coli* B/r (Donachie *et al.*, in preparation) by timing the periods of increase in inducibility of six enzymes whose structural genes have been mapped on the chromosome. For technical reasons it has not been possible to measure all enzymes during a single cell cycle and therefore the timing of inducibility steps has been measured relative to the time of cell division. Such times were estimated for, usually, seven separate cell cycles and the average times for the different enzymes collected together to give an order of events. However, the timing of these events is not exactly the same, *relative to division*, in every experiment and an unambiguous order has not be obtained. The suggested order of events, correlated with the genetic map, is given in Fig. 5.

The reason for the variation in the timing of inducibility changes relative to division is probably that the timing of initiation of rounds of DNA synthesis shows a similar variation relative to division (see Fig. 4). An unambiguous order should therefore be obtainable by relating steps in inducibility to the DNA replication cycle rather than to the somewhat loosely coupled cell division cycle. One such experiment is illustrated in Fig. 6, in which the inducibility of β-galactosidase and the rate of DNA synthesis were both measured in the same experiment. This experiment suggests that the replicative origin is located almost directly opposite the *lac* locus on the circular linkage map. This is consistent with its very approximate location derived from the data in Fig. 5. The accumulated data on timing of inducibility changes make it probable that the direction of replication of the *E. coli* B/r chromosome is "clockwise" (as the map is usually written).

Therefore analysis of enzyme inducibility during the cell cycle makes it virtually certain that many strains of *E. coli* replicate their chromosomes sequentially from a fixed point of origin, but can only approximately indicate the chromosomal locus of this origin and, somewhat more certainly, the order of replication around the chromosome. This question ought to be better answered by a more direct method (e.g., by transduction, Berg and Caro, 1967, 1968; Abe and Tomizawa, 1967).

B. REPLICATION OF EPISOMES

One other interesting conclusion can be drawn from a study of enzyme inducibility. This is that F' episomes, like chromosomes, have a fixed period in the cell cycle during which they replicate. Thus the infection of an F⁻ *lac*⁺ strain with an F *lac*⁺ episome results in a strain in which the inducibility of β-galactosidase increases at two times during the

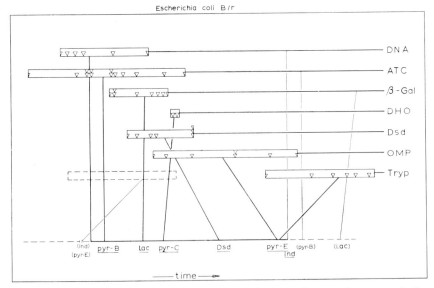

FIG. 5. Correlation between the order of markers on the genetic map and the relative order of events during the cell cycle. The triangles represent the time of an experimentally determined change in rate of induced synthesis of enzymes (or in rate of synthesis of DNA) determined relative to cell division. These have been collected together into one figure by normalizing the cell cycle times for every experiment. The horizontal boxes enclose separate estimates of this time of change in inducibility for particular enzymes. Lines have been drawn from the mean time of an event to the corresponding genetic locus on the linkage map at the bottom. The length of the linkage map has been made equal to the time of a cell division cycle. Therefore a perfect correlation would give vertical lines from the mean time of change in rate of induced synthesis to the corresponding genetic locus. The nature of the timed event is identified on the right of the figure. Dashed lines indicate the continuous succession of cell cycles. (ATC, aspartate transcarbamylase; β-Gal, β-galactosidase; DHO, dihydroorotase; Dsd, D-serine deaminase; OMP, orotidine monophosphate pyrophosphorylase; Tryp, tryptophanase.)

cell cycle, instead of at a single time as in the parent F⁻ cells. The inducibility of enzymes whose locus is not carried on the episome is unaffected (Donachie and Masters, 1966). Similarly a strain having a chromosomal deletion of the *lac* locus and carrying F′ *lac⁺* showed a single period in each cell cycle when the inducibility of β-galactosidase increased (Nishi and Horiuchi, 1966).

Since cells carrying the *lac⁺* gene on both the chromosome and on an episome show two steps in inducibility which are approximately equal,

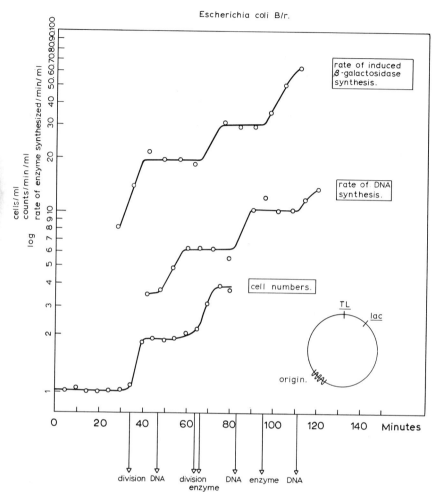

Fig. 6. Change in cell numbers, in rate of DNA synthesis, and in rate of induced β-galactosidase synthesis in a synchronous population of *Escherichia coli* B/r. The circle represents the linkage map with the suggested location of the origin of chromosome replication. (*TL*, threonine and leucine loci; *lac*, lactose locus.) The arrows at the bottom indicate the times of change in the indicated events.

it seems likely that there is only a single F′ factor per chromosome in the population. This supports the argument derived from the quantity of enzyme produced by such partial diploids (Jacob *et al.*, 1963). Both the timing and the number of rounds of episomal replication are therefore regulated.

IV. SYNTHESIS OF ENZYMES IN SYNCHRONOUS POPULATIONS

In the previous two sections we have been considering the response of cells to external stimuli. We have seen that genes can express their

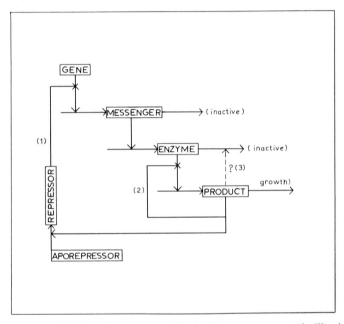

FIG. 7. The regulation of enzyme synthesis. Arrows represent facilitation of a reaction and crosses the inhibition of a reaction. Both messenger RNA and enzyme are represented as unstable, and products are presumed to be removed at a rate depending on the rate of cell growth. (1) Repression of messenger transcription; (2) end-product inhibition of enzyme activity; (3) represents the possibility that enzyme may be inactivated by interaction with end product [see H. Holzer, D. Mecke, K. Wulff, K. Liess, and L. Heilmeyer (1967), *Advan. Enzyme Regulation* **5**, 211].

activities at all times during the cell cycle in response to specific induction or derepression. In this section we will examine the synthesis of enzymes in undisturbed cultures of growing cells, in order to see how the expression of various genes is related to cell growth and division.

The way in which enzyme synthesis is regulated by end products is illustrated in Fig. 7. The degree to which the synthesis of an enzyme is repressed will depend on several factors, including the overall growth rate of the cell and the particular kinetic constants of end-product inhibition and repression. In a cell growing in a medium of a particular composition, different enzymes will be produced at a wide variety of rates. The synthesis of some enzymes may be completely repressed by the presence of corepressor in the medium, while others may be produced at nearly maximal rates in response to the presence of an inducer. In a minimal salts medium however, most of the enzymes concerned with the biosyntheses of amino acids, pyrimidines, purines, etc., will be synthesized at intermediate rates. Thus the rates of synthesis of most enzymes will be higher than their minimum, fully repressed, rates but lower than their maximum, fully derepressed, rates. The range of rates between the minimum and the maximum for a given enzyme in cells growing at a constant growth rate can be very large in bacteria (of the order of one thousandfold, see Pardee and Beckwith, 1963). This range of rates is a continuous one, but for our present purposes we will divide the manner of synthesis of enzymes into three classes, which roughly correspond to minimal, maximal, and intermediate rates of synthesis. These three classes occur in response to different states of repression, and this will be seen to have a profound effect on the manner of production of enzymes during the cell cycle.

These three classes of enzyme synthesis are "basal," "derepressed," and "autogenous." They will be discussed in turn.

A. BASAL SYNTHESIS

The basal rate of synthesis of an enzyme is its minimum rate of production at a particular growth rate. Such a condition is approached when an excess quantity of repressor is present, such that an increase in its concentration will not appreciably lower the rate of enzyme synthesis. The basal rate of synthesis of an enzyme is never quite zero although it may be as low as a few molecules per cell per generation (Hogness, 1959). In actual experiments, basal rates of synthesis are found for inducible enzymes in cells growing in the absence of inducer, and for repressible enzymes in cells growing in the presence of an excess of end products. These rates are often too low to be easily measured, but in a few cases assays sensitive enough to measure basal synthesis in small synchronous populations are available.

The basal synthesis of alkaline phosphatase has been followed in synchronous populations of *B. subtilis* (Donachie, 1965) and *E. coli* (Kuempel *et al.*, 1965) which were growing in high concentrations of inorganic phosphate. The production of the inducible enzymes sucrase (in *B. subtilis*, Masters and Donachie, 1966) and β-galactosidase (in *E. coli*, Kuempel, 1965) has been measured in cultures growing in the

absence of inducers. Ferretti and Gray (1968) have measured the
synthesis of ornithine transcarbamylase in the presence of arginine in
Rhodopseudomonas spheroides.

In each case it has been found that a synchronously dividing population of cells produces a given enzyme *continuously* under basal conditions.

Bacillus subtilis W 23

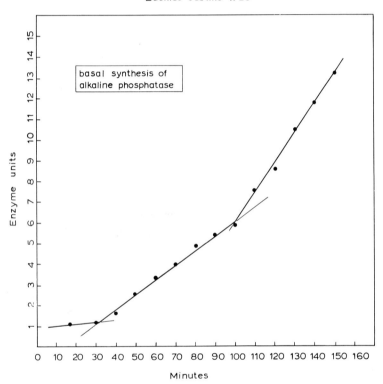

Fig. 8. The synthesis of alkaline phosphatase in a synchronous population of
Bacillus subtilis W23 (generation time 70 minutes) growing in minimal medium
containing sufficient inorganic phosphate to maximally repress the synthesis of
the enzyme.

However, the rate of basal synthesis of these enzymes is so low that
probably no individual cell is synthesizing enzyme continuously. Therefore at any instant of time the population of cells could probably be
divided into two classes, those actively engaged in synthesizing the
enzyme and those which are not. The probability of synthesizing a molecule of the enzyme during any given interval of time, however, is likely

to be the same for all cells. Thus not only can the activity of any gene be "turned on" at any stage in the cell cycle by specific induction or derepression (Sections II and III above), but even under conditions of constant repressor concentration the probability of transcription of a given gene seems to be the same at all times.

Although it is difficult to distinguish experimentally between a continuous exponential curve and a series of straight lines with doublings in slope, in two cases (Donachie, 1965; Kuempel et al., 1965) where assays have been accurate enough, it has been claimed that the rate of basal enzyme production can be best fitted by a series of straight lines, the slope of which increase at the same point in each successive cell cycle. Figure 8 shows alkaline phosphatase synthesis in synchronous B. subtilis. These are the kinetics of synthesis that would be expected if (as we have concluded in previous sections) gene replication is synchronous in synchronously dividing cells. The average rate of enzyme synthesis under basal conditions would then be proportional to gene dosage, as is the rate of synthesis following induction.

B. DEREPRESSED SYNTHESIS

The fully derepressed rate of synthesis of an enzyme is the maximum rate at which the enzyme can be produced under given growth conditions. Such a rate of synthesis can be approached in cells growing in the presence of inducer (although not necessarily; see Section IV,c) and in constitutive mutants. The synthesis of β-galactosidase has been measured during synchronous growth both of a constitutive mutant (an i⁻ strain; Donachie, unpublished) and of wild-type E. coli in the presence of lactose or a gratuitous inducer (Abbo and Pardee, 1960; Cummings, 1965; D. Martin and W. D. Donachie, unpublished). As with basal synthesis the production of enzyme is continuous throughout the cell cycle.

Since fully induced rates of synthesis of β-galactosidase are easy to measure accurately, it is possible to be more certain of the kinetics of synthesis under these conditions. Synthesis takes place at a constant rate with an increase in rate once in each cell cycle. Figure 9 shows the synthesis of the enzyme in E. coli growing on lactose as sole carbon source (D. Martin and W. D. Donachie, unpublished). The data for a constitutive mutant growing in the absence of inducer are similar.

The simplest interpretation of the kinetics of enzyme production under basal or derepressed conditions is that the activity of bacterial genes is regulated by repressor levels and that there is no intrinsic limitation to the time of their action during the cell cycle. These results also support the idea that the rate of enzyme synthesis is proportional to gene dosage under conditions of constant repression (see Donachie, 1964).

Escherichia coli B/r

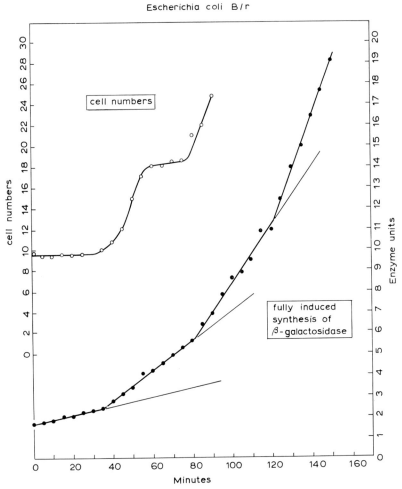

FIG. 9. The synthesis of β-galactosidase in a synchronous population of *Escherichia coli* B/r growing in minimal medium with lactose as sole carbon source.

C. AUTOGENOUS SYNTHESIS

The synthesis of many enzymes takes place under conditions in which repression is neither maximal nor minimal. The rate at which the cell produces the enzyme is proportional to the rate at which the cell utilizes the end products of the enzyme's action. This regulation is achieved by means of the twin control systems of end-product repression of enzyme synthesis and end-product inhibition of enzyme activity. Figure 7 provides a simplified picture of this regulation. Since the degree of repression is in a part a reflection of the enzyme's own catalytic activity,

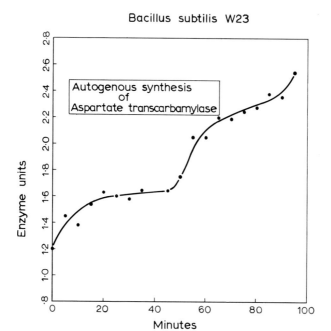

FIG. 10. The synthesis of aspartate transcarbamylase in a synchronous population of *Bacillus subtilis* W23 growing in minimal medium.

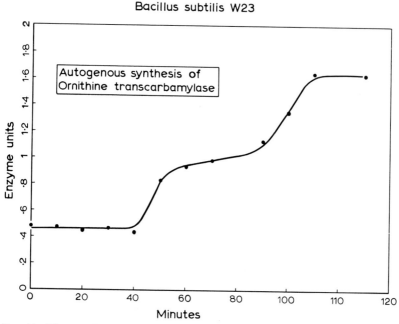

FIG. 11. The synthesis of ornithine transcarbamylase in a synchronous population of *Bacillus subtilis* W23 growing in minimal medium.

TABLE II

Synthesis of Enzymes in Synchronous Bacterial Populations

Enzyme	Organism	Type of synthesis	Conditions of synthesis	Reference
1. Alkaline phosphatase	B. subtilis	Continuous	Basal	Donachie (1965)
	E. coli K12	Continuous	Basal	Kuempel et al. (1965)
	B. cereus	Periodic	Germinating spores	Kobayashi et al. (1965)
	R. spheroides	Periodic	?	Ferretti and Gray (1968)
2. β-Galactosidase	E. coli K12	Continuous	Basal	Kuempel et al. (1965)
	E. coli K12	Continuous	Fully induced	Abbo and Pardee (1960)
	E. coli B/r	Continuous	Fully induced	Cummings (1965)
	E. coli K12	Continuous	Constitutive	Donachie (unpublished)
3. Sucrase[a]	B. subtilis	Continuous	Basal	Masters and Donachie (1966)
4. Aspartate transcarbamylase[a]	B. subtilis	Periodic	Autogenous	Masters and Donachie (1966)
	B. subtilis	Periodic	Autogenous	Masters et al. (1964); Donachie (1965); Masters and Donachie (1966)
5. Ornithine transcarbamylase	E. coli K12	Periodic	Autogenous	Kuempel et al. (1965)
	B. subtilis	Periodic	Autogenous	Donachie (1965); Kuempel et al. (1965)
	E. coli K12	Periodic	Autogenous	Kuempel et al. (1965)
	R. spheroides	Periodic	Autogenous	Ferretti and Gray (1968)
6. Dehydroquinase	B. subtilis	Periodic	Autogenous	Masters and Pardee (1966)
7. Histidase	B. subtilis	Periodic	Autogenous	Masters and Pardee (1966)
8. α-Aminolevulinic acid synthetase[a]	R. spheroides	Periodic	?	Ferretti and Gray (1968)
9. α-Aminolevulinic acid dehydrase	R. spheroides	Periodic	?	Ferretti and Gray (1968)
10. Succinyl-CoA thiokinase	R. spheroides	Periodic	?	Ferretti and Gray (1968)
11. Glycylglycine dipeptidase[a]	E. coli K12	Periodic	?	Nishi and Hirose (1966)
12. Protease[a]	E. coli ML308	Periodic	?	Kogoma and Nishi (1965)
13. Leucine aminopeptidase[a]	E. coli ML308	Periodic		Kogoma and Nishi (1965)
14. α-Glucosidase	B. cereus	Periodic	Germinating spores	Kobayashi et al. (1965); Steinberg et al. (1965)
15. L-Alanine dehydrogenase	B. cereus	Periodic	Germinating spores	Kobayashi et al. (1965)

[a] Unstable in growing cells.

regulation of enzyme synthesis under these conditions is to some degree "self-regulation." Synthesis under conditions where the enzyme determines its own rate of synthesis in this way has therefore been called "autogenous" (Masters and Pardee, 1965).

The formation of enzymes under autogenous conditions is strikingly different from that under basal or derepressed conditions. Figures 10 and 11 give examples of the autogenous synthesis of two biosynthetic enzymes in synchronously dividing cultures of *B. subtilis* (Donachie, 1965 and unpublished). The rate of enzyme production is not constant but periods of rapid synthesis alternate with periods of little net synthesis. Periodic production has been reported for several bacterial enzymes. Table II lists these, together with enzymes whose formation is continuous. Whenever periodic synthesis of enzymes has been reported, the conditions governing the level of enzyme have either been known to be autogenous or were unknown. There is no certain report of periodic enzyme synthesis under basal or derepressed conditions.

The timing of periods of enzyme production is such that there is one period per cell division cycle. Moreover the formation of a given enzyme takes place, on the average, at a particular stage in the cell cycle. These periods of synthesis, which are different for different enzymes, can be arranged in order, to give a temporal sequence of events in the cell cycle of the bacterium (Masters and Pardee, 1965; Ferretti and Gray, 1968).

V. METABOLIC OSCILLATIONS AND THE TEMPORAL CONTROL OF ENZYME SYNTHESIS

In the preceding section we saw that periodic enzyme synthesis takes place only when the negative feedback control system is free to operate. This has led to the suggestion that concentration of end product and rate of synthesis of enzyme oscillate as a consequence of their mutual interdependence under autogenous conditions (Pardee, 1966; Kuempel *et al.*, 1965; Goodwin, 1966; Masters and Donachie, 1966). Enzyme and end product form a system in which the concentration of one determines the rate of synthesis of the other. Since there must be a time lag between a change in the concentration of one component and its effect on the other, it is possible that oscillating changes in the concentrations of both components would ensue (Goodwin, 1963). Periodic synthesis of enzymes would be the observed result.

Mathematical models of such regulatory systems have been made by Goodwin (1963, 1966) and Pardee (1966). Both these workers have studied the behavior of their systems using computers and both have found oscillatory solutions by a suitable choice of numerical constants.

The model proposed by Pardee (1966) assumes that

$$V = V_0 e^{kt}$$

$$\frac{dR}{dt} = \beta E - \gamma V$$

$$\frac{dE}{dt} = \frac{\alpha G}{k_1 + R/V}$$

where V is cell volume, R is number of repressor molecules, G is number of genes, E is number of enzyme molecules, k, k_1, α, β, γ are constants, and $V_0 = V$ at time $t = 0$. This model is given again here to point out that the assumptions contained in it are in accord with the main accepted experimental conclusions concerning the system of enzyme regulation in bacteria. Goodwin's models are formally rather similar (1963 *et seq.*), and the fact that all these sets of equations can have oscillatory solutions is therefore of importance in considering how real bacterial systems may be expected to behave.

The most obvious prediction of these models is that the concentration of repressor in the cell should fluctuate during the cell cycle. However, this is a difficult prediction to test. Pools of end products can be measured in synchronous populations, and it could be seen whether their concentration varied or not. Despite this, since the concentration of biosynthetic enzymes *is* known to oscillate, any variations in concentration of end products could merely be the *result* of the periodic change in catalytic activity and not necessarily a part of the mechanism generating the oscillations in rate of enzyme production. Similarly, if variations in pool size of end product were not detected it would not be possible to exclude the model. This would be for two reasons: that fluctuations in pool size below the level of experimental detection might nevertheless be sufficient to have large effects on the rate of enzyme synthesis (which is the case during derepression of pyrimidine enzymes in *Neurospora crassa*, Donachie, 1964 and unpublished), and second the real possibility that one was not measuring the actual corepressor but a precursor or derivative. Until more is known about the nature of the corepressor and about the kinetics of repression in response to changing concentrations of this corepressor, the measurement of pool sizes of end products offers little prospect of deciding among alternative models of periodic enzyme synthesis.

Experiments can be devised which test the hypothesis that the observed periodic changes in rate of enzyme production are the consequence of oscillations in the feedback control system. A few of these tests have now been carried out with synchronously dividing bacteria.

1. For example, if the oscillations are self-generating then they should

also be to some extent self-maintaining. Therefore, unlike the periodic increases in the inducibility of enzymes (Sections II and III), these cyclic changes in rate of enzyme synthesis should continue in the absence of gene replication. This prediction has been tested (Masters and Donachie, 1966) by inhibiting DNA synthesis in *B. subtilis* and following the formation of ornithine transcarbamylase thereafter. Figure 12 shows

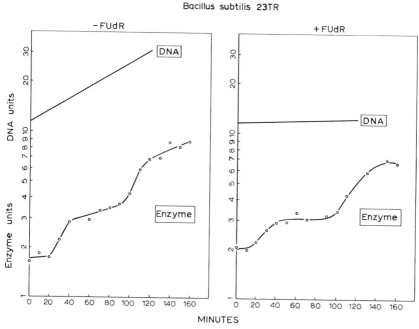

FIG. 12. The synthesis of ornithine transcarbamylase in a synchronous population of *Bacillus subtilis* 23TR. The culture was divided into two at 0 minutes and either uracil alone or uracil + 5-fluorodeoxyuridine (FUdR) was added. DNA synthesis is almost completely inhibited in the presence of FUdR, but the periodic synthesis of the enzyme is not affected.

the results of such an experiment. It is clear that cyclic enzyme synthesis is not dependent on continuing DNA replication. Steinberg *et al.*(1965) have also reported periodic enzyme synthesis in germinating spores of *B. cereus*, during the period when there is no net DNA synthesis.

2. A central prediction of models of oscillating feedback control is that the rate of enzyme synthesis will become constant under conditions of constant repression or derepression. The observations on enzyme synthesis under basal, derepressed, and autogenous conditions reported in the previous section support the hypothesis. However, in only two cases

has it been possible to measure the *same* enzyme under both autogenous and basal conditions. Neither example is perfectly convincing, and it is essential that more observations of this kind be made. The synthesis of the inducible enzyme sucrase in *B. subtilis* is continuous under basal conditions but periodic in cells growing on sucrose as sole carbon source (Masters, 1965; Masters and Donachie, 1965, 1966). The feedback control could be from the catabolite repressors produced as the result of the catabolism of sucrose onto the synthesis of the induced enzyme itself. Whether this explanation is correct or not, the synthesis of this enzyme can clearly be periodic under some conditions and continuous under others. A second example has been reported by Ferretti and Gray (1968), who measured the synthesis of ornithine transcarbamylase in synchronously dividing populations of *Rhodopseudomonas spheroides*. In minimal medium the production of this enzyme was greatly disturbed in the first division following the synchronization treatment (dilution from stationary phase) but showed stepwise synthesis in later division cycles. In the presence of arginine, however, the rate of enzyme synthesis was greatly lowered and its production was continuous.

3. If periodic enzyme synthesis reflects oscillations in pools of end products, then it should be simple to upset the timing of these periods of synthesis by artificially altering the concentration of end products. This has been adequately shown in the experiments on enzyme induction and derepression discussed in previous sections. Inhibition of end-product formation, as for example, by the inhibition of pyrimidine synthesis by 6-azauracil (Kuempel *et al.*, 1965) results in immediate production of the derepressed enzymes at all stages in the cell cycle. Similarly the synthesis of enzymes concerned in the production of bacteriochlorophyll can be repressed by a high light intensity or by aerobic conditions in *Rhodopseudomonas* (Lascelles, 1959, 1960), but their synthesis can be induced at any time in the cell cycle by lowering the light intensity or reverting to anaerobic conditions (Ferretti and Gray, 1967). Under undisturbed conditions of anaerobic growth at low light intensities, the synthesis of these enzymes is nevertheless periodic.

4. The imposition of a period of repression should erase the fluctuations in end-product pool, but these would be expected to build up once more after the removal of external repression. The imposition of a short period of repression could then be expected to alter the timing of enzyme synthesis relative to other events in the cell cycle. This prediction has been borne out in experiments with *B. subtilis* (Masters and Donachie, 1966) in which the synthesis of aspartate transcarbamylase can be repressed by the addition of uracil. Cyclic synthesis of the enzyme stops on addition of uracil and recommences when uracil is removed. The

timing of the periods of enzyme synthesis is however altered following a short period of growth with uracil and remains out of phase with the periodic enzyme synthesis in control cells for several cycles. Figure 13 illustrates such an experiment. The effect of uracil is specific for enzymes of the pyrimidine pathway and does not alter the timing of periods of synthesis of enzymes in other biosynthetic pathways.

We may conclude that there is no fixed time in the bacterial cell cycle when the synthesis of an enzyme must occur and that periodic enzyme synthesis reflects stable oscillations in the regulation of enzyme synthesis through negative feedback.

There are some reports of enzyme synthesis in synchronous cultures which apparently do not fit this model. Steinberg et al. (1965) and Kobayashi et al. (1965) have described the periodic formation of alkaline phosphatase (and other enzymes) during the synchronous division cycles following spore germination in Bacillus cereus. Similarly, Ferretti and Gray (1968) have described periodic synthesis of this enzyme in Rhodopseudomonas spheroides in synchronous cultures after reinoculation from stationary phase. In both cases cells were growing in the presence of inorganic phosphate, which would be expected to repress the synthesis of this enzyme to its basal rate. One may suggest several possible explanations for these observations, but it remains necessary to determine experimentally the causes of the fluctuations in rate of enzyme synthesis. Masters and Donachie (1966) showed that, following transfer of B. subtilis from a low phosphate medium to one lacking phosphate, the synthesis of alkaline phosphatase was at first periodic (rather than continuous as would be expected for synthesis following derepression). At all stages of the cell cycle an initial short period of rapid synthesis was followed by an approximately equal period without synthesis. After this, rapid synthesis of the enzyme resumed. It was suggested, on the basis of these kinetics, that the observed periodic synthesis resulted from the production of inorganic phosphate from an intracellular store of organic phosphate, as a result of the catalytic activity of the enzyme formed initially. Therefore the periodic synthesis of alkaline phosphatase under these conditions would in fact be autogenous. Measurements of intracellular concentrations of inorganic phosphate after the removal of all external supplies of phosphate have indeed demonstrated a pool of inorganic phosphate which fluctuates out of phase with the production of alkaline phosphatase (M. Masters and W. D. Donachie, unpublished). The hypothesis that alkaline phosphatase synthesis is in fact autogenous for a period after the removal of phosphate from the medium is thus a tenable one in this instance. Whether germinating spores utilize internal phosphate to some extent remains to be

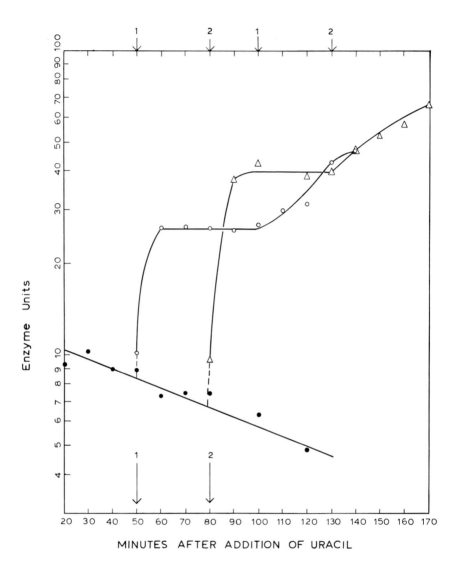

FIG. 13. The synthesis of aspartate transcarbamylase in a synchronous population of *Bacillus subtilis* W23. At 0 minutes uracil was added to repress the synthesis of the enzyme. Because it is unstable, the total quantity of enzyme then decreases (filled circles). At 50 minutes (1) and 80 minutes (2), aliquots were washed free of uracil and the course of enzyme synthesis followed. Enzyme synthesis was stepwise and the times of the first and second bursts of enzyme synthesis are marked at the top of the graph. "1" refers to the first derepressed sample and "2" to the second. The generation time of the cells was 60 minutes.

determined. Periodic synthesis of alkaline phosphatase was first described in a paper by Torriani (1960), who studied the formation of the enzyme during the transition from growth on inorganic phosphate to growth on glycerol phosphate. Torriani was able to demonstrate that the concentration of inorganic phosphate in the medium fluctuated during this transition and that the rate of synthesis of alkaline phosphatase also fluctuated, out of phase with the concentration of inorganic phosphate. It is particularly interesting that this "copy book" example of autogenous enzyme synthesis took place in an asynchronous population of cells (*E. coli*). The periodic synthesis of the enzyme in these cells was in response to fluctuating supplies of inorganic phosphate. It is therefore clear that phased periodic enzyme synthesis in a population of cells may reflect the response of the enzyme-forming system to a common external disturbance, rather than to any internal event in the cell cycle. There is even evidence that cyclic events can build up spontaneously in asynchronous populations of bacteria as the result of the operation of the feedback system involving extracellular components. Thus Sikyta and Slezak (1965) showed that the extracellular concentration of pyruvate showed sharp fluctuations at regular intervals in chemostat cultures of *E. coli*. The intracellular concentrations of enzymes concerned with pyruvate production also varied cyclically in these populations (B. Sikyta, personal communication), and it therefore appears that this is an example of autogenous enzyme and product synthesis phased at the level of the whole population rather than the single cell. Since most pools of end products are intra- rather than intercellular, this is not likely to occur with other enzyme systems but these last few examples do show quite clearly that periodic synthesis of enzymes can arise in a population, synchronous or asynchronous, simply in response to environmental disturbance at some point. Since the procedures used to produce synchronously dividing populations of cells often involve a period of metabolic disturbance, great care is required in distinguishing the results of this disturbance from events which are in fact characteristic of the cell cycle itself.

However, the interpretation of cyclic enzyme synthesis as the consequence of a common metabolic disturbance of the cells is clearly insufficient to explain the fact that the timing of enzyme synthesis under autogenous conditions *is* related to the time of cell division. The time of synthesis of a particular enzyme remains the same, on average, from one experiment to the next. Masters and Pardee (1965) have shown this to be true for four enzymes in *B. subtilis*, and Ferretti and Gray (1968) have done the same for four enzymes of *R. spheroides*. Although in both these cases synchrony of division is achieved by diluting stationary phase cells into fresh medium, thus introducing a large metabolic disturbance, the synchrony of *B. subtilis* populations may persist for up to 7 generations without the cyclic synthesis of the enzymes disappearing. Measurements of the timing of synthetic periods were made some time after the initial dilution of the cells, and it must be considered highly unlikely that these oscillations were the consequence of the dilution. Indeed the fact that the periodicity of enzyme synthesis is always the same as the periodicity of cell division is not to be expected from a freely oscillating feedback system (Goodwin, 1966) and itself indicates a coupling between the cell cycle and the time of enzyme synthesis.

Masters and Pardee (1965) have suggested that the ordered replication of the genes may provide the link between the cell cycle and the times of synthesis of enzymes under autogenous conditions. They have based this proposal on the observation of a correlation between the order of replication of 4 genes in the genome of *B. subtilis* and the relative order of synthesis of the 4 corresponding enzymes. Goodwin (1966) has devised a model by which such a correspondence between order of gene replication and order of enzyme synthesis might come about. According to this model, the synthesis of a particular enzyme is periodic as a result of the oscillation of the feedback control system, but the periodicity of the synthesis periods is "entrained" by the replication of the corresponding structural gene. The entraining signal following gene replication was suggested to be the synthesis of a small amount of messenger RNA following gene replication, but other events could also provide the necessary signal. (For example, the steady-state level of an unstable messenger would rise sharply to a new level following gene doubling, and this could equally well act as an entraining signal.) Since the entrainment of cyclic enzyme synthesis would result only from a *tendency* for some enzyme to be made following gene replication, it would be easy to disturb temporarily the timing of synthetic periods by a period of repression or derepression. Similarly, since the synthesis of enzymes would be inherently oscillatory, synthesis should continue to be periodic following the inhibition of DNA synthesis. Thus this model, with its loose coupling between gene replication and timing of enzyme synthesis behaves very much in accord with the experimental facts. It is, however, difficult to test the model directly in other ways.

In conclusion, the experiments which have attempted to uncover the mechanism of periodic enzyme synthesis in synchronous bacterial populations have suggested that periodic enzyme synthesis is the result of self-generating and self-maintaining oscillations in the feedback system controlling rate of enzyme synthesis but that, in a constant environment, the cyclic formation of enzymes is in some way coupled to the cell cycle. It has been suggested that this coupling is a consequence of the regular sequential replication of the genome.

VI. UNSTABLE ENZYMES AND THE CELL CYCLE

At least six of the fifteen enzymes listed in Table II are unstable in growing cells (Donachie, 1965; Nishi and Hirose, 1966; Ferretti and Gray, 1968, and unpublished observations in this laboratory). The true number is likely to be even higher since not all the enzymes have been tested. At first sight this observation seems to run counter to established experimental facts, since Hogness et al. (1955) showed very clearly that there is almost no turnover of proteins (and of β-galactosidase in particular) in growing E. coli. Even in stationary phase cells this turnover is only 4–5% per hour (Mandelstam, 1958, 1961). In contrast to this stability of protein structure, Nishi and Kogoma (1965) have demonstrated a short period of turnover of proteins during division of E. coli ML308. However, inactivation of enzyme activity is not necessarily the same thing as the breakdown of the protein structure into amino acids, and the observed instability of the enzymes in Table II may represent changes of enzyme configuration from active to inactive forms. Nevertheless it cannot be ruled out that a percentage of bacterial enzymes are turning over in growing cells since this would make little difference to the overall estimates of turnover for total protein.

The question of the behavior of unstable enzymes during the cell cycle becomes important if we consider periodic enzyme synthesis as having a role in the timing of other cellular events. It is possible to make models whereby the simple periodic doubling in quantity of a critical enzyme could trigger other events and so control the timing of, for example, cell division. This is because the accompanying fluctuations in end-product concentration could be large (Pardee, 1966). But clearly a better system for control would exist if the critical enzyme was present in the cell only at the critical time, and not at others. This appears to be the case for several enzymes during the development and differentiation of higher organisms (see, for example, Stern, 1961; Sussman, 1965). The conditions for such a periodic appearance and disappearance of enzyme would appear to exist in the synthesis of an unstable enzyme under autogenous conditions.

Under basal or derepressed conditions of synthesis, the amount of an unstable enzyme would increase in a stepwise manner rather than continuously as in the case of a stable enzyme (Donachie, 1965). This follows because, at a given gene dosage, the rate of synthesis of enzyme would come to be balanced by its rate of inactivation. This constant level of enzyme activity would change to one twice as high following gene duplication, and this would appear as a stepwise increase in total enzyme activity. Under autogenous conditions, where oscillations in the feedback system are superimposed on the effects of gene duplication, the stepwise increase in enzyme activity could become much sharper (Masters and Donachie, 1966). In the case of a very unstable enzyme or when large fluctuations in end product were occurring, the total amount of active enzyme could increase sharply and then disappear again. Thus active enzyme would be found only at one period in the cell cycle. Such behavior has been observed for the unstable enzyme aspartate transcarbamylase in *B. subtilis* following the release of a prolonged period of repression (unpublished observations). However, in undisturbed conditions the amount of this enzyme increases in a simple stepwise manner under autogenous conditions (see Fig. 10). A much more relevant example is therefore that of Nishi and his collaborators (Nishi and Hirose, 1966; Kogoma and Nishi, 1965) that the total amount of protease and dipeptidases rises and falls during the cell cycle of *E. coli*. Since these periods of high activity roughly correspond in time with the periods of division, and since there is evidence of protein turnover at the time of division (Nishi and Kogoma, 1965), these authors propose that the periodic appearance of proteolytic activity is indeed part of the process of cell division.

If the periodic production of these unstable enzymes is determined in a similar way to the periodic production of other enzymes under autogenous conditions, then the timing of their appearance could depend on the time of replication of their structural genes.

VII. DNA REPLICATION AND THE INTEGRATION OF CELL GROWTH AND DIVISION

The size and composition of bacteria are dependent upon the medium in which they are growing. Schaechter *et al.* (1958) have made a detailed investigation of the composition of *Salmonella* growing at a large number of different growth rates in different media. They found that the proportions of a number of cellular constituents changed from one growth rate to the next. In this discussion we will refer to these as "physiological variables." Among these variables we may list the mass per cell, the DNA per cell, the number of ribosomes per cell, and the number of ribosomes per mass (Schaechter *et al.*, 1958).

In contrast to this variation in cell composition, certain other relationships remain constant at different growth rates. Some of these are taken for granted (e.g., that there is always one cell division per mass doubling), but others that are much more unexpected have recently been discovered by Cooper and Helmstetter (1968). As described by Helmstetter in this volume, these "physiological constants" are the time taken for a replication point to traverse a single genome, and the interval between the termination of such a round of replication and cell division. For *Escherichia coli* B/r these constants are very close to 40 and 20 minutes, respectively (at all growth rates between 1 and 3 doublings per hour).

Since these relationships are essential to an understanding of the timing of gene expression in different environments, it is necessary to briefly discuss the relationships between some of the physiological variables and constants. To do this and to gain some insight into the probable mechanism controlling the time of initiation of DNA replication at different growth rates, a more fundamental physiological constant can be derived from the above observations. This constant is the cell mass at which rounds of DNA replications are initiated (Donachie, 1968).

A. TIMING OF CHROMOSOME INITIATION

It has been shown that a period of protein synthesis is necessary between the initiation of one round of DNA replication and the initiation of the next (Maaløe and Hanawalt, 1961; Hanawalt *et al.*, 1961; Maaløe, 1961; Nakada, 1960; Pardee and Prestidge, 1956). The time between cell division and the subsequent initiation of a round of DNA replication is different from one growth rate to the next (Cooper and Helmstetter, 1968), but one may ask whether a constant amount of protein synthesis takes place in these different time intervals. To calculate this, it has been assumed that the behavior of *Salmonella* and *Escherichia* is qualitatively similar at different growth rates. The measurements of Schaechter *et al.* (1958) showed that the mass per cell of *Salmonella* varies with growth rate in such a way that

$$\log (\text{mass/cell}) = k \text{ (growth rate)} \tag{1}$$

From this relationship the relative masses of cells at the time of division can be calculated for various growth rates. The rules for the timing of DNA synthesis relative to division (i.e., that a round of DNA replication takes 40 minutes and that the interval between termination of a round and the following division is 20 minutes in *E. coli* B/r) can then be used to find the time in the cell cycle when rounds of DNA replication are initiated at various growth rates (Cooper and Helmstetter, 1968). The cell mass at this time can then be determined. The result of such calculations is that the mass at which initiation takes place is always an integral multiple of a unique mass, I.

For cells growing at rates between 1 and 2 doublings per hour, initiation of chromosome replication takes place at a mass $= I$, while for faster growing cells with growth rates between 2 and 3 doublings per hour the mass at initiation $= 2I$.

Schaechter *et al.* (1958) found that DNA per cell increased with increase in growth rate for *Salmonella*, and the work of Cooper and Helmstetter (1968) has confirmed this for *E. coli* B/r. It can also be calculated from the data of the latter authors that the number of chromosome origins at which replication is initiated in a cell cycle increase discontinuously as a function of growth rate. In fact for cells with growth rates of between 1 and 2 doublings per hour the number of chromosome origins at which replication is initiated at any time is a constant, Ω, while for cells growing at rates between 2 and 3 doublings per hour the number of origins at the time of initiation is equal to 2Ω.

It is therefore clear from these experiments that the initiation of rounds of DNA replication begins for all cells at all growth rates when the ratio

$$I/\Omega = 1 \qquad (2)$$

As a model to explain this constant proportionality, it is proposed that the interval between one initiation of chromosome replication and the next is determined by the time required to synthesize a fixed quantity of initiator protein. This quantity is sufficient for the initiation of replication at only a limited number, Ω, of chromosome origins. Because of the unitary behavior of the initiator it is proposed that it is a single structure. Because of the time required to complete the synthesis of the initiator it is likely to be a large structure rather than a single molecule. (This seems likely also because no mechanism has yet been proposed whereby the synthesis of a single molecule can be exactly timed. See Section IV in the discussion of basal enzyme synthesis.)

This model provides an explanation for the timing of DNA synthesis at different growth rates and also for many experiments on unbalanced growth during a shift from one medium to another. Thus, during a "shift-up" of bacteria from a poor to a rich medium, the mass doubling time changes at once but the DNA doubling time increases gradually until the mass/DNA ratio characteristic of the new medium has been attained (see Maaløe and Kjeldgaard, 1966). Only then does the DNA doubling time reach its new value. On this model the change would be due to the initiation of extra rounds of DNA synthesis as cells of different ages successively reached the critical mass, I. Similarly Pritchard and Lark (1964) showed that the initial rate of DNA synthesis following the termination of a period of thymineless growth increased in proportion to the length of the period of thymineless growth. According to the present model, initiators would be synthesized during thymineless growth and the subsequent rate of DNA synthesis would be increased due to extra initiations of replication at the chromosome origins. Pritchard and Lark showed that such extra initiations were the cause of the increasing rate of DNA replication. The well-known heterogeneity of resumption of DNA synthesis in a population of bacteria whose chromosome replication has been aligned by amino acid starvation is also explained by this model. During amino acid starvation cell mass remains constant and therefore the number of initiators will also remain constant, however, all rounds of DNA replication will be completed. On restoration of amino acids, few cells will have an initiator but progressively more and more will synthesize one as the average cell mass increases. The rate of overall DNA synthesis will therefore increase until, after one mass doubling time, all cells will have initiated a round of replication. This is a reasonable approximation to what in fact appears to happen (Hanawalt et al., 1961).

This model for the timing of rounds of DNA replication was clearly foreshadowed by Maaløe and Rasmussen (1963), who suggested that "the timing of the initiation process is linked to cell growth in terms of mass and protein rather than to the process of DNA replication itself."

The relations set out in Eqs. (1) and (2), together with the constancy in the time required to replicate a chromosome, necessarily imply that there must be a constant interval between termination of a DNA round and the time of the subsequent cell division. This follows if there is always one division per mass doubling. That the constant interval between DNA rounds and subsequent cell division can be a consequence of the other physiological constants is satisfactory since it is difficult to conceive of a separate timing mechanism that would result in cell division at a constant time after DNA replication at all growth rates. There is no implication in this formulation of any causal connection between DNA replication and cell division; provided cell division is timed by some mechanism and provided initiation of DNA rounds always begins at a fixed mass, then there will necessarily be a fixed interval between the two events. However, there is independent evidence of a causal connection between DNA synthesis and cell division, and this will be reviewed below.

B. TIMING OF CELL DIVISION

It has long been known that there is some kind of causal relationship between DNA replication and division. This comes from the observation that agents that specifically inhibit DNA synthesis often result in the growth of long filamentous cells without septa. It is now equally well known that the converse need not be true, since agents are known that inhibit cell division without apparently interfering with DNA replication. Some recent experiments will now be briefly described which indicate that the time of cell division, like the time of chromosome initiation, depends on the mass/DNA ratio in the cells (Donachie *et al.*, 1968).

The following experiments depend on the observation that growth of a thymine auxotroph in the absence of thymine leads to no irreversible changes in the cells and that growth and division can be restored after any period of thymineless growth (Donachie and Hobbs, 1967). (The irreversible changes previously thought to accompany thymine starvation have now been shown to be the secondary result of the induction of a defective prophage—at least in *E. coli* 15T⁻.)

On removal of thymine from a steadily growing population of *E. coli* 15T⁻, no new septa are formed in the cells (although septa already being synthesized at the commencement of thymine starvation are completed in its absence). There is therefore an almost immediate inhibition of septum initiation.

Cells continue to grow in length and mass in the absence of thymine, and the rate of mass increase is almost identical to that of cells growing with thymine. In minimal medium, growth ceases after about one mass doubling. If thymine is added back to the culture at any time up to the point where growth stops, the exponential rate of mass increase will continue indefinitely. DNA synthesis recommences immediately on addition of thymine (Barner and Cohen, 1956), but cell division does not resume until an additional interval of time has elapsed. Figure 14 illustrates this behavior following one generation time of growth in the absence of thymine. After this period of thymineless growth the exponential rate of resumed DNA synthesis is twice that in control cultures (as predicted by the hypothesis on the control of the initiation of chromosome replication presented above). As a result, the total DNA content of the culture is restored to that of a control culture which has been growing for the whole time in the presence of thymine (see Fig. 14). At this point the rate of DNA synthesis has returned to the control rate (again as the hypothesis predicts) and continues at that rate indefinitely.

No new cell septa can be seen in the cells during this period of rapid DNA synthesis, until the mass/DNA ratio of the culture has been restored almost to that in control cultures. At this point there is a rapid increase in cell numbers (Fig. 14). This first period of moderately synchronous division is followed by a return to the control rate of cell division. The initial period of rapid division results in a doubling in

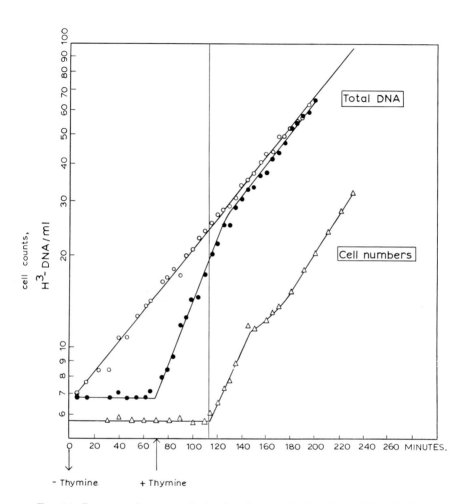

Fig. 14. Recovery from growth in the absence of thymine in *Escherichia coli* 15T⁻ J.G. 151 (see Donachie and Hobbs, 1967). Cells were prelabeled by growth in thymine-³H for 3 generations and then washed free of thymine at 0 minutes. The culture was divided into two equal parts and reinoculated with or without thymine-³H. After 70 minutes (one generation time) thymine-³H was added back to the thymine-starved culture. Total DNA was followed in both cultures by measurement of DNA-³H (open circles, control culture; filled circles, experimental culture). Cell numbers were followed in both cultures, but they are shown only for the experimental culture (triangles). Cell numbers increased logarithmically in the control. The vertical line marks the reinitiation of cell division.

cell numbers, and microscopic examination of the cells during this time shows that every cell in the population has divided once.

If thymine is added back to cultures after various periods of thymineless growth, cell division follows after an interval and always has the same kinetics as in Fig. 14. However, the interval between reinitiation of DNA replication and division is not a constant but depends on the length of the period of thymineless growth. In fact the interval between reinitiation of DNA replication and division is always approximately equal to the period of thymineless growth.

In this way division can be initiated at any time after the initiation of DNA replication (in actual experiments this interval has been varied between 0.25 and 1.0 generation time). Since DNA replication reinitiates at about twice the control rate, the interval between the beginning of DNA synthesis and division is equal to the time required to restore the mass/DNA ratio to the control level.

There appears to be no way in which a cell can "measure" the ratio of any of its components except by the specific stoichiometric combination of one with the other. Formally therefore the control of division time in these experiments can be described by the following model. Repressors of cell septum initiation are formed at a rate proportional to the rate of mass increase. These repressors are inactivated by specific combination with one or more sites on the DNA of the cell. When DNA synthesis is inhibited, excess repressor accumulates during subsequent cell growth and the initiation of new cell septa is repressed. On restoration of DNA replication, DNA is synthesized at twice the rate of repressor synthesis and eventually all repressor is inactivated by combination with DNA. Cell division is then initiated.

There are obvious similarities between this model for the control of septum initiation and the earlier model for the control of chromosome initiation. Both events depend on the mass/DNA ratio. Both models assume the synthesis of a hypothetical entity, initiator or repressor, at a rate proportional to the rate of mass increase. Both initiators and repressors are assumed to combine with specific sites on the DNA. There is, however, as yet no other evidence that the initiator of DNA replication is identical to the repressor of septum initiation, attractive though such an idea may be, and the proof or disproof of these speculations must wait on new experiments.

VIII. SUMMARY AND CONCLUSIONS

In this chapter we have described some of the timed events that take place during the bacterial cell cycle. Although this description is incomplete, it has been possible to see that the continued replication of the

DNA is an essential component in the system of control by which the timing of successive events is regulated.

We now have a reasonable picture of the course in time of the translation of individual genes. At a particular growth rate, the rate of production of an enzyme depends solely on the number of copies of the gene which encodes its structure, and on the concentration of specific repressor molecules. If the concentration of repressor is held constant, for example, by the addition of a large excess of inducer or corepressor (derepressed or basal conditions), then the rate of enzyme synthesis is proportional to the number of these genes. This in turn is determined by the age of the cell, since chromosome replication always begins at a fixed time in the cell cycle (at constant growth rate) and proceeds in a fixed order from a unique point of origin. Consequently enzyme synthesis under conditions of a constant repression takes place at a constant rate until the time when its corresponding gene is replicated. The rate of synthesis then doubles. We have found no evidence of there being any period in the cell cycle when the activity of a gene is inhibited.

When the rate of synthesis of an enzyme is determined by the cell itself, i.e., when the degree of repression is not artificially held constant, then regular oscillations in rate of synthesis can build up. The periodic synthesis of enzymes under these conditions is not *immediately* dependent on gene replication, since it can continue in the absence of DNA synthesis. However the periodicity of synthesis is always the same as the periodicity of the cell cycle, and its timing is therefore in some way coupled to the cell cycle. It has been suggested that this coupling takes place through periodic gene replication. Once in every cell cycle the potential to synthesize a given enzyme will double, and this has been shown to be sufficient to entrain the periodicity of the discontinuous synthesis of enzymes to that of their genes (Goodwin, 1966). Although there is no evidence that the periodic production of biosynthetic enzymes is important in the regulation of cell growth, they do provide a model for the periodic synthesis of other gene products, the timed production of which might be essential for the integration of cellular growth.

The most obvious processes that must be regulated with respect to one another, if the integrity of the cell is to be maintained, are DNA replication and cell division. Both processes show evidence of being regulated in that they each occur once in every mass doubling time. The timing of DNA rounds appears to be determined by cell mass while cell division is dependent on DNA replication. However, in some bacterial mutants cell division is uncoupled from the DNA cycle and can proceed autonomously. In such mutants enucleate cells may be produced during growth (Adler *et al.*, 1967; Hirota and Jacob, 1967; Gross *et al.*,

1968) while DNA synthesis may be either normal or abnormal. Evidently cell septum synthesis is periodic in itself and although normally coupled to DNA replication can continue in its absence in mutant populations. Experiments reported here show that the timing of cell division can be controlled by the DNA/mass ratio in cells growing under special conditions. The actual ratio of total DNA/total mass changes very little during balanced cell growth, but the ratio of specific components could well change regularly. The possible nature of such components, which could control both cell division and initiation of DNA synthesis, has been discussed in the previous section.

ACKNOWLEDGMENTS

We wish to thank Miss Suzanne Armour and Mr. D. G. Hobbs for valuable technical assistance. We are indebted to Drs. Ernest Gray, J. J. Ferretti, and C. E. Helmstetter for sending us data prior to publication.

REFERENCES

Abe, M., and Tomizawa, J. (1967). *Proc. Natl. Acad. Sci. U.S.* **58**, 1911.
Abbo, F. E., and Pardee, A. B. (1960). *Biochim. Biophys. Acta* **39**, 478.
Adler, H. I., Fisher, W. D., Cohen, A., and Hardigree, A. A. (1967). *Proc. Natl. Acad. Sci. U.S.* **57**, 321.
Barner, H. D., and Cohen, S. S. (1956). *J. Bacteriol.* **71**, 149.
Berg, C. M., and Caro, L. G. (1967). *J. Mol. Biol.* **29**, 419.
Berg, C. M., and Caro, L. G. (1968). *Cold Spring Harbor Symp. Quant. Biol.*, in press.
Cairns, J. (1963). *J. Mol. Biol.* **6**, 208.
Clark, D. J., and Maaløe, O. (1967). *J. Mol. Biol.* **23**, 99.
Cooper, S., and Helmstetter, C. (1968). *J. Mol. Biol.* **31**, 519.
Cummings, D. J. (1965). *Biochim. Biophys. Acta* **85**, 341.
Cutler, R. G., and Evans, J. E. (1967). *J. Mol. Biol.* **26**, 81.
Donachie, W. D. (1964). *Biochim. Biophys. Acta* **82**, 293.
Donachie, W. D. (1965). *Nature* **205**, 1084.
Donachie, W. D. (1968). *Nature* **219**, 1077.
Donachie, W. D., and Hobbs, D. G. (1967). *Biochem. Biophys. Res. Commun.* **29**, 172.
Donachie, W. D., and Masters, M. (1966). *Genet. Res.* (Cambridge) **8**, 119.
Donachie, W. D., Masters, M., and Armour, S., in preparation.
Donachie, W. D., Hobbs, D. G., and Masters, M. (1968). *Nature* **219**, 1079.
Ferretti, J. J., and Gray, E. D. (1967). *Biochem. Biophys. Res. Commun.* **29**, 501.
Ferretti, J. J., and Gray, E. D. (1968). *J. Bacteriol.* **95**, 1400.
Goodwin, B. C. (1963). "Temporal Organization in Cells." Academic Press, New York.
Goodwin, B. C. (1966). *Nature* **209**, 476.
Gross, J. D., Karamata, D., and Hempstead, P. (1968). *Cold Spring Harbor Symp. Quant. Biol.*, in press.
Hanawalt, P. C., Maaløe, O., Cummings, D. J., and Schaechter, M. (1961). *J. Mol. Biol.* **3**, 156.

Helmstetter, C. E. (1967). *J. Mol. Biol.* **24**, 417.
Helmstetter, C. E., and Cooper, S. (1968). *J. Mol. Biol.* **31**, 507.
Helmstetter, C. E., and Cummings, D. J. (1964). *Biochim. Biophys. Acta* **82**, 608.
Helmstetter, C. E., and Cummings, D. J. (1963). *Proc. Natl. Acad. Sci. U.S.* **50**, 767.
Hirota, Y., Jacob, F., Ryter, A., Buttin, G., and Nakai, T., (1968). *J. Mol. Biol.* **35**, 175.
Hogness, D. S. (1959). *In* "Biophysical Science" (J. Oncely, ed.), p. 256. Wiley, New York.
Hogness, D. S., Cohn, M., and Monod, J. (1955). *Biochim. Biophys. Acta* **16**, 99.
Jacob, F., Brenner, S., and Cuzin, F. (1963). *Cold Spring Harbor Symp. Quant. Biol.* **28**, 329.
Kobayashi, Y., Steinberg, W., Higa, A., Halvorson, H. O., and Levinthal, C. (1965). *In* "Spores III" (L. L. Campbell and H. O. Halvorson, eds.), p. 200. Am. Soc. Microbiol., Ann Arbor, Michigan.
Kogoma, T., and Nishi, A. (1965). *J. Gen. Appl. Microbiol.* **11**, 321.
Kuempel, P. L. (1965). Doctoral Dissertation, Princeton University, Princeton, New Jersey.
Kuempel, P. L., Masters, M., and Pardee, A. B. (1965). *Biochem. Biophys. Res. Commun.* **15**, 38.
Lark, K. G., Repko, T., and Hoffman, E. J. (1963). *Biochim. Biophys. Acta* **76**, 9.
Lascelles, J. (1959). *Biochem. J.* **72**, 508.
Lascelles, J. (1960). *J. Gen. Appl. Microbiol.* **23**, 487.
Maaløe, O. (1961). *Cold Spring Harbor Symp. Quant. Biol.* **26**, 45.
Maaløe, O. (1963). *J. Cell. Comp. Physiol.* **62** (Suppl. 1), 31.
Maaløe, O., and Hanawalt, P. C. (1961). *J. Mol. Biol.* **3**, 144.
Maaløe, O., and Kjeldgaard, N. O. (1966). "Control of Macromolecular Synthesis; a Study of DNA, RNA and Protein Synthesis in Bacteria." Benjamin, New York.
Maaløe, O., and Rasmussen, K. V. (1963). *Colloq. Intern. Centre Natl. Rech. Sci. (Paris)* **124**, 165.
Mandelstam, J. (1958). *Biochem. J.* **69**, 110.
Mandelstam, J. (1961). *Biochem. J.* **79**, 489.
Masters, M. (1965). Doctoral Dissertation, University of California, Berkeley, California.
Masters, M., and Donachie, W. D. (1965). *Proc. Symp. Mutational Process, Prague*, pp. 285–292. Academia, Prague.
Masters, M., and Donachie, W. D. (1966). *Nature* **209**, 476.
Masters, M., Kuempel, P. L. and Pardee, A. B. (1964). *Biochem. Biophys. Res. Commun.* **15**, 38.
Masters, M., and Pardee, A. B. (1965). *Proc. Natl. Acad. Sci. U.S.* **54**, 64.
Meselson, M., and Stahl, F. (1958). *Proc. Natl. Acad. Sci. U.S.* **44**, 671.
Mitchison, J. M., and Vincent, W. S. (1965). *Nature* **205**, 987.
Mitchison, J. M., and Vincent, W. S. (1966). *In* "Cell Synchrony—Studies in Biosynthetic Regulation" (I. L. Cameron and G. M. Padilla, eds.), p. 328. Academic Press, New York.
Nakada, D. (1960). *Biochim. Biophys. Acta* **44**, 241.
Nishi, A., and Hirose, S. (1966). *J. Gen. Appl. Microbiol.* **12**, 293.
Nishi, A., and Horiuchi, T. (1966). *J. Biochem.* **60**, 338.
Nishi, A., and Kogoma, T. (1965). *J. Bacteriol.* **90**, 884.
Pardee, A. B. (1966). *In* "Metabolic Control Colloquium of the Johnson Research Foundation," p. 239. Academic Press, New York.

Pardee, A. B., and Beckwith, J. (1963). *In* "Informational Macromolecules" (H. J. Vogel, V. Bryson, and J. O. Lampen, eds.), p. 255. Academic Press, New York.

Pardee, A. B., and Prestidge, L. (1956). *J. Bacteriol.* **71**, 677.

Pritchard, R. H., and Lark, K. G. (1964). *J. Mol. Biol.* **9**, 288.

Schaechter, M., Maaløe, O., and Kjeldgaard, N. O. (1958). *J. Gen. Appl. Microbiol.* **19**, 592.

Sikyta, B., and Slezak, J. (1965). *Biochim. Biophys. Acta* **100**, 311.

Stern, H. J. (1961). *J. Biophys. Biochem. Cytol.* **9**, 271.

Steinberg, W., Halvorson, H. O., Keynan, A., and Weinberg, E. (1965). *Nature* **208**, 710.

Sussman, M. (1965). *Brookhaven Symp. Biol.* **18**, 66.

Tevethia, M. J., and Mandel, M. (1967). *Proc. Natl. Acad. Sci. U.S.* **58**, 1174.

Torriani, A. (1960). *Biochim. Biophys. Acta* **38**, 460.

Yanagita, T., and Kaneko, K. (1961). *Plant Cell Physiol.* **2**, 443.

Yoshikawa, H., and Sueoka, N. (1963). *Proc. Natl. Acad. Sci. U.S.* **49**, 806.

CHAPTER 4

Synchrony and the Formation and Germination of Bacterial Spores

Ralph A. Slepecky

I. INTRODUCTION TO GERMINATION AND SPORULATION

The formation from an actively metabolizing vegetative cell of a resistant spore, having a low metabolism provides the bacterial cell with unique survival advantages and represents a distinct metamorphosis. Likewise, germination, the breaking of the spore's dormancy, and subsequent emergence of a vegetative cell represents a true cellular differentiation. Since these processes of differentiation occur in a microorganism, they can be applied to studies of control mechanisms at the cellular and molecular level of organization. Interest in spore formation and germination and awareness that they represent model systems for studying differentiation at a very basic level are indicated by the recent appearance of many reviews in this area (Halvorson, 1965; Halvorson et al., 1966; Vinter, 1967; Murrell, 1967; Kornberg et al., 1968).

Certain genetically competent species of bacteria, mainly *Bacillus* and *Clostridium*, are capable of undergoing the "spore—vegetative cell–spore" metamorphosis. Germination of the spore, the breaking of the highly dormant state, involves a series of degradative reactions. Outgrowth is the period during which the spore gradually becomes a vegetative cell and requires new macromolecular synthesis (Halvorson *et al.*, 1966). The vegetative cell is then capable of undergoing various morphological and biochemical changes which lead to a series of cell divisions and subsequent spore formation or to the production of a spore without intermittent cell division (Vinter and Slepecky, 1965).

Fine structure analyses of spore formation have revealed seven distinct stages. The cytological changes and the approximate time of their appearance during the growth cycle have been well described and documented with electron micrographs (Young and Fitz-James, 1959a; Ryter, 1965; Ohye and Murrell, 1962; Ellar and Lundgren, 1966). These stages are I, preseptation-nuclear material in an axially disposed filament; II, septation; III, protoplast envelopment; IV, cortex formation; V, coat formation; VI, maturation; and VII, free spore (Fitz-James, 1965a; Schaeffer *et al.*, 1965a; Murrell, 1967) (see Fig. 1). Prior to sporulation, the nuclear material is in an axially disposed filament. Segregation of the chromatin material to the poles of the cell occurs concomitantly with the invagination of the plasma membrane and its associated mesosome. The invagination moves concentrically toward the center of the cell and fuses to complete the spore septum. The mode of formation of this septum is similar to the formation of the normal division transverse septum. However, in sporulation the division of the cell is not equal and subsequent proliferation of the larger cell's cell membrane leads to complete engulfment of the "forespore" and liberation of the immature spore, surrounded now by a double unit membrane, into the cytoplasm of the larger cell. Cortex material similar to vegetative cell wall mucopeptide is laid down between the unit membranes, and its deposition corresponds in time to the accumulation of dipicolinic acid and calcium; presumably, these are located here. Protein coats are synthesized around the outside of the spore, and in some species an additional layer, the exosporium, is synthesized. After the spore has matured, the lytic enzymes produced during the sporulation process lyse the sporangial cell and the spore is liberated into the medium. Several of these stages are readily assessed by phase contrast microscopy or by simple staining (Gordon and Murrell, 1967) allowing for routine monitoring of the culture during the spore's differentiation.

In addition, many biochemical and physiological events associated with the formation of a bacterial spore have been recorded and characterized in a number of different sporulation systems (Table I). The events given in the table have not been put in order since the exact sequence is not known. Since recent reviewers (Vinter, 1967; Halvorson, 1965; Murrell, 1967; Kornberg et al., 1968) have chronicled and discussed these important contributions, they will not be reviewed here. A few attempts have been made to relate on a time scale the distinct events or appearance of specific products with the aforementioned observed cytological stages (Hashimoto et al., 1960; Schaeffer et al., 1965a; Fitz-James, 1965a; Halvorson, 1965; Vinter, 1967); however, in most instances, the time of the formation of a particular product has been only approximate. This fact is the strongest argument in favor of attempting to improve growth and sporulation synchrony (Halvorson, 1965; Vinter, 1967; Murrell, 1967).

The genetics of sporulation is quite complex and awaits further clarification. Most of the definitive work has been done on *Bacillus subtilis*, which is amenable to transformation (Spizizen, 1958; Schaeffer and Ionesco, 1960) and transduction (Takahashi, 1961; Thorne, 1962). Two major types of sporulation mutants have been described: Sp^-, which are completely asporogenous, and Osp (oligosporogenic), which can sporulate but at extremely low frequencies. The latter presumably are damaged in some regulatory function (Schaeffer, 1963). In addition it is possible to have mutants with blocks at various cytological stages during sporulation (Schaeffer, 1963). The availability of *B. licheniformis* for transformation studies (Thorne and Stull, 1966) and *B. cereus* for transduction studies (C. B. Thorne, personal communication, 1968) should lead to further expansion of our knowledge of sporulation genetics.

Current evidence suggests that the sporulation process is controlled by a large number of unlinked genes dispersed over the chromosome (Schaeffer et al., 1965a; Takahashi, 1965a,b), and there is additional evidence that there may be a cluster of closely linked genes which regulates the first stage (Spizizen, 1965). The possibility that a plasmid or episome may be involved in sporulation (Jacob et al., 1960; Rogolsky and Slepecky, 1964) awaits further confirmation.

Thus sporulation consists of a series of complex biochemical changes occurring hand in hand with distinct morphological changes, all of which must be determined by distinct genetic loci. Consideration of the many enzymes and new proteins involved in sporulation has led to the speculation that the spore genome is probably quite large and may contain at least 100 structural and regulatory genes (Halvorson, 1965). The ordered appearance of these cytological and biochemical changes implies a sequential reading of the genome. Just what induces the reading and how it is regulated are among the challenging questions.

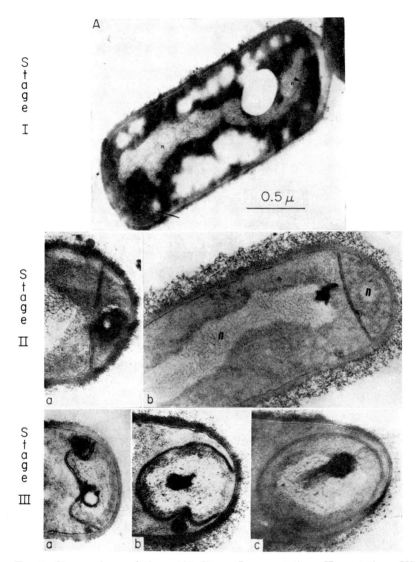

Fig. 1. Stages of sporulation: (A) Stages I, preseptation; II, septation; III, protoplast envelopment. (B) Stages IV, cortex formation; V, coat formation; VI, maturation. (Stage VII is a free spore, not shown.) n = nuclear material; C_1, C_2 = cortex; IM = inner spore membrane; OM = outer spore membrane; IC = inner coat; OC = outer coat; ex = exosporium.

Stages I, IVb, and VIb are of *Bacillus cereus* (from Remsen et al., 1966); stages II, III, and IVa, of *Bacillus megaterium* (from Ellar et al., 1967); stages V and VIa, of *Bacillus cereus* (from Ellar and Lundgren, 1966). I–IV and VIb are longitudinal sections; Va, Vb, and VIa are cross sections.

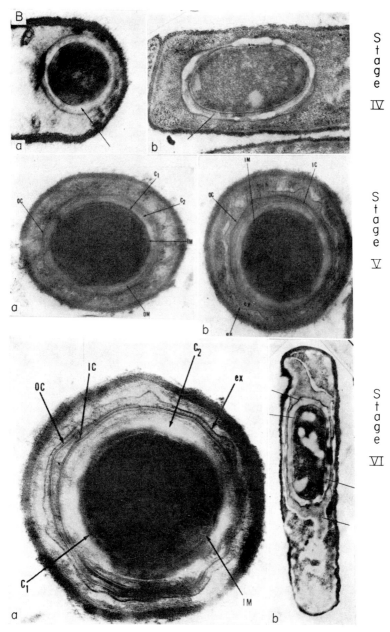

FIG. 1. For legend see opposite page.

TABLE I

PHYSIOLOGICAL AND BIOCHEMICAL CHANGES ASSOCIATED WITH THE FORMATION OF A BACTERIAL SPORE

Event	Representative references	Event	Representative references
I. DNA All fractions synthesized	Young and Fitz-James, 1959a,b; Hodson and Beck, 1960; Canfield and Szulmajster, 1964; Kogoma and Yanagita, 1967	**III. Protein and enzymes** New protein synthesis	Foster and Perry, 1954; Nakata et al., 1956; Fitz-James, 1960; Canfield and Szulmajster, 1964; Vinter, 1960; Halvorson, 1965; Leitzmann and Bernlohr, 1965; Ryter and Szulmajster, 1965
II. RNA All fractions synthesized	Spotts and Szulmajster, 1962; Aronson and Del Valle, 1964	Protein turnover	Foster and Perry, 1954; Urba, 1959; Young and Fitz-James, 1959a,b; Monro, 1961; Spotts and Szulmajster, 1962; Aubert and Millet, 1963a,b; Aubert and Millet, 1965; Spudich and Kornberg, 1968;
Stable t- and rRNA	Balassa, 1965		
Labile mRNA	Del Valle and Aronson, 1962		
New mRNA differs from vegetative cell mRNA			
a. More complex	Doi and Igarashi, 1964a	Protease production	Bernlohr and Novelli, 1959, 1960, 1963; Balassa et al., 1963; Schaeffer et al., 1963; Spizizen et al., 1965a; 1963;
b. Less heterogeneous	Aronson, 1965a		
RNA shifts			
a. Increase in t:r	Doi and Igarashi, 1964b		
b. Decrease in the number of polysomes and ribosomes	Fitz-James, 1965a	Functional TCA cycle	Hanson et al., 1963a, 1964a,b; Megraw and Beers, 1964; Pepper and Costilow, 1964; Szulmajster and Hanson, 1965
c. Stable membrane bound polysomes remain	Aronson, 1965b		
RNA turnover	Young and Fitz-James, 1959a,b; Balassa, 1963a,b, 1964a,b, 1965; Szulmajster, 1964; Fitz-James, 1965a	Formation of toxins	Sebald and Schaeffer, 1965; Lamanna, 1940

(Continued)

III. (Continued)

Formation of antigens	Vennes and Gerhardt, 1957; Tomcsik and Baumann-Grace 1959; Norris and Wolf, 1961; Norris, 1962; Baillie and Norris, 1963, 1964; Cavallo et al., 1963; Walker and Batty, 1965	
Formation of cystine-rich proteins	Vinter, 1957; 1961b	
Changes in enzymatic constitution of the cell	Stewart and Halvorson, 1953; Lawrence and Halvorson, 1954; Powell and Strange, 1956, 1957; Simmons and Costilow, 1962; Halvorson and Srinivasan, 1964	
Different pattern of enzymes	Halvorson, 1962; Simmons and Costilow, 1962	
Glucose dehydrogenase	Bach and Sadoff, 1962	
Ribosidase	Powell and Strange, 1956	
Acetoacetyl-CoA reductase	Halvorson and Srinivasan, 1964	
Adenosine deaminase	Powell and Strange, 1956	
Alanine racemase	Stewart and Halvorson, 1953	
Soluble NADH oxidase	Szulmajster and Schaeffer, 1961	
Inorganic pyrophosphatase	Tono and Kornberg, 1967	
Purine nucleoside phosphorylase	Gardner and Kornberg, 1967	
Spore specific enzymes	Stewart and Halvorson, 1953; Powell and Hunter, 1956; Bach and Sadoff, 1962	
Lytic enzymes	Greenberg and Halvorson, 1955; Powell and Strange, 1956; Strange and Dark, 1957; Tomcsik and Bouille, 1961	
Other enzymes		Halvorson, 1962, 1965
IV. Other components		
Cortical mucopeptide		Warth et al., 1963a,b
Diaminopimelic acid		Murrell and Warth, 1965
Calcium		Vinter, 1956
Dipicolinic acid		Powell, 1953; Perry and Foster, 1956; Kondo et al., 1964; Schaeffer et al., 1965a
N-Succinylglutamic acid		Millet and Pineau, 1960; Aubert et al., 1961
Sporulation factor		Srinivasan and Halvorson, 1963; Srinivasan, 1965
Antibiotics		Schaeffer et al., 1965a
V. Resistance		
Heat		Vinter, 1960, 1962; Ryter and Szulmajster, 1965
Radiation		Vinter, 1961a,b; Bott and Lundgren, 1964; Romig and Wyss, 1957
Chemical		Ryter, 1965
VI. Additional events		
Sporulation commitment		
Forespore stage		Fitz-James, 1965a
Septum completion		Vinter and Chaloupka, 1967
Cessation of vegetative wall synthesis		Vinter, 1967
Degradation of sporangial wall		Vinter, 1963b
Refractility		Knaysi, 1951
Release of free spore		Anonymous

Until recently sporulation was thought to ensue only upon the cessation of vegetative growth. However, that cells are capable of forming spores during exponential growth (Aubert *et al.*, 1961; Kerravala *et al.*, 1964; Schaeffer *et al.*, 1965b) and during outgrowth after germination (Vinter and Slepecky, 1965) implies that derepression of the spore genome can occur at any time. This derepression (or induction) might be brought about in a variety of ways. Depletion of carbon or nitrogen substrates, growth factors or minerals (Grelet, 1957); the action of a sporulation factor (Srinivasan and Halvorson, 1963; Srinivasan, 1965); retardation of the growth rate (Bergère and Hermier, 1965a,b); decrease in protein synthesis (Doi and Igarashi, 1964a); blockage of DNA synthesis (Schaeffer *et al.*, 1965a); derepression of a catabolite repressor (Schaeffer *et al.*, 1965b), or induction of key enzyme systems (Hanson *et al.*, 1963b; Szulmajster and Hanson, 1965; Ramaley and Bernlohr, 1965) have all been implicated.

Once the sporulation genome has been activated, presumably at stage I (Fig. 1), further differentiation is subjected to a regulation which is not well understood. The occurrence of regulatory genes (Lundgren and Beskid, 1960; Spizizen, 1965) and the appearance of specific mRNA during sporulation (Balassa, 1963a,b; Doi and Igarashi, 1964a,b) implies that regulation is at the transcription level. In addition, recent evidence has shown the appearance of specific tRNA (Kaneko and Doi, 1966) which suggests that control is also at the translation level.

Schaeffer (1963; Schaeffer *et al.*, 1965a) has proposed that sporulation may be controlled by several operons. One sporulation-specific operon, possibly responsible for the production of a protease enzyme (Spizizen *et al.*, 1963; Bernlohr, 1964; Balassa, 1964b), is derepressed by nitrogen-containing metabolites; after this the other operons, expressed later, are sequentially induced (Stanier, 1950).

Halvorson (1965) has formulated another model. After derepression of a sporulation initiating factor (S), possibly the sporulation factor of Srinivasan and Halvorson (1963), the spore genome is transcribed. An early product of the transcription (X), possibly the antibiotic of Balassa (1964b), selectively blocks further transcription and translation of the vegetative genome. The order of succession of the gene products then results from either an unidirectional transcription of the genome or by sequential induction and repression of the genome.

Studies on the induction and regulation of sporulation are somewhat hampered by the fact that during the early stages of sporulation the mother cell is apparently still capable of further synthesis of DNA (Young and Fitz-James, 1959a) and some vegetative cell protein (Aubert and Millet, 1965). Because of such difficulties and of the fact that the time of appearance of events and products can be only calculated approximately, elucidation of induction and regulation may require more precise synchrony than has been used in most studies of these problems.

II. ANALYSIS OF SYNCHRONY DURING SPORULATION

A. BATCH CULTURE

Most studies on sporulation have been carried out using conventional batch cultures or the "active culture" technique (see Section II,C). Batch cultures of spore-forming bacteria display growth patterns typical of other procaryotic organisms, with a lag, a logarithmic, and a stationary phase. A spore inoculum is used in preference to vegetative cell inocula since the former are much more stable over prolonged storage. At some time in the late log or early stationary phase of the growth curve, the cell initiates the formation of a spore. It has been common practice to follow the convention (Schaeffer, 1961; Szulmajster and Schaeffer, 1961; Murrell, 1967) of arbitrarily designating the commencement of sporulation at the point at which exponential growth ends (T_0), corresponding to the axial chromatin or preseptation stage (stage I) as determined by examination of thin sections in the electron microscope. Stages I through VI are usually completed in 6–8 hours $(T_6–T_8)$ and represent several generation times in terms of exponential growth.

Currently, on the basis of the distribution of various sporulation cell types at relatively late stages, and of the heterogeneity as determined by density gradient centrifugation of final free spore crops usually harvested at very late times in the stationary phase, the sporulation process occurring in batch culture is believed to be asynchronous. However, the early stages may be more synchronous than we realize. Batch culture studies are usually carried out in minimal defined media or dilute complex media, and the beginning of the sporulation process has been associated with nutrient depletion (Grelet, 1951). A similar carbon limitation in cells genetically incapable of sporulation leads to cessation of growth and usually induces lysis of the cells (Monod, 1942); many asporogenic mutants usually exhibit the same behavior upon nutrient exhaustion. In batch cultures, two distinct physiological states exist (Foster, 1956; Schaechter et al., 1958)—one associated with the exponential phase of growth during which growth is balanced (Campbell, 1957) and the other associated with postlogarithmic phases during which the growth is unbalanced (Cohen and Barner, 1954). Both states are characterized by different growth rates. It is interesting that sporulation commences in batch cultures during the period of transition from one state to the other—a period which is ill understood (Maaløe and Kjelgaard, 1966) and is probably analogous to the shift-down transitions described in the classic studies of Kjelgaard et al. (1958) (Chaloupka, 1967). Most synchronization procedures have used cells taken from the exponential phase of growth (Cameron and Padilla, 1966), but there are cases where synchrony was obtained after dilution of cells from the stationary phase (Yoshikawa and Sueoka, 1963; Masters et al., 1964; Cutler and Evans, 1966; Tevithia and Mandel, 1967). Cutler and Evans (1966), cognizant of some earlier work (Henrici, 1928; Hershey and Bronfenbrenner, 1937; Hershey, 1938; Winslow and Walker, 1939; Huntington and Winslow, 1937) indicating a homogeneous physiological and morphological state in the early stationary phase, particularly in minimal media, have reexamined the ability of cells to synchronize at this stage and have hypothesized that cells have a tendency to synchronize themselves when approaching the stationary phase. Furthermore, it has been noted that *Saccharomyces cerevisiae* has a tendency to enter a potentially synchronous condition upon aging in the culture (Williamson, 1964). The mechanism underlying the attainment of synchronous division after dilution is not understood, however. If cells do become naturally synchronous during the transition state or in early stationary phase, then perhaps the early stages of sporulation in the batch culture have been fortuitously more synchronous than is realized, and it is only during the ensuing relatively long period of time required for the completion of the differentiation that the process of sporulation becomes increasingly asynchronous. Quantitative measurements of the degree of synchrony during development stages I and II will have to be performed. Such studies will be enhanced by the relatively simple and rapid staining method now available (Gordon and Murrell, 1967) to relate stage II and stage III development. Stage IV development can readily be estimated by the counts of the brighter-than-white forespores as seen in the phase contrast microscope. Such studies will require more uniform criteria for estimating the degree of synchrony. The various cell stages can be expressed as a percentage of the total cell units recognizable at the time of sampling, and the developmental rates can be expressed as the percentage of increase per minute of the various forms (W. G. Murrell, personal communication, 1968).

Free spore populations formed in batch cultures, harvested at late times of culture and thus representing "old" cells, have been shown to be discontinuously heterogeneous with regard to density when centrifuged in gradients of sucrose (Church and Halvorson, 1959; Howitt, 1960), a lead chelate of N,N'-(dihydroxyethyl)-N,N'-(decarboxymethyl)-ethylenediamine (Lewis et al., 1965) and urografin or N,N'-diacetyl-3,5-diamino-2,4,6-triiodobenzoate, an X-ray contrasting agent (Tamir and Gilvarg, 1966). The germination properties of light spores differed from those of heavy spores (Tamir and Gilvarg, 1966). Whether the differences in density reflect a relationship between the time of formation of the spore and the synchronous fraction of the population to which the cell which formed it belonged, is not known. However this seems an interesting possibility, as it would offer a means of improving spore germination synchrony (see Section III). In fact synchronously formed spores (by the procedure given in Section II,D) of *B. megaterium* are known to initiate germination faster than asynchronously produced spores (R. A. Slepecky, unpublished data, 1965). The ratio of the sizes of the different density classes would also offer a method of quantitating the degree of synchrony during synchronous sporulation.

The fact that synchronous fractions of yeast and bacterial cell populations can be obtained by density gradient (isopycnic) centrifugation (Mitchison and Vincent, 1965, 1966) suggests that it might be possible to separate cells in different stages of sporulation by the same method. Tamir and Gilvarg (1966) initiated work along these lines when they separated spores from vegetative cells. Spudich and Kornberg (1968) have recently used that method to isolate homogeneous populations of sporulating cells. They obtained four discrete density stages and have determined protein turnover in each stage.

B. "ENDOTROPHIC" SPORULATION

Mention should be made of another method which has been used to study sporulation, namely, "endotrophic" sporulation (reviewed by Black and Gerhardt, 1963). This technique involves the replacement of the actively growing late log phase cells from a batch culture into water or preferably water plus calcium (Black *et al.*, 1960). It results in the cells being able to complete the formation of refractile spores. Modification of this technique, wherein additional substances have been added, is usually referred to as a "replacement technique." While the techniques have been useful in defining the stage of commitment of cells to sporulation (the point of no return) and for suggesting that the sporulation process is largely endogenous after a certain stage (Foster and Perry, 1954; Foster, 1956), the degree of synchrony of spore development depends on the synchrony obtained in the batch culture and needs more consideration.

C. THE ACTIVE CULTURE TECHNIQUE

The second important method used in the study of sporulation is the "active culture" technique. It is a modification of the batch culture and originated in the laboratory of Professor H. Orin Halvorson; it has been used by them (Halvorson, 1957) and others (Collier, 1957; Young and Fitz-James, 1959a; Hashimoto et al., 1960; Lundgren and Beskid, 1960) in an attempt to obtain more synchronous sporulation. A spore suspension is subcultured three times; in each culture a 10% inoculum is used—that is, the initial population is approximately 10% of the possible maximum growth attainable. Sporulation ensues after the growth of the last actively growing culture. Reference has been made to this synchronous or near-synchronous sporulation (Collier, 1957; Halvorson, 1957; Young and Fitz-James, 1959a; Hashimoto et al., 1960), but it must be noted that the "sporulation synchrony" obtained is based only on the more rapid appearance of refractile (stage IV) or heat-resistant (stages V–VI) spores. Thus only late stages of sporulation are monitored, and synchrony of the cell division prior to sporulation is not considered.

This method has been quite useful in defining many of the events during the late stages of sporulation. However, if the induction and regulation of sporulation are to be understood, one must consider the early stages of sporulation as well as the actual degree of synchrony during the late stages of this system.

D. Division Synchrony followed by Sporulation

Although there has been much success in obtaining synchronous cell division in some protists (Campbell, 1957; Scherbaum, 1960; Zeuthen, 1964; Cameron and Padilla, 1966), relatively little use has been made of the available methods with regard to sporulation. Imanaka et al. (1967) have described an easily performed "selective synchrony" technique which gives good division synchrony followed by a high degree of sporulation synchrony. Synchrony was achieved by a modification of the filtration technique of Maruyama and Yanagita (1956). Filtration of the late log-phase batch cultures of Bacillus megaterium, grown on defined sucrose salts medium (SS) or SS plus glutamate (SSG), through nine layers of Whatman No. 40 filter paper in a fritted-glass disk Büchner funnel resulted in filtrates containing cells which showed synchronous division(s) and proceeded to sporulation. SS-grown cells completed one synchronous division after filtration, and sporulation ensued after the cessation of growth. SSG-grown cells completed two synchronous divisions, and sporulation occurred during or just after the second division. The stepwise pattern of growth and the doubling of cell numbers at each division fit the criteria for synchronous division as described in other protistan systems (Scott et al., 1955; Barner and Cohen, 1956; Lark and Maaløe, 1956; Burns, 1959; Halvorson et al., 1964; Donachie, 1965; Bostock et al., 1966).

The degree of division synchrony was further substantiated by a high division index (80%); the division index in these cells is analogous to the division index in synchronous eucaryotic organisms (Zeuthen, 1964; Bostock *et al.*, 1966; Padilla *et al.*, 1966) and is considered to be a good indication of overall synchrony (Zeuthen, 1964; Cameron and Padilla, 1966). Sporulation synchrony was improved greatly over that obtained with batch cultures grown under the same conditions. There was rapid formation of sporulation types, and, more importantly, there was an increase in the homogeneity of sporulation cell types at any particular time. Because the described system gives both good division and sporulation synchrony, the method should be useful in delineating early events in sporulation and their regulation.

The method proved useful in a fine structure study of synchronously dividing *Bacillus megaterium* (Ellar *et al.*, 1967); attention was focused mainly on the first synchronous division. The sequence of events resulting in the division of the cell was revealed, as well as the role of the mesosome associated with the developing cell wall septum in the deposition of new cell wall material. Examination of sporulation stages indicated a pattern similar to that found in other studies. The fact that during the early stages of sporulation the cells were found to contain relatively few ribosomes as compared with vegetative cells agrees with findings of others (Balassa, 1965; Doi, 1965; Fitz-James, 1965a), who found, by chemical analyses, rRNA turnover and a decrease in ribosomes during sporulation. It is of interest that Maaløe and Kjelgaard (1966) described a similar decrease in ribosomes in electron photomicrographs in cells grown under conditions to induce low growth rates.

III. GERMINATION SYNCHRONY

The process of spore germination consists of an activation stage, a germination stage during which degradative changes break the dormant state, and an outgrowth stage during which the spore elongates and becomes a new vegetative cell capable of dividing.

Activation is a reversible process which conditions the spore for germination and increases the number of spores undergoing germination as well as the rate of germination. Spores can be activated by a variety of treatments (Sussman and Halvorson, 1966; Holmes and Levinson, 1967a), notably exposure to heat (Curran and Evans, 1945).

Germination is triggered by simple compounds like certain amino acids and ribosides or their mixtures and is monitored by the loss of spore refractility as seen by phase contrast microscopy and by the decrease in optical density. In addition, the spores lose their resistance to heat, radiation, injurious chemicals, and dilute stains. Concomitantly with "phase" darkening, the spores swell, break out of their coats, and exude up to 30% of their dry weight; about half of the exudate consists of a calcium chelate of the spore-specific substance dipicolinic acid and the rest of it is mucopeptide fragments and amino acids.

Germination rates vary with the species and strain and also within each strain. As described previously, free spores formed in batch cultures have been found to be heterogeneous with respect to density, and the germination properties of the different density spores differed. Thus a population of spores may germinate heterogeneously.

It has been estimated that the average lag time necessary for an individual spore to become "phase" dark is 3.6 minutes (Vary and Mc-Cormick, 1965). Thus a homogeneous population of rapid strains may germinate with a high degree of synchrony; if there are available additional precursors, metabolites, and O_2, these germinated spores are capable of DNA replication and subsequent growth and division.

Synchrony of the germination process was first recognized by Fitz-James (1955), who found that spores of B. cereus, at high densities to retard later growth, germinated synchronously when placed in heart infusion broth. Germination synchrony was also obtained by harvesting the spores from a chilled culture and then placing them into a special germination medium (Young and Fitz-James, 1959b). The germinated cells changed uniformly, and at 120 minutes after germination DNA synthesis was first observed. Both of these studies indicated that during the two subsequent synchronous cell divisions, DNA synthesis was continuous throughout the growth cycle.

However, using similar techniques, Woese and Forro (1960) found that during synchronous spore germination and outgrowth RNA synthesis occurred before DNA synthesis, and others found that in synchronously germinated B. subtilis spores, chromosome replication was synchronous (Wake, 1963; Oishi et al., 1963). Yoshikawa et al. (1964), combining this system with genetic transformation, found that DNA replication must start at one point on the chromosome and proceed in the same direction along its length. Synchronous replication of B. subtilis DNA during outgrowth has been reviewed in an earlier book on synchrony (Sueoka, 1966).

The synchrony system arising from initiation of spore germination and carried over to the first cell division has aided the study of the initiation of transcription and translation. During a stepwise doubling of cell numbers for several generations, initiation of three enzymes occurred at a specific time during each division cycle and resulted in each enzyme doubling during only a fraction of the cycle in B. cereus (Kobayashi et al., 1965). These data and those of others (Steinberg et al., 1965; Donellan et al., 1965; Torriani and Levinthal, 1967; Steinberg and Halvorson, 1968) demonstrate that sequential transcription and translation take place during the stage of outgrowth.

Studies are not available on events occurring after a few divisions in this naturally synchronous system. Presumably the culture's divisions become asynchronous and thus it appears to be ill suited for synchronous sporulation studies. However, it is highly likely that the system could be modified easily so that induction and regulation of sporulation could be studied also.

IV. MICROCYCLE SPORULATION

This is sporulation without intermediate cell division. Sporogenesis in the aforementioned cultivation systems is preceded by a number of cell divisions. That it is possible to have a direct transition of outgrowing bacterial spores to new sporangia without intermediate cell division has been demonstrated by Vinter and Slepecky (1965). They termed this phenomenon "microcycle sporogenesis." The technique involved the replacement of germinated spores into a medium deficient in nutrients. New sporangia were formed without intermediate cell division only when the cells had previously developed to the stage of elongation and initiated DNA synthesis. The beginning of the first DNA replication in the outgrowing spore at the moment of transfer was found to be most advantageous for the highest yield of new sporangia (Vinter and Chaloupka, 1967). Subsequently, other investigators (Holmes and Levinson, 1967b; MacKechnie and Hanson, 1968) have shown microcycle sporogenesis to occur in a chemically defined medium without using a replacement technique. Fine structure analysis of *B. megaterium* (Freer and Levinson, 1967) during microcycle sporogenesis showed that the second-stage spores were formed in a manner similar to those formed with other methods of cultivation. Other evaluations indicate that the microcycle-formed spores appear to be normal with regard to colony-forming ability, heat resistance, DPA content, and other attributes (Vinter and Slepecky, 1965; Holmes and Levinson, 1967b; MacKechnie and Hanson, 1968). Perhaps the most striking dividend of the microcycle investment has been the indication that a new round of DNA replication occurs during outgrowth prior to spore development (Holmes and Levinson, 1967b; MacKechnie and Hanson, 1968). It is tempting to speculate that since DNA replication is associated with division and formation of two cells, the round of DNA replication occurring during microcycle may represent further evidence of the kinship between cell division and the "atypical cell division" of sporulation. However the finding of new DNA synthesis needs confirmation and further elucidation as to its occurrence during the stage of sporulation. Extensive studies on DNA synthesis in a number of different species during normal sporulation have indicated a cessation of DNA synthesis during the axial filament stage (Young and Fitz-James, 1959a,b) or at least prior to stage II (Canfield and Szul-majster, 1964).

The degree of synchrony during microcycle sporogenesis is not known. Presumably early stages of the process would be relatively synchronous due to initial germination synchrony. The time course for the complete development and the pattern of sporulation types, however, is approximately that found in batch cultures. Further studies on the degree of synchrony seem to be warranted.

V. FUTURE POSSIBILITIES OR PROBLEMS POSSIBLY SOLVABLE BY A SYNCHRONOUS SYSTEM OF SPORULATION

Although much is known about the sporulation process many problems remain to be solved. It is obvious that one of the requisites to any further studies is more precise synchrony. Once this is established it may be possible to clarify the sporulation induction process, to expand understanding of the regulation process, and to understand other aspects of the differentiation. Reference has been made previously to the similarity of stage II septation formation to the "normal" division transverse septum formation. This may be one area that is elucidated further with synchronous techniques. A number of authors (Murrell, 1961; Robinow, 1960; Vinter, 1963a, 1967; Starka and Caslavska, 1965; Fitz-James, 1962, 1965b; Ellar and Lundgren, 1966; Freer and Levinson, 1967) have noted this similarity, and Ellar et al. (1967), have studied the fine structure of both processes in a synchronously dividing and sporulating culture of B. megaterium. In that study and in the extensive fine structure investigations noted earlier (Young and Fitz-James, 1959a; Ryter, 1965; Ohye and Murrell, 1962; Ellar and Lundgren, 1966), it was found that although the initiation of the invagination of the plasma membrane and its subsequent fusion near the center of the cell to form a complete septum are similar in division and sporulation, the processes differ in two respects: (a) in the normal division process, the transverse septum formation occurs in the center of the cell whereas in sporulation, spore septum formation occurs closer to one of the poles; (b) no new cell wall material is deposited at the base of the spore septum as it is in the division septum.

Occasionally cells have been found in which abnormal spore septum formation occurs (Ryter et al., 1961; Young, 1964; Ryter, 1965) leading to the formation of a septum at each pole. Both of these "forespores" were initially formed by the invagination of the plasma membrane, but in one forespore maturation was incomplete and there was deposition of cell wall material between the layers of the membrane septa; the other forespore developed into a mature spore. Remsen and Lundgren (1965) have described abnormal nonsporulating cells in a normal sporulating culture in which both membrane and cell wall invaginations were distinct from normal transverse septum and forespore formation. The multiple septation variants contained as many as eight invaginations (septa) at either pole. That the cell has the capacity for cell wall synthesis at the forespore septum under normal conditions has been shown further by rejuvenation experiments. Rejuvenation, or the ability of the sporulating cell to return to normal growth in the proper media, occurred in cells just prior to the end of stage II and thin sections indicated that completed forespore septa were the sites of cell wall formation after rejuvenation (Fitz-James, 1965a). Vinter and Chaloupka (1967) have extended these findings and have shown further that after stage II the forespore part is unable to be rejuvenated but the mother cell is still capable of further division. It has been proposed that these findings indicate that the "commitment of the cell to sporulation" occurs at the time of the completion of the forespore septum formation. It appears then that the processes of cell division and spore septation may be similar and may be related in some manner; when the normal process of sporulation is upset, the spore septum contains cell wall material.

The asymmetrical "sporulation division" results in a daughter cell which is of smaller size. Balanced growth implies a doubling in the size of the cell prior to division with the resulting cells of equal size; however, in unbalanced growth the cell may fail to double its size and the resulting daughter cell is smaller than the mother cell. This reasoning is largely based on conjecture in the case of bacterial sporulation, but such an occurrence has been shown to be the case in dividing yeasts (Williamson, 1964).

The relationship between transverse septum and spore septum can be extended to include the effects of balanced and unbalanced growth on DNA replication and sporulation, and later, the effect of growth rate on DNA replication and sporulation. Prior replication of the DNA is a mandatory requirement for normal cell division. The exact relationship between DNA replication and the time of septum formation is not clear (Clark and Maaløe, 1967). Prior DNA synthesis in the "sporulation division" is occurring during stage I (Young and Fitz-James, 1959a,b) and is concluded by stage II (Canfield and Szulmajster, 1964), the time of the completion of the spore septum. It is interesting to note that during shift-down transitions, previously suggested (Section II,A) as a physiological transition analogous to the transition which occurs

from log to stationary phase in batch cultures, Kjelgaard *et al.* (1958)
noted a rapid and "reductive" division. There was a last doubling in
the number of cells along with a reduction in the number of nuclei
per cell. In the "sporulation division" there would be the formation
of two "cells" leading to each having one nucleus per cell. Whether
the observed "sporulation division" is analogous to that "reductive" division will have to await further experiment.

Studies on the mechanism controlling initiation of DNA synthesis
have been enhanced by the use of phenethyl alcohol (Berrah and
Konetzka, 1962). At the appropriate concentration of this compound,
cell division but not protein or RNA synthesis will be inhibited, and
it has been shown that the addition of phenethyl alcohol inhibits a
new round of DNA replication (Treick and Konetzka, 1964; Lark, 1966;
Lark and Lark, 1966). Phenethyl alcohol has been demonstrated to inhibit sporulation at substantially lower concentrations than that required
for cell division inhibition (Slepecky, 1963). Results obtained by
the addition of phenethyl alcohol to cultures at various times during
growth suggested that forespore formation was prevented (Slepecky and
Celkis, 1964). Fine structure studies indicated that phenethyl alcohol
completely inhibited the formation of spore septa (Remsen *et al.*, 1966).
giving support to the idea of Treick and Konetzka (1964) that the
primary site of phenethyl alcohol inhibition may be upon bacterial membranes. Others (Lester, 1965; Silver and Wendt, 1967) have examined
the effects of phenethyl alcohol on cell permeability and have concluded
that there is a breakdown of cellular permeability barriers. However,
although the action of phenethyl alcohol in inhibiting DNA replication
may be due to the prevention of the attachment of the chromosome
to the cell membrane as proposed by Treick and Konetzka (1964), it
could also be a secondary consequence of the initial change in cell permeability (Silver and Wendt, 1967). The former suggestion implies that
control of DNA replication is regulated by a structural site on the membrane for the attachment of the DNA, as per the replicon model (Jacob
and Brenner, 1963; Jacob *et al.*, 1963). This now has some experimental
backing from electron microscopic (Jacob *et al.*, 1966) and chemical
(Ganesan and Lederberg, 1965) data. The number of replication points
per chromosome, available within the bacterial cell, is believed to vary
with the cell growth rate (Lark, 1966; Sueoka, 1966; Helmstetter, Chapter 2 in this volume). It has been implied that in rapidly growing cells
sufficient structural sites would be available for the extra chromosome
origins (Lark, 1966); however, it is not known whether the number
of structural sites would vary with the growth rate. It is interesting
in this regard that the required effective concentration of phenethyl

alcohol for inhibition of DNA replication increases with the growth rate (Lark and Lark, 1966), and that as mentioned earlier, less phenethyl alcohol is required for inhibition of the "sporulation division" than was required to inhibit normal division (Slepecky, 1963). Perhaps during the low growth rate that prevails during sporulation, fewer structural sites would be available.

The consideration of spore septation as a "sporulation division" has many implications with reference to the relationship between replication and normal division (see Donachie and Masters, Chapter 3 and Helmstetter, Chapter 2). The formation of a model applicable to sporulation will have to await further developments in that exciting area. However, the similarities briefly outlined above are worth noting and they may stimulate thinking and lead to fruitful experiments.

ACKNOWLEDGMENTS

The author is grateful to his wife, Norma, for her help, to Tony Hitchins and Professor Donald Lundgren for reading and criticizing the manuscript, and to Mary Lasky for typing the manuscript. Some of the work reported here was supported by grants from the National Science Foundation (GB-6433) and from the National Aeronautics and Space Administration (NSG-693).

REFERENCES

Aronson, A. I. (1965a). *J. Mol. Biol.* **11**, 576.
Aronson, A. I. (1965b). *J. Mol. Biol.* **13**, 92.
Aronson, A. I., and Del Valle, M. R. (1964). *Biochim. Biophys. Acta* **87**, 267.
Aubert J. P., and Millet, J. (1963a). *Compt. Rend. Acad. Sci.* **256**, 1866.
Aubert, J. P., and Millet, J. (1963b). *Compt. Rend. Acad. Sci.* **256**, 5442.
Aubert, J. P., and Millet, J. (1965). *In* "Méchanismes de Régulation des Activités Cellulaires chez les Microorganismes" (J. Senez, ed.), p. 545. Gordon and Breach, New York.
Aubert, J. P., Millet, J., Pineau, E., and Milhaud, G. (1961). *Biochim. Biophys. Acta* **51**, 529.
Bach, J. A., and Sadoff, H. L. (1962). *J. Bacteriol.* **83**, 699.
Baillie, A., and Norris, J. R. (1963). *J. Appl. Bacteriol.* **26**, 102.
Baillie, A., and Norris, J. R. (1964). *J. Bacteriol.* **87**, 1221.
Balassa, G. (1963a). *Biochim. Biophys. Acta* **72**, 497.
Balassa, G. (1963b). *Biochim. Biophys. Acta* **76**, 410.
Balassa, G. (1964a). *Biochem. Biophys. Res. Commun.* **15**, 236.
Balassa, G. (1964b). *Biochem. Biophys. Res. Commun.* **15**, 240.
Balassa, G. (1965). *In* "Méchanismes de Régulation des Activités Cellulaires chez les Microorganismes" (J. Senez, ed.), p. 565. Gordon and Breach, New York.
Balassa, G., Ionesco, H., and Schaeffer, P. (1963). *Compt. Rend. Acad. Sci.* **257**, 986.
Barner, H. D., and Cohen, S. S. (1956). *J. Bacteriol.* **72**, 115.
Bergère, J. L., and Hermier, J. (1965a). *Ann. Inst. Pasteur* **109**, 80.
Bergère, J. L., and Hermier, J. (1965b). *Ann. Inst. Pasteur* **109**, 391.
Bernlohr, R. W. (1964). *J. Biol. Chem.* **239**, 538.

Bernlohr, R. W., and Novelli, G. D. (1959). *Nature* **184**, 1256.
Bernlohr, R. W., and Novelli, G. D. (1960). *Biochim. Biophys. Acta* **41**, 541.
Bernlohr, R. W., and Novelli, G. D. (1963). *Arch. Biochem. Biophys.* **103**, 94.
Berrah, G., and Konetzka, W. A. (1962). *J. Bacteriol.* **83**, 738.
Black, S. H., and Gerhardt, P. (1963). *Ann. N.Y. Acad. Sci.* **102**, 755.
Black, S. H., Hashimoto, T., and Gerhardt, P. (1960). *Canad. J. Microbiol.* **6**, 213.
Bostock, C. J., Donachie, W. D., Masters, M., and Mitchison, J. M. (1966). *Nature* **210**, 808.
Bott, K. F., and Lundgren, D. G. (1964). *Radiation Res.* **21**, 195.
Burns, V. W. (1959). *Science* **129**, 566.
Cameron, I. L., and Padilla, G. M. (eds.) (1966). "Cell Synchrony—Studies in Biosynthetic Regulation." Academic Press, New York.
Campbell, A. (1957). *Bacteriol. Rev.* **21**, 263.
Canfield, R. A., and Szulmajster, J. (1964). *Nature* **203**, 496.
Cavallo, G., Falcone, G., and Imperato, S. (1963). *Bacteriol. Proc.*, p. 25.
Chaloupka, J. (1967). *Folia Microbiol.* **12**, 75.
Church, B. D., and Halvorson, H. (1959). *Nature* **183**, 124.
Clark, D. J., and Maaløe, O. (1967). *J. Mol. Biol.* **23**, 99.
Cohen, S. S., and Barner, H. D. (1954). *Proc. Natl. Acad. Sci. U.S.* **40**, 885.
Collier, R. E. (1957). *In* "Spores." (H. O. Halvorson, ed.), p. 10. Am. Inst. Biol. Sci., Washington, D.C.
Curran, H. R., and Evans, F. R. (1945). *J. Bacteriol.* **49**, 335.
Cutler, R. G., and Evans, J. E. (1966). *J. Bacteriol.* **91**, 469.
Del Valle, M. R., and Aronson, A. I. (1962). *Biochem. Biophys. Res. Commun.* **9**, 421.
Doi, R. (1965). *In* "Spores III" (L. L. Campbell and H. O. Halvorson, eds.), p. 111. Am. Soc. Microbiol., Ann Arbor, Michigan.
Doi, R., and Igarashi, R. T. (1964a). *J. Bacteriol.* **87**, 323.
Doi, R., and Igarashi, R. T. (1964b). *Proc. Natl. Acad. Sci. U.S.* **52**, 755.
Donachie, W. D. (1965). *Nature* **205**, 1084.
Donellan, J. E., Nags, E. H., and Levinson, H. S. (1965). *In* "Spores III" (L. L. Campbell and H. O. Halvorson, eds.), p. 152. Am. Soc. Microbiol., Ann Arbor, Michigan.
Ellar, D. J., and Lundgren, D. G. (1966). *J. Bacteriol.* **92**, 1748.
Ellar, D. J., Lundgren, D. G., and Slepecky, R. A. (1967). *J. Bacteriol.* **94**, 1189.
Fitz-James, P. C. (1955). *Canad. J. Microbiol.* **1**, 525.
Fitz-James, P. C. (1960). *Biochem. Biophys. Res. Commun.* **8**, 507.
Fitz-James, P. C. (1962). *J. Bacteriol.* **84**, 104.
Fitz-James, P. C. (1965a). *In* "Méchanismes de Régulation des Activités Cellulaires chez les Microorganismes" (J. Senez, ed.), p. 529. Gordon and Breach, New York.
Fitz-James, P. C. (1965b). *Symp. Soc. Gen. Microbiol.* **15**, 369.
Foster, J. W. (1956). *Quart. Rev. Biol.* **31**, 102.
Foster, J. W., and Perry, J. J. (1954). *J. Bacteriol.* **87**, 295.
Freer, J. H., and Levinson, H. S. (1967). *J. Bacteriol.* **94**, 441.
Ganesan, A. T., and Lederberg, J. (1965). *Biochem. Biophys. Res. Commun.* **18**, 824.
Gardner, R., and Kornberg, A. (1967). *J. Biol. Chem.* **242**, 2383.
Gordon, R. A., and Murrell, W. G. (1967). *J. Bacteriol.* **93**, 495.
Greenberg, R. A. and Halvorson, H. O. (1955). *J. Bacteriol.* **69**, 45.
Grelet, N. (1951). *Ann. Inst. Pasteur* **81**, 430.

4. SYNCHRONY AND BACTERIAL SPORES

Grelet, N. (1957). *J. Appl. Bacteriol.* **20**, 315.
Halvorson, H. O. (1957). *J. Appl. Bacteriol.* **20**, 305.
Halvorson, H. O. (1962). In "The Bacteria" (I. C. Gunsalus and R. Y. Stanier, eds.), Vol. V, p. 223. Academic Press, New York.
Halvorson, H. O. (1965). *Symp. Soc. Gen. Microbiol.* **15**, 343.
Halvorson, H. O., and Srinivasan, V. R. (1964). *Natick Symp. Food Sci.,* p. 1.
Halvorson, H. O., Gorman, J., Tauro, P., Epstein, R., and LaBerge, M. (1964). *Federation Proc.* **23**, 1002.
Halvorson, H. O., Vary, J. C., and Steinberg, W. (1966). *Ann. Rev. Microbiol.* **20**, 169.
Hanson, R. S., Srinivasan, V. R., and Halvorson, H. O. (1963a). *J. Bacteriol.* **85**, 451.
Hanson, R. S., Srinivasan, V. R., and Halvorson, H. O. (1963b). *J. Bacteriol.* **86**, 45.
Hanson, R. S., Blicharska, J., Arnaud, M., and Szulmajster, J. (1964a). *Biochem. Biophys. Res. Commun.* **17**, 690.
Hanson, R. S., Blicharska, J., and Szulmajster, J. (1964b). *Biochem. Biophys. Res. Commun.* **17**, 1.
Hashimoto, T., Black, S. H., and Gerhardt, P. (1960). *Canad. J. Microbiol.* **6**, 203.
Henrici, A. (1928). "Morphological Variation and the Rate of Growth of Bacteria." Baillière, Tindall and Cox, London.
Hershey, A. (1938). *Proc. Soc. Exptl. Biol. Med.* **38**, 127.
Hershey, A., and Bronfenbrenner, J. (1937). *J. Gen Physiol.* **21**, 721.
Hodson, P. H., and Beck, J. V. (1960). *J. Bacteriol.* **79**, 661.
Holmes, P. K., and Levinson, H. S. (1967a). *Currents Modern Biol.* **1**, 256.
Holmes, P. K., and Levinson, H. S. (1967b). *J. Bacteriol.* **94**, 434.
Howitt, C. J. (1960). M. Sc. Thesis, Univ. of Wisconsin, Madison, Wisconsin.
Huntington, E., and Winslow, C.-E.A. (1937). *J. Bacteriol.* **33**, 123.
Imanaka, H., Gillis, J. R., and Slepecky, R. A. (1967). *J. Bacteriol.* **93**, 1624.
Jacob, F., and Brenner, S. (1963). *Compt. Rend. Acad. Sci.* **256**, 298.
Jacob, F., Schaeffer, P., and Wollman, E. L. (1960). *Symp. Soc. Gen. Microbiol.* **10**, 67.
Jacob, F., Brenner, S., and Cuzin, F. (1963). *Cold Spring Harbor Symp. Quant. Biol.* **28**, 329.
Jacob, F., Ryter, A., and Cuzin, F. (1966). *Proc. Royal Soc.* **B164**, 267.
Kaneko, I., and Doi, R. H. (1966). *Proc. Natl. Acad. Sci. U.S.* **55**, 564.
Kerravala, Z. J., Srinivasan, V. R., and Halvorson, H. O. (1964). *J. Bacteriol.* **88**, 374.
Kjelgaard, N. O., Maaløe, O., and Schaechter, M. (1958). *J. Gen. Microbiol.* **19**, 607.
Knaysi, G. (1951). "Elements of Bacterial Cytology," 2nd ed. Comstock Pub. Assoc., Ithaca, New York.
Kobayashi, Y., Steinberg, W., Higa, A., Halvorson, H. O., and Levinthal, C. (1965). In "Spores III" (L. L. Campbell and H. O. Halvorson, eds.), p. 200. Am. Soc. Microbiol., Ann Arbor, Michigan.
Kogoma, T., and Yanagita, T. (1967). *J. Bacteriol.* **94**, 1715.
Kondo, M., Takeda, Y., and Yoneda, M. (1964). *Bikens J.* **7**, 153.
Kornberg, A., Spudich, J. A., Nelson, D. L., and Deutscher, M. P. (1968). *Ann. Rev. Biochem.* **37**, 51.
Lamanna, C. (1940). *J. Infect. Diseases* **67**, 193.
Lark, K. G. (1966). In "Cell Synchrony–Studies in Biosynthetic Regulation" (I. L. Cameron and G. M. Padilla, eds.), p. 54. Academic Press, New York.

Lark, K. G., and Lark, C. (1966). *J. Mol. Biol.* **20,** 9.

Lark, K. G., and Maaløe, O. (1956). *Biochim. Biophys. Acta* **21,** 448.

Lawrence, N. L., and Halvorson, H. O. (1954). *J. Bacteriol.* **68,** 334.

Leitzmann, C., and Bernlohr, R. W. (1965). *J. Bacteriol.* **89,** 1506.

Lester, G. (1965). *J. Bacteriol.* **90,** 29.

Lewis, J. C., Snell, N. S., and Alderton, G. (1965). In "Spores III" (L. L. Campbell and H. O. Halvorson, eds.), p. 47. Am. Soc. Microbiol., Ann Arbor, Michigan.

Lundgren, D. G., and Beskid, G. (1960). *Canad. J. Microbiol.* **6,** 135.

Maaløe, O., and Kjeldgaard, N. O. (1966). "Control of Macromolecular Synthesis; a Study of DNA, RNA, and Protein Synthesis in Bacteria." Benjamin, New York.

MacKechnie, T. J., and Hanson, R. S. (1968). *J. Bacteriol.* **95,** 355.

Maruyama, Y., and Yanagita, T. (1956). *J. Bacteriol.* **71,** 542.

Masters, M., Kuempel, P., and Pardee, A. (1964). *Biochem. Biophys. Res. Commun.* **8,** 348.

Megraw, R. E., and Beers, R. J. (1964). *J. Bacteriol.* **87,** 1087.

Millet, J., and Pineau, E. (1960). *Compt. Rend. Acad. Sci.* **250,** 1363.

Mitchison, J. M., and Vincent, W. S. (1965). *Nature* **205,** 987.

Mitchison, J. M., and Vincent, W. S. (1966). In "Cell Synchrony-Studies in Biosynthetic Regulation" (I. L. Cameron and G. M. Padilla, eds.), p. 328. Academic Press, New York.

Monod, J. (1942). "Recherches sur la Croissance des Cultures Bacteriennes," Actualites Scientifiques et Industrielles No. 911, Hermann, Paris.

Monro, R. E. (1961). *Biochem. J.* **81,** 225.

Murrell, W. G. (1961). *Symp. Soc. Gen. Microbiol.* **11,** 100.

Murrell, W. G., (1967). *Advan. Microbiol. Phys.* **1,** 133.

Murrell, W. G., and Warth, A. D. (1965). In "Spores III" (L. L. Campbell and H. O. Halvorson, eds.), p. 1. Am. Soc. Microbiol., Ann Arbor, Michigan.

Nakata, D., Matsushiro, A., and Miwatoni, T. (1956). *Med. J. Osaka. Univ.* **6,** 1047.

Norris, J. R. (1962). *J. Gen. Microbiol.* **28,** 393.

Norris, J. R., and Wolf, J. (1961). *J. Appl. Bacteriol.* **24,** 42.

Ohye, D. F., and Murrell, W. G. (1962). *J. Cell Biol.* **14,** 111.

Oishi, M., Yoshikawa, H., and Sueoka, N. (1963). *Nature* **204,** 1069.

Padilla, G. M., Cameron, I. L., and Elrod, L. H. (1966). In "Cell Synchrony–Studies in Biosynthetic Regulation" (I. L. Cameron and G. M. Padilla, eds.), p. 269. Academic Press, New York.

Pepper, R. E., and Costilow, R. N. (1964). *J. Bacteriol.* **87,** 303.

Perry, J. J., and Foster, J. W. (1955). *J. Bacteriol.* **69,** 337.

Powell, J. F. (1953). *Biochem. J.* **54,** 210.

Powell, J. F., and Hunter, J. R. (1956). *Biochem. J.* **62,** 381.

Powell, J. F., and Strange, R. E. (1956). *Biochem. J.* **63,** 661.

Powell, J. F., and Strange, R. E. (1957). *Biochem. J.* **65,** 700.

Ramaley, R. F., and Bernlohr, R. W. (1965). *J. Mol. Biol.* **11,** 842.

Remsen, C. C., and Lundgren, D. G. (1965). *J. Bacteriol.* **90,** 1426.

Remsen, C. C., Lundgren, D. G., and Slepecky, R. A. (1966). *J. Bacteriol.* **91,** 324.

Robinow, C. F. (1960). In "The Bacteria" (I. C. Gunsalus and R. Y. Stanier, eds.), Vol. 1, p. 207. Academic Press, New York.

Rogolsky, M., and Slepecky, R. A. (1964). *Biochem. Biophys. Res. Commun.* **16,** 204.

Romig, W. R., and Wyss, O. (1957). *J. Bacteriol.* **74,** 386.

Ryter, A. (1965). *Ann. Inst. Pasteur* **108**, 40.

Ryter, A., and Szulmajster, J. (1965). *Ann. Inst. Pasteur* **108**, 640.

Ryter, A., Ionesco, H., and Schaeffer, P. (1961). *Compt. Rend. Acad. Sci.* **252**, 3675.

Schaechter, M., Maaløe, O., and Kjelgaard, N. O. (1958). *J. Gen. Microbiol.* **19**, 592.

Schaeffer, P. (1961). Ph.D. Thesis, Paris.

Schaeffer, P. (1963). *Symp. Biol. Hung.* **6**, 123.

Schaeffer, P., and Ionesco, H. (1960). *Compt. Rend. Acad. Sci.* **251**, 3125.

Schaeffer, P., Ionesco, H., Ryter, A., and Balassa, G. (1965a). In "Mechanismes de Régulation des Activités Cellulaires Chez les Microorganismes" (J. Senez, ed.), p. 553. Gordon and Breach, New York.

Schaeffer, P., Millet, J., and Aubert, J. P. (1965b). *Proc. Natl. Acad. Sci. U.S.* **54**, 704.

Scherbaum, O. H. (1960). *Ann. Rev. Microbiol.* **14**, 283.

Scott, D. B. M., Delamater, E. D., Minsavage, E. J., and Chu, E. E. (1955). *Science* **123**, 1036.

Sebald, M., and Schaeffer, P. (1965). *Compt. Rend. Acad. Sci.* **260**, 5398.

Silver, S., and Wendt, L. (1967). *J. Bacteriol.* **93**, 560.

Simmons, P. J., and Costilow, R. N. (1962). *J. Bacteriol.* **84**, 1274.

Slepecky, R. A. (1963). *Biochem. Biophys. Res. Commun.* **12**, 369.

Slepecky, R. A., and Celkis, Z. (1964). *Bacteriol. Proc.*, p. 14.

Spizizen, J. (1958). *Proc. Natl. Acad. Sci. U.S.* **44**, 1072.

Spizizen, J. (1965). In "Spores III" (L. L. Campbell and H. O. Halvorson, eds.), p. 64. Am. Soc. Microbiol., Ann Arbor, Michigan.

Spizizen, J., Reilly, B., and Dahl, B. (1963). *Proc. 11th Intern. Congr. Genet., The Hague*, p. 31.

Spotts, C. R., and Szulmajster, J. (1962). *Biochim. Biophys. Acta* **61**, 635.

Spudich, J. A., and Kornberg, A. (1968). *J. Biol. Chem.*, in press.

Srinivasan, V. R. (1965). In "Spores III" (L. L. Campbell and H. O. Halvorson, eds.), p. 64. Am. Soc. Microbiol., Ann Arbor, Michigan.

Srinivasan, V. R., and Halvorson, H. O. (1963). *Nature* **197**, 100.

Stanier, R. Y. (1950). *Bacteriol. Rev.* **14**, 179.

Starka, J., and Caslavska, J. (1965). In "Méchanismes de Régulation des Activités Cellulaires Chez les Microorganismes" (J. Senez, ed.), p. 583. Gordon and Breach, New York.

Stewart, B. J., and Halvorson, H. O. (1953). *J. Bacteriol.* **65**, 160.

Steinberg, W., and Halvorson, H. O. (1968). *J. Bacteriol.* **95**, 469.

Steinberg, W., Halvorson, H. O., Keynan, A., and Weinberg, E. (1965). *Nature* **208**, 710.

Strange, R. E., and Dark, F. A. (1957). *J. Gen. Microbiol.* **17**, 525.

Sueoka, N. (1966). In "Cell Synchrony—Studies in Biosynthetic Regulation" (I. L. Cameron, and G. M. Padilla, eds.), p. 38. Academic Press, New York.

Sussman, A. S., and Halvorson, H. O. (1966). "Spores: Their Dormancy and Germination." Harper & Row, New York.

Szulmajster, J. (1964). *Bull. Soc. Chim. Biol.* **46**, 443.

Szulmajster, J., and Hanson, R. S. (1965). In "Spores III" (L. L. Campbell and H. O. Halvorson, eds.), p. 162. Am. Soc. Microbiol., Ann Arbor, Michigan.

Szulmajster, J., and Schaeffer, P. (1961). *Compt. Rend. Acad. Sci.* **252**, 220.

Takahashi, I. (1961). *Biochem. Biophys. Res. Commun.* **5**, 171.

Takahashi, I. (1965a). *J. Bacteriol.* **89**, 294.

Takahashi, I. (1965b). *J. Bacteriol.* **89**, 1065.

Tamir, H., and Gilvarg, C. (1966). *J. Biol. Chem.* **241**, 1085.
Tevithia, M. J., and Mandel, M. (1967). *Proc. Natl. Acad. Sci. U.S.* **58**, 1174.
Thorne, C. B. (1962). *J. Bacteriol.* **83**, 106.
Thorne, C. B., and Stull, H. B. (1966). *J. Bacteriol.* **94**, 1012.
Tomcsik, J., and Baumann-Grace, J. B. (1959). *J. Gen. Microbiol.* **21**, 666.
Tomcsik, J., and Bouille, M. (1961). *Ann. Inst. Pasteur* **100**, 25.
Tono, H., and Kornberg, A. (1967). *J. Bacteriol.* **93**, 1819.
Torriani, A., and Levinthal, C. (1967). *J. Bacteriol.* **94**, 176.
Treick, W., and Konetzka, W. A. (1964). *J. Bacteriol.* **88**, 1580.
Urba, R. C. (1959). *Biochem. J.* **71**, 513.
Vary, J. C., and McCormick, N. G. (1965). In "Spores III" (L. L. Campbell and H. O. Halvorson, eds.), p. 188. Am. Soc. Microbiol., Ann Arbor, Michigan.
Vennes, J. W., and Gerhardt, P. (1957). *Bacteriol. Proc.*, p. 105.
Vinter, V. (1956). *Folia Biol.* **2**, 216.
Vinter, V. (1957). *Folia Biol.* **3**, 193.
Vinter, V. (1960). *Folia Microbiol.* **5**, 217.
Vinter, V. (1961a). *Nature* **189**, 589.
Vinter, V. (1961b). In "Spores II" (H. O. Halvorson, ed.), p. 127. Burgess Publ., Minneapolis, Minnesota.
Vinter, V. (1962). *Folia Microbiol.* **7**, 120.
Vinter, V. (1963a). *Experientia* **19**, 307.
Vinter, V. (1963b). *Folia Microbiol.* **8**, 147.
Vinter, V. (1967). *Folia Microbiol.* **12**, 89.
Vinter, V., and Chaloupka, J. (1967). *Acta Faculty Med. Univ. Brunenis* **29**, 63.
Vinter, V., and Slepecky, R. A. (1965). *J. Bacteriol.* **90**, 803.
Wake, R. G. (1963). *Biochem. Biophys. Res. Commun.* **13**, 67.
Walker, P. D., and Batty, I. (1965). *J. Appl. Bacteriol.* **28**, 194.
Warth, A. D., Ohye, D. F., and Murrell, W. G. (1963a). *J. Cell. Biol.* **16**, 579.
Warth, A. D., Ohye, D. F., and Murrell, W. G. (1963b). *J. Cell. Biol.* **16**, 593.
Williamson, D. (1964). In "Synchrony in Cell Division and Growth" (E. Zeuthen, ed.), p. 589. Wiley (Interscience), New York.
Winslow, C. E. A. and Walker, H. (1939). *Bacteriol. Rev.* **3**, 147.
Woese, C. R., and Forro, J. R. (1960). *J. Bacteriol.* **80**, 811.
Yoshikawa, H., and Sueoka, N. (1963). *Proc. Natl. Acad. Sci. U.S.* **49**, 559.
Yoshikawa, H., O'Sullivan, A., and Sueoka. N. (1964). *Proc. Natl. Acad. Sci. U.S.* **52**, 973.
Young, I. E. (1964). *J. Bacteriol.* **88**, 242.
Young, I. E., and Fitz-James, P. C. (1959a). *J. Biophys. Biochem. Cytol.* **6**, 467.
Young, I. E., and Fitz-James, P. C. (1959b). *J. Biophys. Biochem. Cytol.* **6**, 483.
Zeuthen, E. (1964). "Synchrony in Cell Division and Growth." Wiley (Interscience), New York.

CHAPTER 5

Synthesis of Macromolecules during the Cell Cycle in Yeast

Patric Tauro, Eckhart Schweizer, Ray Epstein, and
Harlyn O. Halvorson*

I. INTRODUCTION

Yeast provides one of the simplest eucaryotic cellular systems in which
the order of biosyntheses of macromolecules can be examined during
the division cycle and in which a sufficient amount of genetic information is available to examine the relationship between gene order and
the order of macromolecular synthesis. Over one hundred genes have
been mapped on the eighteen chromosomes of the yeast Saccharomyces
cerevisiae (Mortimer and Hawthorne, 1966), but only a few of these
genes have been clearly demonstrated to be structural genes for known
enzymes.

It was first observed by Sylvén and co-workers (1959) that in synchronously dividing yeast several enzymes were synthesized during a

* National Institutes of Health Research Career Professor.

brief period of the cell cycle. The subsequent findings in our laboratory by Gorman et al. (1964) that in interspecific yeast hybrid(s) the parental genes for the same enzyme are expressed at different periods of the cell cycle, and that neither various methods of inducing cell synchrony (Tauro and Halvorson, 1966) nor the presence or absence of inducers influenced the time of enzyme synthesis (Halvorson et al. 1966), suggested that the entire genome was not continuously accessible for transcription during the cell cycle but was transcribed in an ordered manner.

The periodic synthesis of α-glucosidase in S. cerevisiae is, however, determined by gene position. Different strains of yeast were constructed, each of which contained a varying number of unlinked structural genes for the enzyme α-glucosidase. These cultures were synchronously grown, and the timing of α-glucosidase synthesis was determined during the cell cycle (Tauro and Halvorson, 1966). The number of periods of enzyme synthesis was proportional to the number of structural genes present, and each of these structural genes was expressed at a characteristic period of the cell cycle. Thus, we may conclude that in the case of the identical structural genes for α-glucosidase responding to a common regulatory system, the time of enzyme formation is influenced by the position of the genes on the yeast chromosome, but not by regulatory controls. With regard to these enzymes, the genome is apparently accessible for transcription only over a brief period of the cell cycle. Similar phenomena have been observed by Tingle (1967) for β-galactosidase synthesis in Saccharomyces lactis.

II. GENE POSITION AND ENZYME TIMING

The above findings have led us to the following question: What is the relationship between the position of the structural gene and the time of expression during the division cycle? Ordered enzyme synthesis during the cell cycle influenced by gene position implies that the cell contains a clock mechanism for ordering the expression for various segments of the genome. The simplest, but not the only, model to provide such order is the yeast chromosome itself. Consequently, we have attempted to test whether the yeast genome is transcribed in a sequential polarized manner.

This possibility was tested by examining the time of expression during the cell cycle of several structural genes on different chromosomes of S. cerevisiae. Our current understanding of the timing-distance relationship in S. cerevisiae is summarized in Table I. The time of expression of a number of genes on the same chromosome as well as on different chromosomes at varying distances from the centromere has been determined. As seen (Table I) there is no direct relationship between the

TABLE I

ENZYMES, STRUCTURAL GENES, AND THE TIME OF EXPRESSION IN *Saccharomyces cerevisiae*

Chromosome number	Structural gene	Distance from centromere[a]	Enzyme	Approximate timing (fraction of a generation)	Reference
III	$M_3(MA_2)$	35 R	α-Glucosidase	0.1	Tauro and Halvorson (1966)
	hi$_4$	24 L	Histidinol dehydrogenase	0.3	Tauro et al. (1968)
V	ur$_3$	6 L	Orotidine 5'-decarboxylase	0.1	Tauro et al. (1968)
	thr$_3$	40 R	Aspartokinase	0.25	Tauro et al. (1968)
	hi$_1$	44 R	PR-ATP-PPase	0.3	Tauro et al. (1968)
	is$_1$	60 R	Threonine deaminase	0.45	Tauro et al. (1968)
VII	$M_1(MA_1)$	80 R	α-Glucosidase	0.5	Tauro and Halvorson (1966)
VIII	ar$_4$	8 R	Arginosuccinase	0.7	Tauro et al. (1968)
IX	ly$_1$	30 R	Saccharopine dehydrogenase	0.45	Tauro et al. (1968)
XIV	ly$_9$	32	Saccharopine reductase	0.65	Tauro et al. (1968)
Segments:					
5	$M_2(MA_3)$	—	α-Glucosidase	0.75	Tauro and Halvorson (1966)
7	$M_4(MA_4)$	—	α-Glucosidase	0.25	Tauro and Halvorson (1966)

[a] R and L refer to the two sides of the centromere on the chromosome as published by Mortimer and Hawthorne (1966).

104   TAURO, SCHWEIZER, EPSTEIN, AND HALVORSON

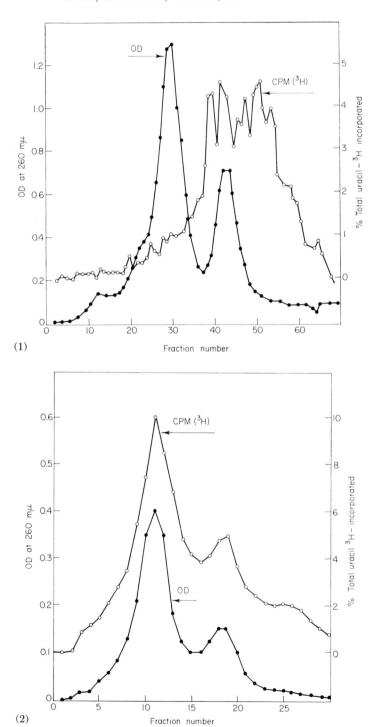

gene distance from the centromere and the time of expression during the cell cycle. The time of expression, on the other hand, is dependent on the direction in which the chromosome is transcribed. For example, the time of expression of hi$_4$ and M$_3$ on chromosome III occurs at widely differing intervals of the cell cycle, M$_3$ is expressed earlier than hi$_4$. On chromosome V, ur$_3$ which is close to the centromere, is expressed earlier in the cell cycle than are three genes on the opposite arm of the same chromosome but at greater distance from the centromere.

Table I also provides a comparison of the time of expression of genes which are either centromere linked or which are relatively equidistant from their respective centromeres. The expression of ur$_3$ on chromosome V and of ar$_4$ on VIII occurs at different times of the cell cycle although both are closely centromere linked. Similarly ly$_1$ and ly$_9$ which are equidistant from the centromere, are expressed at different times during the cell cycle. These results could be summarized as follows:

1. Linked genes are transcribed at close intervals of the cell cycle and in the order of their location on the yeast chromosome.
2. Unlinked genes on the same and different chromosomes are expressed at different times.
3. The time of expression of a given structural gene is dependent on the direction in which the chromosome is transcribed.

These findings are consistent with the view, but do not necessarily prove, that the chromosomes in yeast are transcribed sequentially begin-

FIG. 1. (*top*). Sucrose density gradient analysis of pulse-labeled RNA. A culture (200 ml) of *Saccharomyces cerevisiae* growing exponentially in glucose succinate medium (Tauro and Halvorson, 1966) was labeled with uracil-^{3}H (5 μc/5μg/ml) for 7.5 minutes. The cells were collected by filtration over a 1.2-μ pore size Millipore filter and were disrupted by passing twice through a French pressure cell at 4°C. The broken cell mass was centrifuged in a Sorvall centrifuge at 10,000 g for 20 minutes. To the supernatant solution an equal volume of 4 M LiCl was added to solubilize proteins and to precipitate RNA (Barlow, *et al.*, 1963). After storage for 16 hours at 4°C, the suspension was centrifuged and the rRNA precipitate was washed two times with 2M LiCl. The final precipitate was dissolved in 0.02 M Tris-0.005 M Mg^{2+} buffer at pH 7.2. This solution had a 260 mμ:280 mμ absorbancy ratio greater than 2. The solution of labeled RNA was layered on top of a 5–20% linear sucrose gradient and centrifuged at 78,700 g for 17 hours. After centrifugation, 0.5-ml fractions were collected and analyzed for radioactivity and for optical density at 260 mμ.

FIG. 2. (*bottom*). Sucrose density gradient analysis of pulse-labeled and chased RNA. See legend of Fig. 1 for details. The culture was labeled for 7.5 minutes with uracil-^{3}H (5 μc/5 μg/ml) and then chased with an excess of unlabeled uracil (760 μg/ml) for 15 minutes. The cells were collected, and RNA was isolated and analyzed on sucrose density gradient as described in Fig. 1.

ning from one end of the chromosome and progressing through the centromere to the other end.

III. SYNTHESIS OF RIBOSOMAL AND TRANSFER RNA DURING THE CELL CYCLE

If the transcriptional products of the yeast genome are ordered during the division cycle, then one might expect variations in the appearance of primary products (RNA) of the yeast genome during synchronous growth. Since ribosomal RNA (rRNA) and transfer RNA (tRNA) are stable and comprise nearly 95% of the total cellular RNA, these represent prime objects for analysis. Total cellular RNA increases continuously during the cell cycle in several yeasts (Williamson and Scopes, 1960; Mitchison, 1963; Halvorson et al., 1964). From studies of single cells of S. pombe, Mitchison and Lark (1962) observed that the rate of RNA synthesis in single cells is constant over most of the cell cycle.

To test whether rRNA and tRNA cistrons in yeast are expressed at any unique period of the cell cycle, the patterns of newly synthesized and of stable RNA were examined at intervals. Since there are limited quantities of cells available from synchronous cultures of yeast, methods were developed to permit the isolation of rRNA and tRNA from such small samples (see legend of Fig. 1).

The distribution of pulse-labeled RNA, analyzed by centrifugation in a linear sucrose gradient, is shown in Fig. 1. Large molecular weight RNA was precipitated by LiCl. It can be seen that the RNA synthesized during the 7.5-minute pulse is highly heterogeneous. The smaller molecular weight regions with high specific activity presumably represent messenger RNA. This possibility is supported by the fractionation of cells which were first pulsed for 7.5 minutes with radioactive uracil and then grown for 15 minutes in the presence of excess unlabeled uracil to displace most of the isotope from the low molecular weight mRNA (messenger RNA) into ribosomal RNA. As seen in Fig. 2. only 18 S and 26 S RNA are identified on the sucrose gradients. The radioactivity closely tracks the optical density, indicating that after a pulse-chase experiment one can readily measure newly synthesized rRNA.

To determine whether the two species of rRNA (18 S, 26 S) were made at the same time or at different periods of the cell cycle, synchronous cultures of yeast were labeled early in the cell cycle with uracil-^{14}C and late in the cell cycle with uracil-^{3}H. The two cultures were combined, then the RNA was extracted and analyzed on a sucrose density gradient (Fig. 3). It can be seen that both the ^{14}C and ^{3}H profiles parallel closely the optical density profiles of the 18 S and 26 S RNA. In parallel experiments, covering the entire division cycle, the ratios of incorporation

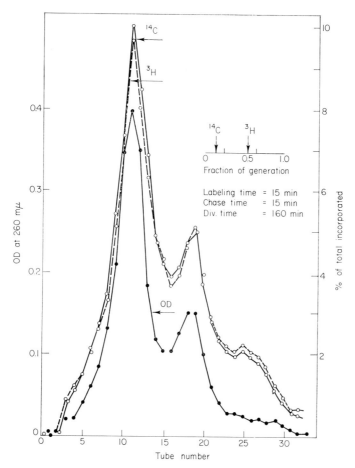

Fig. 3. Synthesis of 18 S and 26 S RNA during the cell cycle of *Saccharomyces cerevisiae*. Cells were grown synchronously as previously described (Tauro and Halvorson, 1966). At two different intervals of the cell cycle aliquots (200 ml) were labeled with either uracil-^{14}C or uracil-^{3}H (5 μc/5 μg/ml) for 15 minutes and later chased with an excess of unlabeled uracil for 15 minutes. The cells were collected and combined; ribosomal RNA was isolated and analyzed in a sucrose density gradient as described in Fig. 1. The radioactivity (^{14}C and ^{3}H) and optical density at 260 mμ of each fraction was determined.

into these two components remain constant. These results suggest that within the limits of this procedure both 18 S and 26 S rRNA are synthesized during the same time intervals of the cell cycle.

Bulk RNA is known to be synthesized throughout the division cycle (Williamson and Scopes, 1960; Mitchison, 1963; Halvorson *et al.* 1964). To determine whether rRNA and tRNA were made continuously or discontinuously during the cell cycle, similar pulse-labeling and chasing experiments were carried out at 15 points during a division cycle. The results (Fig. 4) show that the rate of incorporation of uracil into rRNA and tRNA remains constant throughout the cell cycle.

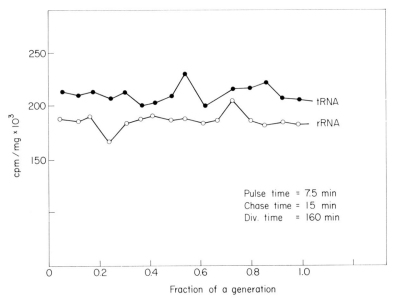

Fig. 4. Syntheses of ribosomal and tRNA during the cell cycle of *Saccharomyces cerevisiae*. Fifty-milliliter aliquots of a synchronously growing culture of *S. cerevisiae* were labeled with uracil-^{14}C (5 μc/5 μg/ml) for 7.5 minutes and chased for 15 minutes with excess of uracil-^{12}C. The cells were collected, and ribosomal RNA was isolated by LiCl precipitation as described in Fig. 1. For tRNA, the supernatant solutions from the LiCl precipitation were deproteinized and the tRNA was isolated by ethanol precipitation and DEAE-cellulose column chromatography (Bock and Cherayil, 1967). The fraction of RNA that eluted between 0.1 and 1 *M* sodium chloride was retained as the tRNA fraction. Radioactivity and optical density at 260 mμ were determined. The specific activities of the RNA samples are plotted as a function of a fraction of a generation time.

IV. THE NUMBER OF RIBOSOMAL RNA CISTRONS IN NUCLEAR DNA OF *Saccharomyces cerevisiae*

The continuous syntheses of rRNA and tRNA during the cell cycle suggest that either the control of transcription of these cistrons is different from that observed for the structural genes of enzymes or that there are a large number of cistrons for rRNA and tRNA scattered over the genome of yeast. In the latter case, each of them could be subject to the normal, ordered transcription. To measure the number of rRNA cistrons, the following DNA-rRNA hybridization studies were conducted.

DNA-^{3}H was isolated from *S. cerevisiae* according to the method of Smith and Halvorson (1967) and separated from mitochondrial DNA by Cs_2SO_4 density gradient centrifugation in the presence of $HgCl_2$ (Nandi *et al.*, 1965). Figure 5A shows the banding of DNA in a preparative ultracentrifugation run. The mitochondrial component had been enriched by previous centrifugation in a CsCl density gradient. The purity of the resulting nuclear and mitochondrial DNA's was tested by equilibrium density gradient centrifugation in CsCl (Fig. 5B).

Bulk rRNA-^{32}P was used for hybridization. This fraction contains mostly cytoplasmic rRNA and a negligible proportion of RNA from mitochondrial particles. Cells of *S. cerevisiae* were first grown in a synthetic medium containing phosphate-^{32}P. The culture was then chased with phosphate-^{31}P for 2.5 generations to displace radioactivity from the unstable RNA fraction. The cells were broken in a French pressure cell at 4°C, the homogenate was deproteinized with phenol, and the rRNA was precipitated as the Li$^+$ salt (Barlow *et al.* 1963). The precipitate was solubilized and the 18 S and 26 S rRNA species separated by repeated centrifugation in a 5–20% linear sucrose gradient. The rRNA species were hybridized with both nuclear and mitochondrial DNA. On the basis of the results (Fig. 6), approximately 0.8% of the nuclear yeast genome was homologous to 18 S RNA, and 1.6% to 26 S RNA. Both subunits together occupied 2.4% of the nuclear genome. In contrast, hybridization could not be detected between bulk rRNA and mitochondrial DNA. From these results we draw the following conclusions:

1. The two ribosomal subunits are coded by distinct cistrons, as can be seen from the additivity of their hybridization values. Similar results have been found in other organisms (Yankofsky and Spiegelman, 1963; Ritossa and Spiegelman, 1965).

2. Both 18 S and 26 S rRNA are coded by the same number of cistrons, since 26 S RNA has twice the molecular weight of 18 S RNA and hybridizes with double the amount of DNA. It is interesting that in HeLa cells 18 S and 26 S RNA arise by controlled degradation of a 45 S molecule, thereby regulating a 1:1 stoichiometry (Scharff and Robbins, 1965; Penman, 1966). This possibility has not been tested in yeast, although large molecular weight RNA species have been reported.

3. From the saturation plateau of the RNA-DNA hybridization, the number of cistrons for each of 18 S and 26 S RNA per haploid nucleus

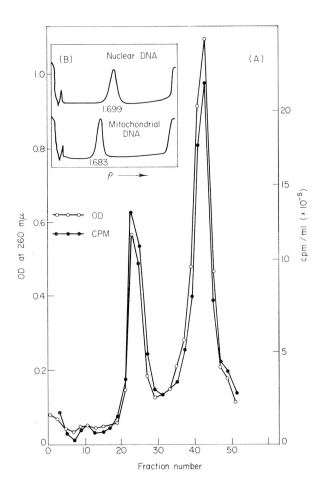

Fig. 5. (A) Separation of nuclear and mitochondrial DNA by density gradient centrifugation. DNA from *Saccharomyces cerevisiae* (350 μg) was mixed with HgCl₂ (145 mμmoles) and centrifuged to equilibrium in Cs₂SO₄ (1.554 gm/ml) at 68,000 *g* for 48 hours. After centrifugation, 10-drop fractions were collected and diluted with 0.5 ml of water. Optical density at 260 mμ and radioactivity of each sample was determined.

(B) Densitometer tracings of ultraviolet photographs of purified nuclear and mitochondrial DNA of *S. cerevisiae*. After centrifugation to equilibrium in CsCl (1.705 gm/ml, pH 8.5) at 44,770 rpm for 17 hours in a Beckman model E analytical centrifuge, ultraviolet photographs were taken and scanned on a Joyce-Loebl microdensitometer.

were calculated to be about 140 (Table II). A similar high multiplicity of rRNA genes has been found in other eucaryotes (Ritossa and Spiegelman, 1965; Attardi *et al.* 1965b; Matsuda and Siegel, 1967) and to a lesser extent in bacteria (Attardi *et al.* 1965a).

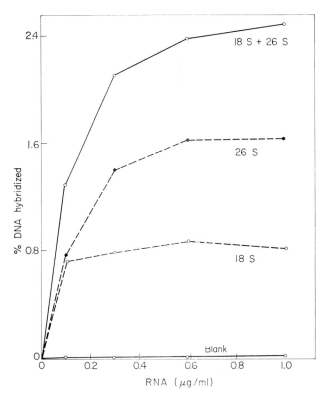

FIG. 6. Hybridization of nuclear DNA with 18 S and 26 S ribosomal RNA from *Saccharomyces cerevisiae*. DNA-RNA hybridization was carried out by the Millipore membrane technique of Gillespie and Spiegelman (1965). Membrane filters containing 10 μg of heat-denatured DNA (650 cpm/μg) were incubated with indicated amounts of RNA-^{32}P (15,000 cpm/μg) in vials containing 5 ml of double-strength sodium–saline citrate (2× SSC) and 0.2% sodium dodecyl sulfate for 12 hours at 62°C. After incubation, each filter was washed with 250 ml of 2× SSC and nonspecific hybrids were removed by treating the filters with pancreatic and T₁ RNase for 2 hours at 22°C. The papers were then washed with 250 ml of 2× SSC and dried, and the radioactivity (^{3}H and ^{32}P) determined.

4. There was no hybridization detected (<0.1%) between mitochondrial DNA and cytoplasmic rRNA. Since one rRNA cistron (18 S and 26 S) would constitute about 25% of the mitochondrial DNA (Fukuhara,

1967), this would have been readily detected. These experiments do not eliminate the possibility that mitochondria contain their own specific ribosomes and that these were excluded in the extraction procedures employed.

From these results it is clear that the extensive production of rRNA in yeast is a consequence of the large number of rRNA genes present. The continuous synthesis of rRNA during the cell cycle would therefore be expected whether the rRNA genes are either randomly scattered or

TABLE II

Estimation of Ribosomal Cistrons in Nuclear DNA of *Saccharomyces cerevisiae*

Size of nuclear genome	1.25×10^{10} daltons
Percent DNA hybridized with	
18 S rRNA	0.8
26 S rRNA	1.6
DNA homologous to 18 S rRNA	1.0×10^8 daltons
DNA homologous to 26 S rRNA	2.0×10^8 daltons
Molecular weight 18 S rRNA	0.6×10^6 daltons
Molecular weight 26 S rRNA	1.2×10^6 daltons
18 S rRNA cistrons	140
26 S rRNA cistrons	140

distributed in several clusters along the genome on the same or on different chromosomes.

V. SYNTHESIS OF MITOCHONDRIAL DNA DURING THE CELL CYCLE OF *Saccharomyces lactis*

The existence of a system in yeast endowed with cytoplasmic inheritance, the mitochondrion (Ephrussi, 1953), raises the question whether the control of these elements is independent of nuclear control during the cell cycle. Yeast mitochondria contain DNA that differs in size and buoyant density from nuclear DNA (Schatz *et al.* 1964; Tewari *et al.* 1965; Corneo *et al.* 1966) and possess a protein-synthesizing system that differs in several respects from the protein-synthesizing system present in the cytoplasm (Clark-Walker and Linnane, 1966). Recently Fukuhara (1967) identified in aerobically grown yeast an RNA fraction that hybridizes specifically with mitochondrial DNA. Until the structural genes of the mitochondria and their products are identified, it will be difficult to examine directly the nature of the transcriptional controls. The synthesis of several mitochondrial components, whose structural genes are determined by the nucleus, are periodic during the cell cycle (Pretlow, unpublished results).

One question that can be explored at the present time is whether

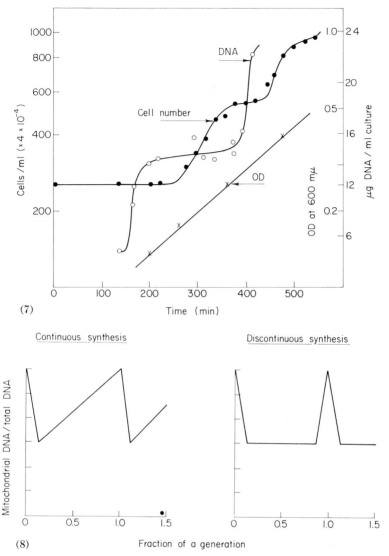

(7)

Time (min)

(8)   Fraction of a generation

FIG. 7. (*top*). Synthesis of bulk DNA in a synchronous culture of *Saccharomyces lactis*. At intervals during synchronous growth in 2% succinate synthetic medium (Tauro and Halvorson, 1966), samples were removed for DNA determination (50 ml), cell counts, and optical density determinations (1 ml). Cell numbers were determined using an electronic Coulter counter after appropriate dilutions with 0.9% NaCl. Optical density was measured at 600 mμ on a Beckman DU spectrophotometer. DNA was determined in perchloric acid extracts of the cells by the diphenylamine reaction (Burton, 1956).

FIG. 8. (*bottom*). Models for the synthesis of mitochondrial DNA during the cell cycle of *Saccharomyces lactis*.

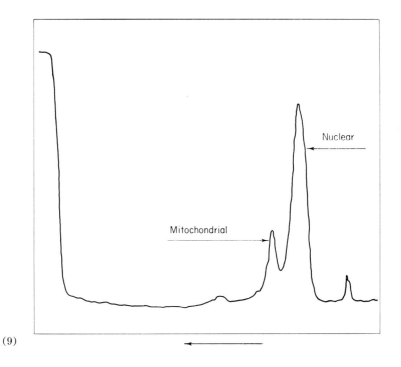

(9)

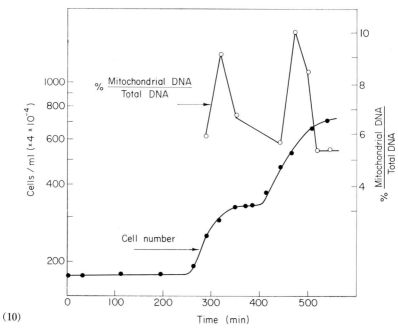

(10)

the replication of the mitochondrial DNA is random or coordinated with the cell cycle, e.g., nuclear DNA synthesis. There is evidence in *Tetrahymena* (Parsons, 1965) that the synthesis of mitochondrial DNA occurs at a time different from that of nuclear DNA. The observations of Guttes *et al.* (1967) in *Physarum*, however, suggest a continuous synthesis of mitochondrial DNA during the cell cycle. In yeast there have been reports suggesting that the biogenesis of mitochondria may occur in the absence of nuclear DNA replication, thus implying an independent mitochondrial DNA replication mechanism (Slonimski, 1953; Moustacchi and Williamson, 1966). Furthermore, the amount of mitochondrial DNA per cell can be considerably varied without changes in the amount of nuclear DNA (Smith, 1967).

To follow the synthesis of mitochondrial DNA during the division cycle, *S. lactis* was chosen as the biological material. This organism can be grown on organic acids, thereby avoiding the repression of mitochondrial synthesis by glucose during vegetative growth. The increase in bulk DNA during the cell cycle is shown in Fig. 7. DNA synthesis is periodic and occupies only a short period of the cell cycle in agreement with the findings with other synchronous cultures of yeast (Halvorson *et al.* 1964; Bostock *et al.*, 1966; Williamson, 1965).

Although bulk DNA is synthesized periodically, mitochondrial DNA which comprises several percent of total DNA could still be synthesized continuously and not be detected by the method employed to determine total DNA.

One could imagine two extreme possibilities for mitochondrial DNA synthesis during the cell cycle: (1) it is continuous over the cell cycle; or (2) it is periodic. In the first case, one would predict that a continuous synthesis of mitochondrial DNA and the discontinuous synthesis of nuclear DNA during the cell cycle would obey the kinetics described in Fig. 8. It is already known that the synthesis of nuclear DNA is periodic. If mitochondrial DNA is synthesized throughout the cell cycle,

---

FIG. 9. (*top*). Microdensitometer tracing of an ultraviolet photograph of *Saccharomyces lactis* DNA. DNA was extracted from *S. lactis* by the method of Smith and Halvorson (1967). Each sample of DNA (10 μg) was complexed with 4.4 mμmoles of $HgCl_2$ and centrifuged to equilibrium in Cs_2SO_4 (1.554 gm/ml) at 44,770 rpm for 17 hours at 25°C. The curve represents a Joyce-Loebl microdensitometer tracing of an ultraviolet photograph.

FIG. 10. (*bottom*). Synthesis of mitochondrial DNA in a synchronous culture of *Saccharomyces lactis*. At intervals during synchronous growth of *S. lactis,* 100-ml samples were removed for DNA extraction and 1 ml for cell number determination. See Fig. 9 for details. The relative amount of nuclear and mitochondrial DNA was calculated by measuring the areas under each of the optical density peaks.

this would lead to a continuous increase in the ratio of mitochondrial DNA to total DNA. On the other hand, if the synthesis of mitochondrial DNA occurs discontinuously, and at a time close to that of nuclear DNA synthesis, one could predict the kinetics of mitochondrial DNA to total DNA ratio to fluctuate as also indicated in Fig. 8. If mitochondrial DNA synthesis occurs periodically but at a widely different time than the nuclear DNA synthesis, then corresponding variations in the ratio would occur at different times of the cell cycle. An experimental test of these predictions has been made with synchronous cultures of *S. lactis* growing in 2% succinate as the carbon source.

Equilibrium density gradient centrifugation was used to separate mito-chondrial DNA and nuclear DNA. Although the buoyant densities differ, considerable overlap of the two components occurs when they are centri-fuged in CsCl. To overcome this difficulty, the DNA extracted from the cells was centrifuged to equilibrium in Cs_2SO_4 after complexing with Hg^{2+} as described by Nandi *et al.* (1965). As shown in Fig. 9, this procedure enables an easy and accurate measurement of both nuclear and mitochondrial DNA of *S. lactis*.

By means of these procedures, the synthesis of mitochondrial DNA during the cell cycle of *S. lactis* was examined. As shown in Fig. 10, the ratio of mitochondrial DNA to total DNA varies discontinuously over a narrow interval of the cell cycle. During the remainder of the period this ratio remains constant. From the results it is evident that the two DNA's are not synthesized at precisely the same time, since in that case the ratio would have remained constant. It is therefore clear that mitochondrial DNA is synthesized discontinuously during the cell cycle and is made at a time very close to the time of nuclear DNA synthesis.

The finding that the population of mitochondrial DNA's, and presum-ably of the mitochondrion itself, shows a cyclic replication paralleling the division cycle opens the possibility for using the latter to examine the possible order of expression of mitochondrial genes.

VI. CONCLUSIONS

The results presented in this paper provide further evidence that in *S. cerevisiae* enzyme synthesis is ordered during the cell cycle. Although the number of enzymes examined thus far is limited, the present data suggest that the genome is expressed in a sequential order.

Such ordering of enzyme expression as is dependent upon gene position, is most easily understood if the control is exercised at the level of transcription of the genome. Thus far it has not been possible directly to test this assumption. No unique periods were found during the cell cycle for rRNA synthesis. However, the finding of over a hundred cistrons for rRNA readily explains the continuous synthesis of this gene product. To directly test the hypothesis of transcriptional control it will be necessary to examine the timing of synthesis of specific tRNA's or of specific mRNA's.

Finally, the periodic synthesis of mitochondrial DNA during the cell cycle of yeast extends the spectrum of genetic determinants whose order of expression can be subjected to biochemical analysis. Of particular interest in the case of mitochondrial DNA is the limited genome size (approximately 10^7 daltons). If the mitochondria are all identical, this material may well provide the experimental material of choice for testing directly whether the temporal control of gene expression in yeast is at the level of transcription or translation.

ACKNOWLEDGMENTS

This investigation was supported by grants from the National Science Foundation (B-1750) and by USPH research grant AI-1459. The technical assistance of Mrs. Christine MacKechnie is gratefully acknowledged.

REFERENCES

Attardi, G., Huang, P. C., and Kabat, S. (1965a). *Proc. Natl. Acad. Sci. U.S.* **53**, 1490.
Attardi, G., Huang, P. C., and Kabat, S. (1965b). *Proc. Natl. Acad. Sci. U.S.* **54**, 185.
Barlow, J. J., Mathias, A. P., and Williamson, R. (1963). *Biochem. Biophys. Res. Commun.* **13**, 61.
Bock, R. M., and Cherayil, J. D. (1967). *In* "Methods in Enzymology: Nucleic Acids" (S. P. Colowick and N. O. Kaplan, eds.), Vol. 12, Part A, pp. 638-644. Academic Press, New York.
Bostock, C. J., Donachie, W. D., Masters, M., and Mitchison, J. M. (1966). *Nature* **210**, 808.
Burton, K. (1956). *Biochem. J.* **62**, 315.
Clark-Walker, G. D., and Linnane, A. W. (1966). *Biochem. Biophys. Res. Commun.* **25**, 8.
Corneo, G., Moore, C., Sanadi, D. R., Grossman, S. I., and Marmur, J. (1966). *Science* **151**, 687.
Ephrussi, B. (1953). "Nucleo-Cytoplasmic Relations in Microorganisms." Oxford Univ. Press, New York.
Fukuhara, H. (1967). *Proc. Natl. Acad. Sci. U.S.* **58**, 1065.
Gillespie, D., and Spiegelman, S. (1965). *J. Mol. Biol.* **12**, 829.
Gorman, J., Tauro, P., Laberge, M., and Halvorson, H. O. (1964). *Biochem. Biophys. Res. Commun.* **15**, 43.
Guttes, E. W., Hanawalt, P. C. and Guttes, S. (1967). *Biochim. Biophys. Acta* **142**, 181.
Halvorson, H. O., Gorman, J., Tauro, P., Laberge, M., and Epstein, R. (1964). *Federation Proc.* **23**, 1002.
Halvorson, H. O., Bock, R. M., Tauro, P., Epstein, R., and Laberge, M. (1966).

In "Cell Synchrony—Studies in Biosynthetic Regulations" (I. L. Cameron and G. M. Padilla, eds.), pp. 102–116. Academic Press, New York.

Matsuda, K., and Siegel, A. (1967). *Proc. Natl. Acad. Sci. U.S.* **58**, 673.

Mitchison, J. M. (1963). *J. Cell. Comp. Physiol.* **62**, 1.

Mitchison, J. M., and Lark, K. G. (1962). *Exptl. Cell. Res.* **28**, 452.

Mortimer, R. K., and Hawthorne, D. C. (1966). *Genetics* **53**, 165.

Moustacchi, E., and Williamson, D. H. (1966). *Biochem. Biophys. Res. Commun.* **23**, 56.

Nandi, U. S., Wang, J. C., and Davidson, N. (1965). *Biochemistry* **41**, 1687.

Parsons, J. A. (1965). *J. Cell Biol.* **25**, 641.

Penman, S. (1966). *J. Mol. Biol.* **17**, 117.

Ritossa, F., and Spiegelman, S. (1965). *Proc. Natl. Acad. Sci. U.S.* **53**, 737.

Scharff, M. D., and Robbins, E. (1965). *Nature* **208**, 464.

Schatz, G., Haselbrunner, H. and Tuppy, H. (1964). *Biochem. Biophys. Res. Commun.* **15**, 127.

Slonimski, P. O. (1953). "Formation des Enzymes Respiratoires chez la Levure." Masson, Paris.

Smith, D. (1967). Ph.D. Thesis, University of Wisconsin, Madison, Wisconsin.

Smith, D., and Halvorson, H. O. (1967). *In* "Methods in Enzymology: Nucleic Acids" (S. P. Colowick and N. O. Kaplan, eds.), Vol. 12, Part A, pp. 538–541. Academic Press, New York.

Sylvén, B., Tobias, C. A., Malmgren, H., Ottoson, R. and Thorell, B. (1959). *Exptl. Cell. Res.* **16**, 77.

Tauro, P., and Halvorson, H. O. (1966). *J. Bacteriol.* **92**, 652.

Tauro, P., Halvorson, H. O., and Epstein, R. L. (1968). *Proc. Natl. Acad. Sci. U.S.* **59**, 277.

Tewari, K. K., Jayaraman, J., and Mahler, H. R. (1965). *Biochem. Biophys. Res. Commun.* **21**, 141.

Tingle, M. (1967). Ph.D. Thesis, University of Wisconsin, Madison, Wisconsin.

Williamson, D. H. (1965). *J. Cell. Biol.* **25**, 517.

Williamson, D. H., and Scopes, A. W. (1960). *Exptl. Cell Res.* **20**, 338.

Yankofsky, S., and Spiegelman, S. (1963). *Proc. Natl. Acad. Sci. U.S.* **49**, 538.

CHAPTER 6

Investigations during Phases of Synchronous Development and Differentiation in *Neurospora crassa**

G. J. Stine

I. INTRODUCTION

Synchrony as defined by conventional use is the simultaneous division or doubling of a cell population. Methods are now available for obtaining synchronous cellular division in bacteria, protozoa, and plant and animal cells in tissue culture (Zeuthen, 1964). A recent survey of the literature reveals that the only synchrony (simultaneous cell division) known in fungi occurs during germination of *Aspergillus niger* spores and division

* Research sponsored by the U.S. Atomic Energy Commission under contract with the Union Carbide Corporation.
  This work was in part supported by a NASA Predoctoral Fellowship and by the Public Health Service under grant No. GM 07607 to the University of Delaware and by the University of Tennessee.

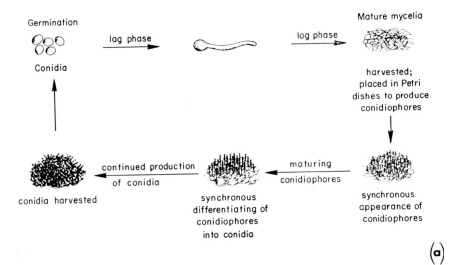

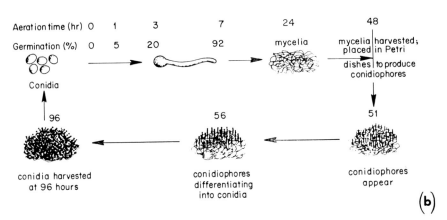

FIG. 1. Asexual life cycle of *Neurospora crassa*. Lag phase represents a 3-hour time period when 0 to 20% of the conidia have germinated. Log phase represents the rapid germination of the remaining conidia (approximately 4 hours) and the rapid growth of the young mycelia (approximately 17 hours). Mature mycelia, 24 to 48 hours are harvested at 48 hours and set at 25°C to develop, conidiophores. Conidiophores synchronously appear over the mycelial mat at 51 hours and elongate to maturity to 56 hours. At 56 hours the conidiophores synchronously differentiate their first conidia. They continue to produce conidia for the next 8 hours. Conidia were harvested at various times from the initial production in synchrony to 96 hours for experimental purposes (Fig. 3).

or budding in *Saccharomyces cerevisiae* (Zeuthen, 1964; Padilla *et al.*, 1965). Many of the fungi, in particular those that maintain a coenocytic multinucleate vegetative state, do not lend themselves to the development of synchrony in cell division. However, besides cell division, there are other events such as morphological differentiation which occur during development and which can be synchronized. A better understanding of how development and differentiation occur in a coenocytic organism might be achieved if one could obtain a system in which the changes in morphology were synchronous in time for a given culture. To accomplish the synchronization of morphological change with time, the asexual cycle of *Neurospora* was phased in a manner which permitted the sampling of growth products and visual observation of morphological change from semidormant asexual conidia through germination, maturation of mycelia, development and differentiation of conidiophores into new conidia (Fig. 1a,b). Another possibility of achieving synchronously produced products for analysis during development is an investigation of the mutants of *Neurospora* which display a definite rhythm in their growth pattern (Pittendrigh *et al.*, 1959; Berliner and Neurath, 1965). A brief review of these strains will be presented following the presentation of synchronization in the phased asexual cycle of *Neurospora crassa*.

II. PHASING OF THE ASEXUAL CYCLE—SYNCHRONIZATION

Neurospora crassa has been studied extensively since its introduction into genetics by Dodge (1929). *Neurospora* exhibits both the sexual and asexual life cycles (Fig. 2). Specific portions of these cycles have been subjected to extensive genetic and biochemical analysis. Only recently, however, have attempts been made to follow a series of particular events throughout an entire life cycle (Stine and Clark, 1967; Stine, 1967).

The phasing of the complete asexual cycle offers one the advantage of observing a large population of a given genotype synchronously undergoing physiological and morphological changes with time.

Neurospora was cultured in minimal medium, and after 48 hours of incubation the mycelial suspension was harvested on a Büchner funnel. Sufficient mycelial suspension was added to the funnel to make a mycelial mat 2–3 mm thick. This mat was washed with cold water and filtered under vacuum to produce a moist mat. The mat was treated with phosphate buffer, pH 6.0, and a mixture of penicillin–streptomycin. Petri dishes containing the mats were inverted and incubated at 25°C. After 3 hours, conidiophores appeared simultaneously over the mycelial mat (Stine and Clark, 1967). The conidiophores continued to elongate for the succeeding 5 hours, at the end of which the conidiophores synchronously produced conidia.

At the time of the initial burst of conidial differentiation, each conidiophore contained a single conidium. With time, however, each conid-

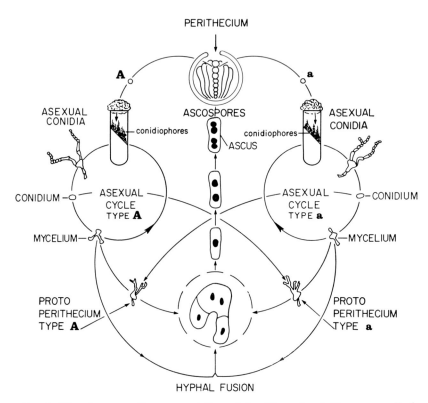

Fig. 2. The diagrammatic representation of the life cycle of *Neurospora* displays a sexual cycle that can occur as a result of hyphal fusion of separate mating types (A, a) or as a result of cross fertilization of the protoperithecium with conidia of the separate mating types (A, a). The asexual cycle as diagrammed in Fig. 1 is represented above in the circles as conidium, germination, mycelium, conidiophores, conidia. (Diagram adapted from the life cycle of *Neurospora crassa*, Srb *et al.*, 1965.)

iophore became a chain of individual conidia; this accounts for the increase in numbers of conidia produced with time. In retrospect, the conidiophore with its single conidium gives an excellent reference to the minimum number of conidiophores produced at the third hour of parent mycelial incubation. An initial burst of conidial differentiation at 1×10^7 with each conidium located on an individual conidiophore of uniform length means that no less than 1×10^7 conidiophores appear synchronously on the parent mycelial mat in a very short time. These conidiophores differentiate their first conidium in synchrony. Thus we have the formation of completely new morphological structures, the conidiophore and conidium, being produced in synchrony and in quantity which allows for experimental investigation (Stine, 1967).

III. UTILIZATION OF THE PHASED ASEXUAL CYCLE OF *Neurospora Crassa*

A. EFFECT OF AERATION AND CHEMICALS ON THE DEVELOPMENT OF CONIDIOPHORES AND CONIDIA

Cultures of *Neurospora* were aerated through 120 hours (Table I). Although 82% of the dry weight of a 72-hour culture was present at

TABLE I

EFFECT OF THE DURATION OF AERATION ON THE PRODUCTION OF CONIDIOPHORES AND CONIDIA

Aeration (hours)	Mycelia (gm dry weight)	Synchronous[a] conidiophores (at 3 hours)	Conidia present[b] at:	
			8 Hours	14 Hours
24	0.205	No	$<10^6$	$<10^6$
48	0.240	No	1×10^8	1×10^9
72	0.250	Yes	1×10^7	1×10^9
96	0.237	Yes	1×10^7	1×10^9
120	0.225	No	$<10^6$	1×10^8

[a] Synchronous means the simultaneous appearance of conidiophores over the mycelial mat.

[b] Production of conidia starts on or about the eighth hour of incubation and continues for approximately 8 hours.

24 hours, synchrony was not observed until the cultures were 72–96 hours old. The ability to produce conidiophores in synchrony was lost at 120 hours. The optimum time for the production of conidiophores and their differentiation into conidia appears to be dependent on the age of the mycelial culture rather than on the dry-weight content.

Benzaldehyde, acetate, malonate, and glycerol alter the production of *Neurospora* conidia (Turian, 1962; Stine, 1967). These compounds were added to the minimal media to ascertain their effect on the development of conidiophores and later conidial differentiation. Only malonate, when added to the medium in varying concentrations, appeared to induce synchrony in the production of conidiophores and increase the total yield of conidia (Table II). When used in concentrations up to 0.10 M, malonate had no effect on the percentage germination and allowed for the synchronous production of conidiophores at 51 hours similar to that achieved in cultures aerated to 72 hours without malonate. It appears that the effect of malonate is to slow or inhibit conidia formation during the elongation of the conidiophores. This inhibition, if it is in fact an inhibition, is released simultaneously in the synchronously growing conidiophores, giving rise to, first, a more synchronous production of conidia

TABLE II

EFFECT OF MALONATE ON GERMINATION, GROWTH, AND PRODUCTION
OF CONIDIOPHORES AND CONIDIA

Malonate concentration[a]	Percent germination		Mycelium (gm dry weight)	Synchronous conidiophores (at 3 hours)	Conidia present at:	
	1 Hour	7 Hours			8 Hours	14 Hours
Control	2	97	0.205	No	1×10^8	1×10^9
0.01	3	97	0.190	Yes	1×10^7	1×10^9
0.05	2	95	0.195	Yes	1×10^7	2×10^9
0.10	0	95	0.205	Yes	1×10^7	2×10^9
0.50	0	9	0.225	No	$<10^6$	$<10^6$

[a] Malonate was added to medium N before autoclaving (10 minutes). Cultures were aerated for 48 hours.

beyond the initial burst of conidial differentiation and, second, a larger total population of conidia.

B. INFLUENCE OF TEMPERATURE AND pH ON GROWTH AND DEVELOPMENT

Various combinations of heat and cold treatments had no effect on synchrony with the following exceptions. Mycelial mats were frozen ($-20°C$) for 48 hours without losing their ability for uniform production of conidiophores or synchronous production of conidia. The effect of a cold shock simulated the effects of using malonate. That is, if one aerates the culture without malonate for 48 hours and then places the mycelial mat at $-20°C$ for 3 hours, one will obtain a similar degree of synchrony in the production of the conidiophores and their differentiation into conidia. However, the total number of conidia produced will be less than in the malonate cultures.

The influence of pH on the production of conidiophores is evident (Table III). Below pH of 5.5 the formation of conidiophores is uneven

TABLE III

EFFECT OF MOISTENING 48-HOUR MYCELIA WITH BUFFERS OF DIFFERENT
pH ON THE PRODUCTION OF CONIDIOPHORES AND CONIDIA

Treatment[a,b] of mycelia	pH of buffer	Synchronous conidiophores (at 3 hours)	Conidia present at:	
			8 Hours	14 Hours
Na phosphate	4.0	No	$<10^6$	$<10^6$
Na phosphate	4.5	No	$<10^6$	1×10^6
Na phosphate	5.0	No	$<10^6$	1×10^6
Na phosphate	5.5	Yes	1×10^6	1×10^8
Na phosphate	6.0	Yes	1×10^7	1×10^9
Na phosphate	6.5	No	$<10^6$	$<10^6$
Tris	7.0			
K phosphate	7.5	Conidiophore and conidia production		
K phosphate	8.0	absent between pH 7.0 and 8.5		
Tris	8.5			
Distilled H_2O	6.9	No	$<10^6$	$<10^6$
Distilled deionized H_2O	6.8	No	$<10^6$	$<10^6$
Supernatant (medium after harvest)	6.2	No	$<10^6$	1×10^6

[a] Mycelia were washed before the treatments listed. Two milliliters of each reagent was spread evenly over the appropriate dish of mycelia before incubation at 25°C.

[b] The growth medium contained 0.05 M malonate.

and the formation of conidia is inhibited. The optimal pH for the uniform production of conidiophores and synchronous production of conidia is approximately 6.0. Either the buffer ingredients used in obtaining pH's above 6.0 or the pH itself completely inhibits the formation of conidiophores.

C. GERMINATION OF AGING CONIDIA

Conidia were obtained at the beginning of their formation on the conidiophores and tested for percentage germination with time (Fig. 3). The 1-hour harvest of conidia includes all those formed from the onset of formation through 1 hour. The conidia tested at 1, 3, 6, 12, 24, and 60 hours likewise contained all the conidia formed from the onset of their formation. The curve of "30 hours" (Fig. 3) represents the germination rate of conidia produced between 18 and 48 hours from conidiophores of mycelia grown on solidified media, a standard nonsynchronous method for growing conidia.

Apparently conidia, regardless of age or culture technique, retain a random capacity to germinate. This may mean that the events necessary for germination are divorced from those events that are necessary for the development and differentiation of conidia from conidiophores. Based on the slopes of the curves in Fig. 3, conidia produced on the solidified medium would not appear to germinate faster than the conidia of the same age produced from the conidiophores of the harvested mycelial mats. There is, however, a difference in the rate of germination for conidia of different ages produced from conidiophores of harvested mycelial mats. These data are in keeping with those of Ryan (1948) and

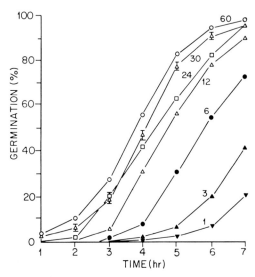

Fig. 3. Germination of aging conidia. Conidia were harvested from separate cultures of conidiophores on filter-formed mats of mycelia at 60, 24, 12, 6, 3, and 1 hour from the onset of conidia formation. The points are average values of duplicate determinations which differed by no more than 6% in germination at 2, 4, and 6 hours. The remaining curve, "30 hours," represents the germination rate of conidia produced between 18 and 48 hours from mycelia grown on solidified medium N. The horizontal bars about and relating to the 30-hour curve represent the standard error calculated from the results of seven similar experiments.

G. J. Stine (1967, unpublished results), which showed that, as the conidia age, they display a faster rate of germination.

Another study (Bianchi and Turian, 1967), using asynchronously produced conidia, showed that heat shocks reduced the average time required for conidial germination; however, the heat shocks did not significantly synchronize the germination process. When asynchronously produced conidia are aerated, they reach a maximum germination in 8 hours. However, if these conidia are kept oxygen free for 5 hours and then aerated, a more rapid conidial germination is obtained. Although there was no apparent lag period in the 5-hour nitrogen-flushed culture, there was no synchrony in the germination process (Bradford and Gibgot, 1963).

D. ENZYME FLUCTUATION DURING DEVELOPMENT OF WILD-TYPE AND MUTANT ACONIDIAL *Neurospora*

The levels of activities for nicotinamide adenine diphosphate (NAD)- and nicotinamide adenine triphosphate (NADP)-glutamic dehydrogenases, nicotinamide adenine diphosphate nucleotidase (NADase), and succinic dehydrogenase were determined with respect to the percentage of conidial germination and mycelial growth in medium N (Vogel and Bonner, 1956) and in altered medium N. The medium was altered either by the addition of 0.05 M glutamate (Sanwal and Lata, 1962a) or by the omission of the trace element zinc. The separate media were used in trying to correlate noticeable enzyme fluctuation with morphological change.

The selection of enzymes to be studied in the investigation was based on their wide distribution in plants, animals, and protists. Two different glutamic dehydrogenases are known as catalysts in the reversible deamination of L-glutamate to α-ketoglutaric acid via a reduction or oxidation of either NAD and/or NADP in the presence of ammonium ions. The reversibility of this oxidative deamination provides a major enzymatic mechanism for the interconversion of α-keto acids and amino acids coupled with the formation or utilization of ammonia. In *Neurospora* the reductive amination of α-ketoglutarate is the only major path for the entry of ammonia (Sanwal and Lata, 1961). The activity of succinic dehydrogenase has been shown in plant (Shepard, 1951; Zalokar, 1959a,b; Fisher, 1960; Stine, 1963) and animal (Kuff and Schneider, 1954) mitochondria. Its presence suggests a functioning tricarboxylic acid cycle, a major energy-producing system in aerobic metabolism. The properties and purification of diphosphopyridine nucleotidase (Kaplan *et al.*, 1951) of *N. crassa* have been described. The *Neurospora* enzyme has been compared to a diphosphopyridine nucleotidase of brain tissue (McIllwain and Rodnight, 1949). Both enzymes were found to be similar except for the inhibitory effects of nicotinamide and a sharp pH optimum for enzyme activity in the animal tissue. A particular biological role for NADase in *Neurospora* has not been determined. Kaplan *et al.* (1951) devised a method for the estimation of the relative activity of NADase in *N. crassa*. Zalokar and Cochrane (1956) concluded that NADase was present in high concentrations on or near the surface of the conidia whereas the mycelia synthesized appreciable amounts of NADase only under conditions favorable to the formation of conidia.

1. Levels of Succinic Dehydrogenase

During the lag phase of germination (Fig. 1), succinic dehydrogenase was not measurable, but during the differentiation of conidiophores from the parent mycelia succinic dehydrogenase activity was very pronounced (Table IV). The fluctuation in enzyme levels during these similar morphological states of differentiation [Weiss (1965) reported that the cell wall of the conidium appears to be identical with that of the hyphae] may indicate significant changes in the underlying physiology of *Neurospora* during these periods of development. It has already been reported that the use of 0.1 M malonate does not inhibit germination of *Neurospora* conidia (Stine and Clark, 1967) and that conidia can actively ferment glucose (Cochrane, 1966). This suggests that the citric acid cycle is functioning during the formation of conidiophores and a different metabolic pathway is operational during conidial germination. Earlier studies by Owens (1954) indicated that the citric acid cycle was not a principal pathway for oxidation of glucose or acetate in the conidia of *Neurospora sitophila*.

The patterns of enzyme fluctuation versus soluble protein found in the medium N and medium N + glutamate cultures were similar except for the apparent inducement of higher levels of succinic dehydrogenase found in the glutamate culture during a period of logarithmic mycelial growth. Although the enzyme level was increased by the addition of glutamate to medium N, the total yield per culture of wet-weight mycelia, dry-weight mycelia, and the time of morphological differentiation (i.e., germination, production of conidiophores, and conidia) were not significantly different from the medium N culture. The medium N glutamate-influenced and zinc-deficient cultures showed significant increases of succinic dehydrogenase during the time of conidiophore development without a significant rise in protein.

In each separate culture, the level of enzyme decreased at a time just preceding the formation of conidia. Thus synthesis of succinic dehydrogenase apparently occurred during the production of conidiophores but not during the formation of conidia. This suggests that the enzyme may be necessary for the production of the conidiophore but not for the differentiation of the conidiophore into conidia. Similar fluctuations in the enzyme levels at these periods of morphological development were also observed for the aconidial strain (Table V), supporting the suggestion that succinic dehydrogenase is required during the production of conidiophores. The data from both the wild-type and the aconidial strain agree with the results of Turian (1960, 1962), which suggest a reciprocal relationship between the presence of succinic dehydrogenase and the

TABLE IV

RELATIVE AMOUNTS OF SUCCINIC DEHYDROGENASE CORRELATED TO THE MORPHOLOGICAL DEVELOPMENT OF WILD-TYPE *Neurospora*[a]

Culture medium		Conidia	Lag phase	Log phase	Mature mycelia	Synchronous appearance of conidiophores	Synchronous conidiophore differentiation into conidia	Conidia
Medium N	Enzyme	1	0	55	50	100	60	1
	Protein	1	2	25	30	30	30	1
N + glutamate	Enzyme	1	0	130	80	125	60	4
	Protein	1	2	35	40	30	30	1
N deficient in zinc	Enzyme	1	0	0	0	10[b]	0[c]	0
	Protein	1	1	11	22	10[d]	10	1

[a] See Fig. 1 for diagrammatic representation of the asexual life cycle events.
[b] Synchronous production of conidiophores could not be achieved in the zinc-deficient culture.
[c] Conidia were not produced in synchrony.
[d] Conidiophores were not produced in synchrony.

TABLE V

RELATIVE AMOUNTS OF SUCCINIC DEHYDROGENASE, NAD-, AND NADP-DEPENDENT GLUTAMIC DEHYDROGENASES, AND NADASE CORRELATED TO MORPHOLOGICAL DEVELOPMENT OF AN ACONIDIAL MUTANT *Neurospora*

Culture medium	Log phase	Mature mycelia	Synchronous appearance of conidiophores	Synchronous[a] conidiophore differentiation		Conidiophores only[b]
				Mycelia	Conidiophores	
Medium N						
Succ. D	40	30	50	5	15	0
NAD-enzyme	80	70	300	20	420	600
NADP-enzyme	1740	1125	1375	1375	550	350
NADase	900	2825	3125	2175	5840	6200
Protein	10	10	15	4	6	6

[a] The conidiophores were separated from the parent mycelia at the time of conidia formation. Both fractions, mycelia and conidiophores, were assayed for enzymes and soluble protein.

[b] The conidiophores were harvested at a time comparable to that when conidia were harvested from the wild-type strain. Since the aconidial mutant does not produce conidia, only the conidiophores were assayed for enzyme protein.

ability of wild-type conidiophores to form conidia. However, the loss of this enzyme cannot be solely responsible for the mechanism of conidia formation, as the enzyme was unmeasurable in aged conidiophores, of the aconidial strain (Table V), which do not undergo conidial differentiation. This does not rule out, however, the possibility that a drop in the level of succinic dehydrogenase, at a specific time in the developed conidiophore, may act as an initiator to the formation of conidia. New conidia differentiated from the medium N + glutamate culture contained 4 times more succinic dehydrogenase than did the conidia which were differentiated from conidiophores of the medium N culture. In the zinc-deficient culture, conidiophores containing little if any succinic dehydrogenase produced conidia with *no measurable* enzyme, yet these conidia germinated as efficiently as conidia that did contain succinic dehydrogenase (the initial conidial inoculum and the conidia produced from conidiophores of the medium N and medium N + glutamate cultures). These data plus the fact that succinic dehydrogenase was not measurable in any culture during the lag phase (Table IV) suggest that the enzyme succinic dehydrogenase is not essential for the germination of conidia.

2. *Levels of NAD- and NADP-Dependent Glutamic Dehydrogenases*

Changes in the levels of the NAD- and NADP-dependent glutamic dehydrogenases were correlated to the asexual cycle (Table VI, Fig. 1).

In the medium N culture, the NAD-enzyme increased slowly through the lag phase, reached a plateau at the end of the log phase, and declined slightly through the time of conidiophore development and differentiation. In the medium N + glutamate culture, except for the 3-fold increase between log phase and mature mycelia, the levels of enzyme were similar to those of the medium N culture.

During the time of conidiophore development in the aconidial strain (Table V), there was a 4-fold increase in the NAD-enzyme. On separation of the mature conidiophores from the parent mycelia, 21 times more NAD-enzyme was found in the conidiophores than in the mycelia from which they were differentiated.

The conidia from the medium N + glutamate culture contained more NAD-enzyme than that found in the conidia from the medium N culture, yet the percentages of germination were the same.

There was an increase in the NADP-enzyme during conidiophore development in the medium N, medium N + glutamate (Table VI), and the aconidial cultures (Table V). A large drop in the NADP-enzyme level occurred during the time of conidiophore differentiation into new conidia (Fig. 1). When the conidiophores of the aconidial culture were separated from the parent mycelia (at a time when the wild-type cultures

TABLE VI
Relative Amounts of NAD- and NADP-Dependent Glutamic Dehydrogenases Correlated to the Morphological Development of Wild-Type *Neurospora*

Culture medium		Conidial inoculum	Lag phase	Log phase	Mature mycelia	Synchronous appearance of conidiophores	Synchronous conidiophore differentiation into conidia	Conidia
Medium N	NAD-enzyme	20	25	400	425	425	350	20
	NADP-enzyme	55	250	5700	2500	3300	1100	30
	Protein	1	2	25	30	30	30	1
N + glutamate	NAD-enzyme	20	25	360	1125	1050	850	35
	NADP-enzyme	55	250	3600	4500	4750	2000	5
	Protein	1	2	35	40	30	30	1

would begin differentiating conidia), the conidiophores contained less than half the amount of the NADP-enzyme found in the parent mycelia. These results are unlike those found for the NAD-enzyme. The positive relationship existing between the increase in NADP-enzyme and dry weight in the medium N culture between lag phase and logarithmic germination and mycelial growth strongly suggests that the NADP-enzyme is essential for normal mycelial metabolism during this period of development. These data are in keeping with the fact that NADP-enzyme mutants grow very slowly in minimal medium; near-normal wild-type growth occurs only if transaminable amino nitrogen is added to the medium (Sanwal and Lata, 1961).

The addition of glutamate had a pronounced effect on the synthesis of the NAD-enzyme between the log phase and the maturation of the culture. During this time, the NAD-enzyme increased two and a half times over the level of the NAD-enzyme in the medium N culture. Sanwal and Lata (1962b) have reported a concurrent regulation of the NAD-enzyme and NADP-enzyme based on the presence of glutamate. They state that in terms of specific activities the NADP-enzyme was repressed between 24 and 48 hours in cultures containing 0.05 M glutamate, whereas the NAD-enzyme was correspondingly derepressed. In terms of specific activities and levels of the respective enzymes in the medium N + glutamate culture, glutamate did influence the greater synthesis of the NAD-enzyme between the log phase and mature mycelia (Stine, 1968), but it did not correspondingly repress the synthesis of the NADP-enzyme (Table VI). Actually, the NADP-enzyme appeared to be repressed during this time interval in the medium N culture and derepressed in the medium N + glutamate culture.

3. Levels of Nicotinamide Adenine Dinucleotidase (NADase)

The level of the NADase-enzyme rapidly decreased during germination and the log phase in the medium N and medium N + glutamate cultures (Table VII). NADase was not measurable again until the appearance of conidiophores. Thus, NADase activity is associated with the appearance of the conidiophores (Fig. 1). The level of NADase continued to increase during a time of conidiophore elongation. The conidia which were differentiated from the conidiophores, in turn, contained quantities of NADase similar to those found associated with the initial conidial inoculum. The lack of NADase activities during log phase and maturation of the mycelia from medium N and medium N + glutamate cultures can be readily explained. NADase was found to be very water soluble, with its production dependent on the age of the mycelia. It was discovered that the gentle washing procedure on log phase and mature

TABLE VII

RELATIVE AMOUNTS OF NADase CORRELATED TO THE MORPHOLOGICAL DEVELOPMENT OF WILD-TYPE *Neurospora*

Culture medium		Conidia	Lag phase	Log phase	Mature mycelia	Synchronous appearance of conidiophores	Synchronous conidiophore differentiation into conidia	Conidia
Medium N	Enzyme	1200	900	0	0	60	1800	1100
	Protein	1	2	25	30	30	30	1
N + glutamate	Enzyme	1200	900	0	0	60	1300	1000
	Protein	1	2	35	40	30	30	1

mycelia reduced the NADase content to unmeasurable quantities (Stine, 1967). Washing of the aconidial mycelia when harvested during these periods of development did not result in a complete loss of NADase activity (Table V). The difference in measurable NADase activity after washing of the log phase and mature mycelia can be considered a distinguishing feature between wild-type and the mutant aconidial mycelia.

During the log phase and the maturation phase of the aconidial mycelia (Table V), the total units of activity increased better than 3-fold. During the time of conidiophore development, NADase activity remained constant. At the time when wild-type conidiophores would differentiate into conidia, the aconidial conidiophores were very long and were easily separated from the parent mycelia. The aged aconidial conidiophores contained 2.6 times the total units of NADase found in the mycelia from which they were produced. Considering that NADase is produced in increasing quantity as the mycelia age, i.e., become older with time (Stine, 1967), it can be stated that the conidiophores appear to age very quickly. Is it possible that this enzyme is an "aging enzyme?"

IV. SUMMARY AND CONCLUSIONS

In fungi exhibiting complete septae and in all higher organisms, the nuclear morphogenetic substances necessary for specific differentiation are stored in defined areas of cytoplasm at a particular site (Weijer *et al.*, 1965). But, in a multinucleate coenocytic organism like *Neurospora*, where there is cytoplasmic continuity throughout the mycelium, one would envision certain restrictions upon the fungi's nuclear behavior regarding physiological and morphological differentiation. Any morphogenetic substance produced by the nucleus (which itself is in continuous migration) or by other DNA-containing structures in a coenocyte, e.g., mitochondria (Luck and Reich, 1964), should be diluted throughout the entire vegetative mycelium. Regardless of this fact, predictable morphological and physiological differentiation does occur throughout the *Neurospora* life cycle.

Since there are a large number of biochemical and morphological mutants of *Neurospora* available from various stock centers in the United States (e.g., Fungal Genetic Stock Center, Dartmouth College, Hanover, New Hampshire), the development of synchronous populations with morphologically differentiated structures, such as the conidiophores and conidia of *N. crassa* in particular, would be very useful for studying factors associated with vegetative development.

Essential in the production of conidiophores in synchrony is the proper care administered to the mycelial mat during harvest. If the mycelia are filtered to a dry state, the subsequent addition of buffer has little effect and synchrony is completely lost. If the mycelia are not filtered enough, the resulting "wet" mat will not produce conidiophores until sufficient evaporation has occurred to expose the surface of the mycelia. In this event, the higher uneven portion of the mycelial mat will be exposed and thus will form conidiophores before other portions of the mycelia, again causing loss of synchrony. Once the correct amount of filtration has been determined, treatment of the mycelial mat with sodium phosphate buffer of pH 6.0 is essential (Stine, 1967).

There is one other study (Bianchi and Turian, 1967) in which the authors tried to obtain hyphae that would differentiate conidia in synchrony. Cultures of *Neurospora* were grown for 3 days in M medium, which permitted only vegetative growth (Turian, 1964). On the third day M medium was replaced with 0.1 M phosphate buffer. However, 15 hours was required for the formation of aerial hyphae and production of conidia. From the data presented, one could not evaluate whether or not the conidiophores or the conidia were produced close enough in time to be thought of as being produced synchronously.

Recently, synchronous mitochondrial division was reported to occur behind growing hyphal tips of *Neurospora* (Hawley and Wagner, 1967).

The asexual cycle of *Neurospora* as described in this paper offers a unique advantage for studies concerned with developmental and aging phenomena. With this system one has the outstanding advantage of using mass conidial populations of a given genotype. Equally important is the fact that enough material can be obtained at any given time for thorough biochemical analysis. Another positive factor is the demonstrable morphological and physiological change that occurs in synchrony giving distinct populations of a given molecular species. Further, one can investigate the functioning of special biochemical pathways and the production of specific enzymes, cellular structures, and organelles such as physiologically similar mitochondria and their production of DNA, RNA, and ribosomes. The relationship of fungi to man necessitates that these types of studies be done using fungi instead of the lower life forms. The introduction of synchrony in an organism such as *Neurospora* opens the door for such investigations. Another aspect to be considered is the possibility of obtaining synchronously produced products during development from mutants of *Neurospora* which display a definite rhythm in their growth and conidiation patterns. A number of such strains currently are being used in investigations to clarify the causes of "biological rhythm." Of the strains available, some display a circadian rhythm (Brandt, 1953; Pittendrigh *et al.*, 1959; Bianchi, 1964; Sargent *et al.*, 1966; Sargent and Briggs, 1968) and some noncircadian

(Neurath and Berliner, 1964; Berliner and Neurath, 1965; Sussman *et al.*, 1964; Sussman *et al.*, 1965; Lowry and Sussman, 1966; Durkee *et al.*, 1966; Namboodiri and Lowry, 1967). Circadian rhythm *circa diem* (approximately a day, Halberg, 1959) pertains to endogenous arrangements and processes which, when entrained or released by a *Zeitgeber* (Aschoff, 1963), continue to oscillate with a periodicity of about 24 hours regardless of environmental conditions. Although rhythmic patterns of growth imply a change in time, mere observations cannot tell us what previous experiences have entrained the underlying circadian oscillations. Normally, as fungus growth progresses over a medium there is little change in the density of the mycelium (Sussman *et al.*, 1964). This pattern of morphological extension is in sharp contrast to the rhythmic mutants of *Neurospora* in which the normal growth pattern is maintained for only a short time after initiation. For example, "clock" mutants of *Neurospora* [noncircadian mutants of *Neurospora* (Sussman *et al.*, 1964)] display a banded type of growth, due to dichotomous branching, which varies in size and regularity depending on the kinds of media used. And "patch" mutants of *Neurospora* (periodic formation of tufts of conidia) display a periodicity in their growth and formation of conidia (Brandt, 1953; Stadler, 1959). A rhythmic pattern is the end product of a morphogenetic phenomenon probably many steps removed from the series of chemical reactions that control cell wall formation or mycelial branching. It is probably the oscillations in the regulatory mechanism of biochemical syntheses that are responsible for periodicity resulting in ring or bandlike growth patterns. "Timex," another rhythmic mutant of *Neurospora* (Sargent *et al.*, 1966), displays rhythms closely analogous to the biological rhythms studied in higher organisms and contains a biochemical mutation whereby the enzyme invertase segregates with the character responsible for the expression of rhythmicity. The strain "Timex" in particular offers one a distinct advantage in studying biochemical-genetic control systems involved in understanding some of the basic mechanisms in the display of rhythm in *Neurospora* and the underlying mechanism of circadian rhythm per se.

Rhythmic mutants of *Neurospora* are now available in which the manifestation of rhythm and length of periodicity can be controlled (Berliner and Neurath, 1966). Thus rhythmic mutants of *Neurospora* present a system that is valuable for investigations in the basic problems of biological rhythms, to perhaps establish a connection between circadian and noncircadian rhythm, and as a source from which we can gain knowledge on the processes of molecular morphogenesis through extraction and analysis of families of similar molecules (species of molecules produced in synchrony in the rhythm mutants) produced in time.

To what degree either of the new approaches, synchrony or rhythmicity, is used in furthering our understanding of the regulation of development and differentiation depends on man's future genius.

REFERENCES

Aschoff, J. (1963). *Ann. Rev. Physiol.* **25**, 581.
Berliner, M. D., and Neurath, P. W. (1965). *J. Cell. Comp. Physiol.* **65**, 183.
Berliner, M. D., and Neurath, P. W. (1966). *Can. J. Microbiol.* **12**, 1068.
Bianchi, D. E. (1964). *J. Gen. Microbiol.* **35**, 437.
Bianchi, D. E., and Turian, G. (1967). *Experientia* **192**, 1.
Bradford, S. W., and Gibgot, B. I. (1963). *Neurospora Newsletter* **4**, 17.
Brandt, W. H. (1953). *Mycologia* **45**, 194.
Cochrane, V. W. (1966). *Colston Papers* **18**, 201–215.
Dodge, B. O. (1929). *Mycologia* **21**, 222.
Durkee, T. L., Sussman, A. S., and Lowry, R. J. (1966). *Genetics* **53**, 1167.
Fisher, H. F. (1960). *J. Biochem.* **235**, 1830.
Halberg, F. (1959). *In* "Photoperiodism and Related Phenomena in Plants and Animals" (R. B. Withrow, ed.), p. 803. Am. Assoc. Advan. Sci., Washington, D.C.
Hawley, E. S., and Wagner, R. P. (1967) *J. Cell Biol.* **35**, 489.
Kaplan, O., Colowick, S. P., and Nason, A. (1951). *J. Biol. Chem.* **191**, 473.
Kuff, E. L., and Schneider, W. C. (1954). *J Biochem.* **206**, 677.
Lowry, R. J., and Sussman, A. S. (1966). *Mycologia* **58**, 541.
Luck, D. J. L., and Reich, E. (1964). *Proc. Natl. Acad. Sci. U.S.* **52**, 931.
McIllwain, H., and Rodnight, R. (1949). *J. Biochem.* **44**, 470.
Namboodiri, A. N., and Lowry, R. J. (1967). *Am. J. Botany* **54**, 735.
Neurath, P. W., and Berliner, M. D. (1964). *Science* **146**, 646.
Owens, R. G. (1954). *Phytopathology* **44**, 501.
Padilla, G. M., Cameron, I. L., and Whitson, G. L. (1965). *Science* **147**, 175.
Pittendrigh, L. S., Bruce, V. G., Rosensweig, N. S., and Rubin, M. L. (1959). *Nature* **184**, 169.
Ryan, F. J. (1948). *Am. J. Botany* **33**, 497.
Sanwal, B. D., and Lata, M. (1961). *Nature* **190**, 286.
Sanwal, B. D., and Lata, M. (1962a). *Biochem. Biophys. Res. Commun.* **6**, 404.
Sanwal, B. D., and Lata, M. (1962b). *Arch. Biochem. Biophys.* **57**, 582.
Sargent, M. L., and Briggs, W. R. (1968). *Plant Physiol.*, in press.
Sargent, M. L., Briggs, W. R., and Woodward, D. O. (1966). *Plant Physiol.* **41**, 1343.
Shepherd, C. J. (1951). *Biochem. J.* **48**, 483.
Srb, A. M., Owen, R. D., and Edgar, R. S. (1965). "Genetics." Freeman, San Francisco, California.
Stadler, D. R. (1959). *Nature* **184**, 170.
Stine, G. J. (1963). Masters thesis, Dartmouth College, Hanover, New Hampshire.
Stine, G. J. (1967). *Can. J. Microbiol.* **13**, 1203.
Stine, G. J. (1968). *J. Cell Biol.* **36**, 81.
Stine, G. J., and Clark, A. M. (1967). *Can. J. Microbiol.* **13**, 447.
Sussman, A. S., Lowry, R. J., and Durkee, T. (1964). *Am. J. Botany* **51**, 243.
Sussman, A. S., Durkee, T. L., and Lowry, R. J. (1965). *Mycopathol. Mycol. Appl.* **25**, 381.
Turian, G. (1960). *Pathol. Microbiol.* **23**, 687.
Turian, G. (1962). *Neurospora Newsletter* **2**, 15.

Turian, G. (1964). *Nature* **202,** 1240.

Vogel, H. J., and Bonner, D. M. (1956). *Microbial Genet. Bull.* **13,** 42.

Weijer, J., Koopmans, A., and Weijer, D. L. (1965). *Can. J. Genet. Cytol.* **7,** 140.

Weiss, B. (1965). *J. Gen. Microbiol.* **39,** 85.

Zalokar, M. (1959a). *Am. J. Botany* **46,** 555.

Zalokar, M. (1959b). *Am. J. Botany* **46,** 602.

Zalokar, M., and Cochrane, V. W. (1956). *Am. J. Botany* **43,** 107.

Zeuthen, E. (1964). *In* "Synchrony in Cell Division and Growth" (E. Zeuthen, ed.). Wiley (Interscience), New York.

CHAPTER 7

Nuclear DNA Replication and Transcription during the Cell Cycle of *Physarum*

Joseph E. Cummins

I. INTRODUCTION

If, as students of the cell cycle, we were able to design organisms fulfilling ideal requirements for cell cycle investigations, we would probably begin by listing the properties of these perfect organisms. Most of us would certainly maintain that the organism should easily be synchronized and that this synchrony be unaffected by statistical considerations—in other words the entire culture should undergo mitosis at precisely the same instant. The synchrony should be maintained over several nuclear division periods and its precision should not decay. Synchronization should not alter the metabolic properties of the synchronized organism. For those whose interest is mitosis, it would be very desirable that the ideal organism have a typical eucaryotic nucleus whose mitotic phases are easily resolvable by light microscopy. Finally it would be desirable that the organism be easily and inexpensively cultivated in a defined medium. At least one organism has properties that come close to fitting the entire description given above, and the present article deals with aspects of the nuclear behavior of one of these remarkable creatures.

Physarum polycephalum is a plasmodial slime mold (myxomycete) whose nuclear divisions are naturally synchronous during vegetative growth. The method for cultivating this organism is given by Daniel and Baldwin (1964) and by Guttes and Guttes (1964). Briefly, the organism is maintained in a semidefined broth medium as an agitated culture of microplasmodia (an ameba-like form in which the individuals contain 2–100 diploid nuclei). To initiate an experiment, microplasmodia are inoculated onto a flat surface where they fuse together to form a macroplasmodium containing as many as several million nuclei. Each nucleus contains about 100 times more DNA than the chromosome of a bacterium, and this DNA is distributed between 20 and 50 small chromosomes (Guttes et al., 1961). The first synchronous mitosis is observed about 5 hours after the microplasmodia are fused together, and synchronous mitoses occur at regular intervals of 8–10 hours after the first nuclear division. One of these synchronous plasmodia will yield about 10 mg of protein, 1 mg of RNA, and 100 μg of DNA. Thus a perfectly synchronous sample can be analyzed without using either single-cell methods or artificial means of synchronizing the cultures. The control of synchronous spore formation in this species was discussed earlier by Daniel (1966). The present article will consider work on DNA replication and transcription during the cell cycle. This report indicates that the processes of DNA replication and transcription are synchronous at the molecular level and also presents data that bear on the factors regulating and maintaining the molecular synchrony.

II. DNA REPLICATION

A. The S Phase

The duration of the DNA synthetic period was originally delineated by Nygaard et al. (1960). They found that nuclear DNA was replicated immediately after mitosis and that the duration of synthesis was 2.5–3.0 hours when the time between nuclear divisions is 8–10 hours. Braun et al. (1965) verified the earlier delineation of the S phase and observed that the maximum rate of DNA synthesis as judged by thymidine incorporation was reached within 5 minutes of the uncoiling of telophase chromosomes. This maximum rate of synthesis was maintained

for about 1.5 hours and subsequently decreased, reaching the low pre-
mitotic level about 4 hours after mitosis. Autoradiographic experiments
by the latter authors showed that DNA synthesis starts simultaneously
in all the nuclei of a plasmodium; but the duration of synthesis varied
somewhat between nuclei. A few of these nuclei continued to incorporate
tritiated thymidine for 2 hours after the majority of the nuclei had
completed replication. These observations indicate that the signal to
initiate replication following mitosis affects all the nuclei precisely and
simultaneously but the duration of replication is less precisely controlled.

Guttes and Guttes (1963) have observed that in newly formed plasmodia a few nuclei replicate DNA throughout their nuclear division
cycle during the cell cycle prior to the first postfusion nuclear division.
Apparently these nuclei originated from those microplasmodia that had
entered mitosis just before they fused to form a macroplasmodium. It
is not known whether these late-replicating nuclei completed replication
before they entered the first nuclear division.

The absence of a G_1 period is not particularly uncommon among a
number of different kinds of rapidly dividing cell types. Some examples
of cells without G_1 periods are the early embryonic cells of grasshoppers
(Gaulden, 1956), sea urchins (Hinegardner *et al.*, 1964), and frogs,
(Graham and Morgan, 1966). Graham and Morgan, (1966) observed
that frog embryo cells develop G_1 periods as the generation times of
cell lines increase and the cell differentiates. (See also Chapter 2 for
Escherichia coli.)

B. SEQUENTIAL REPLICATION

Are individual DNA molecules replicated randomly during the *S* phase,
or is there an organized molecular pattern of replication among the
DNA molecules? Braun *et al.* (1965) performed experiments demonstrating that DNA replication is synchronous at the molecular level. Thus
DNA molecules do not replicate randomly during the *S* phase. These
investigators labeled DNA with thymidine-^{3}H during the last half of
the *S* phase of an interphase and then added bromodeoxyuridine to the
same cultures during the first half of the *S* phase of the following interphase. Bromodeoxyuridine replaced thymidine in newly synthesized DNA
and therefore makes the newly synthesized DNA more dense. The DNA
was then isolated from the plasmodium and centrifuged isopycnically
in a cesium chloride gradient to separate the light unreplicated DNA
from newly replicated dense bromodeoxyuridine containing DNA. When
the light and dense DNA fractions were analyzed, only the light unreplicated DNA contained ^{3}H-radioactivity. This experiment shows that those
DNA molecules which are replicated during the first half of one *S* phase

are again replicated during the first half of the S phase in the following interphase. Therefore, at a molecular level, DNA replication is synchronous.

Mueller and Kajiwara (1966a) have reported experiments and results concerning the replication of a tissue culture cell line that are similar to the replication pattern in the slime mold. This similarity in the mode of DNA replication in a mammalian cell and the slime mold suggests that an organized molecular pattern of replication may be a fundamental property of eucaryotic cells. It will be interesting to learn whether this organized sequential pattern of DNA replication can be observed during the rapid, early embryonic cleavage divisions of metazoa or whether it originates later in the development of the organism.

C. Protein Synthesis and the Control of Replication

The role of protein synthesis in initiating and maintaining DNA replication during the S phase was studied by Cummins and Rusch (1966), who showed that the antibiotic cycloheximide instantly and completely, yet reversibly, inhibited protein synthesis at levels of 10 μg or more per milliliter of culture medium. When inhibitor treatment was initiated during late prophase or metaphase, both mitosis and nuclear reconstruction were observed even though protein synthesis was inhibited. During the period of nuclear reconstruction, DNA synthesis continued. About 20% of the unreplicated nuclear complement was synthesized. Inhibitor treatment during the S period permitted a similar partial replication of nuclear DNA, and the proportion of total nuclear DNA replicated was about the same regardless of when inhibitor treatment was begun. The results of an experiment in which cycloheximide was employed to inhibit protein synthesis during the S phase are shown in Fig. 1.

Employing a sequential labeling procedure in which a pulse of thymidine-^{3}H was given during the last half of an S period followed by a bromodeoxyuridine and cycloheximide treatment initiated at the beginning of the following S period and continued for 3–4 hours, J. E. Cummins and H. P. Rusch (unpublished) observed that DNA synthesized in the presence of antibiotic was replicated semiconservatively and sequentially. The semiconservative mode of replication was obvious from a dense hybrid band formed during isopycnic centrifugation. This dense band had, however, no tritium label; thus it appears that the later replicating parts of the nuclear DNA require complete protein synthesis for the initiation of their DNA synthesis and that the early replicating parts of the genome may terminate synthesis before replication of the later portion is initiated. Other experiments by these authors (Cummins and Rusch, 1966) showed that the DNA replicated in the presence of cycloheximide was stable upon recovery of the plasmodium following removal of the inhibitor, provided that the DNA was not previously labeled with bromodeoxyuridine. It is, therefore, reasonable to assume that the newly replicated DNA was functionally normal.

Factors that may limit DNA replication following cycloheximide treatment include inhibition of the activity or synthesis of the enzymes of deoxynucleotide metabolism and polymerization. However current results make it appear unlikely that nuclear replication is limited by these

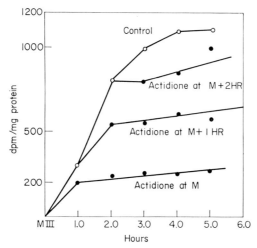

Fig. 1. The effect of cycloheximide (Actidione) on DNA replication. These incubations were in broth medium containing per milliliter 50 μg of fluorodeoxyuridine (*FUdR*), 200 μg of uridine, and 50 μg of cycloheximide. The thymidine-^{3}H level was 100 μg/ml and the specific activity was adjusted to 4.9 μc/μmole. The thymidine counts are normalized regarding protein content. *M* is metaphase of mitosis. (From Cummins and Rusch, 1966.)

factors. Some pertinent observations include those of Brewer and Rusch (1966) that DNA polymerase was highly active in isolated nuclei at least 1 hour prior to the beginning of the *S* phase and this activity did not decrease until the cultures were well into the G_2 phase. Since cycloheximide, even at high levels, has no marked effect on DNA polymerase activity, nuclear replication is probably not prevented by a direct action of the antibiotic on DNA polymerization (Bennett *et al.* 1964). Other pertinent data indicate that the enzyme thymidine kinase is completely synthesized prior to the completion of mitosis and thereafter it is insensitive to cycloheximide inhibition (Sachsenmaier *et al.*, 1967).

The absence of evidence indicating that cycloheximide treatment limits the availability of deoxyuncleotides or DNA polymerase lends credibility to the suggestion of Cummins and Rusch (1966) that nuclear replication is maintained by the synthesis of specific initiator proteins that activate replication in particular chromosomal regions. The inhibition of DNA replication by cycloheximide likely operates by preventing synthesis of these specific proteins.

The following experiment was designed to find out whether or not the replication proteins accumulate during a period when DNA synthesis is prevented by a chemical that does not directly interfere with protein

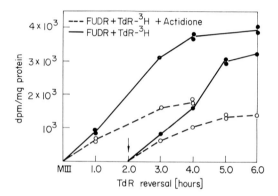

Fig. 2. Failure of replication proteins to accumulate during fluorodeoxyuridine (*FUdR*) blockage. The slime mold was incubated in a medium containing 10 μg of fluorodeoxyuridine and 200 μg of uridine per milliliter. Reversal was accomplished by adding thymidine-^{3}H (*TdR-H*^{3}) 50 μg/ml with the specific activity adjusted to 8.9 μc/μmole in the presence or absence of 50 μg of cycloheximide (Actidione) per milliliter. *MIII* designates that the third mitosis after plasmodial fusion was observed at that point. Reversal of the fluorodeoxyuridine block was initiated at mitosis and 2 hours after mitosis. (From Cummins and Rusch, 1967.)

synthesis. Cummins and Rusch (1966) employed an experiment in which fluorodeoxyuridine inhibition was followed by thymidine reversal in the presence of cycloheximide. This was done to find out whether or not the proteins of replication accumulate during a period when DNA replication is prevented by limiting the availability of thymidine triphosphate. The results of this experiment are shown in Fig. 2. These results show that nuclear DNA is partially, but not completely, replicated following reversal of the FUdR block in the presence of cycloheximide. Based on this experiment it was concluded that completion of an early period of DNA replication is necessary to trigger synthesis of proteins that act to initiate subsequent replication during an *S* phase.

D. REPLICATION AND MITOSIS

Sachsenmaier and Rusch (1964) employed fluorodeoxyuridine as a tool to study the role of DNA synthesis in mitosis. These investigators observed that mitosis cannot occur, in *Physarum*, unless DNA replication is complete. Contact with fluorodeoxyuridine before the end of the S period delayed the onset of the next mitosis. The length of the delay corresponded to the length of inhibition of DNA synthesis, suggesting that a minimum G_2 period several hours in duration precedes mitosis in normal and inhibited cultures. These results appear to be at variance with the results of Guttes and Guttes (1963) indicating that a few nuclei have little or no G_2 period just prior to the first postfusion nuclear division, as mentioned earlier. Nevertheless, the results are most likely not at variance. They suggest that the state of the cytoplasm determines whether or not a plasmodium will complete a long G_2 period, because in the experiments of Guttes and Guttes (1963) the bulk of plasmodial cytoplasm originated from microplasmodia that were in the G_2 phase of the cell cycle prior to fusion. The results of Sachsenmaier and Rusch (1964) suggest that products concerned with the initiation of mitosis begin to accumulate when S phase is terminated in any of the nuclei of the plasmodium.

The next question for consideration is, is mitosis necessary for the initiation of DNA replication? The experiments of Cummins and Rusch (1966) showed that when mitosis was prevented by inhibiting protein synthesis in early prophase, DNA replication was also prevented. When protein synthesis was inhibited in late prophase, mitosis was not prevented and limited DNA replication was observed during the period of nuclear reconstruction. These experiments suggest that mitosis normally plays some necessary role in the initiation of DNA replication. The initiation of DNA replication may, however, not always depend upon completion of mitosis. For example, Brewer and Rusch (1968) have observed that after a plasmodium was subjected to a short heat shock treatment, DNA replication was prematurely initiated late in the G_2 phase of the cell cycle. Heat shock treatment earlier in G_2 phase delayed both mitosis and the onset of DNA replication. Apparently a heat shock during late G_2 phase prevented the normal organization of the mitotic spindle and the nuclei regressed to their interphase state, but activation of replication proteins was not prevented. Consequently, the heat shock treatment induced an endopolyploid replication of the nuclear DNA.

One hypothesis that is consistent with all the observations on protein synthesis, mitosis, and DNA replication in *Physarum*, is that the initiators for an early replicating part of the nuclear DNA complement are synthesized at some time in late G_2 phase or early prophase but are in an inactive form that is activated when the chromosomes uncoil in telophase. A short heat shock during prophase may activate these initiators in the absence of a complete chromosome coiling cycle. The proteins that allow complete replication of nuclear DNA are, however, not synthesized until the first period of replication has been terminated. Complete synthesis of the early replicating portion of nuclear DNA appears to be essential for synthesis of initiators for the later replicating parts of the nuclear DNA. DNA replication may be activated by mitosis in many of the cell types where there is little or no G_1 period. DNA replication may be controlled in a similar manner in cells with a distinct G_1 period, but the initiators for the first period of DNA replication are probably not synthesized until after mitosis is completed but probably before the point at which cells accumulate when they are starved for deoxynucleotides (Mueller *et al.*, 1962; Taylor, 1965, Littlefield and Jacobs, 1965).

III. DNA TRANSCRIPTION

A. RNA Synthesis *in Vivo*

Work on DNA transcription in *Physarum* was begun by Nygaard *et al.* (1960), who followed orotic acid incorporation into RNA pyrimidines during interphase and observed that incorporation was greatly reduced at mitosis. Mittermayer *et al.* (1964) measured changes in the rate of RNA synthesis by determining the amount of radioactive RNA after exposing cultures to labeled uridine for 10 minutes. Their results are shown in Fig. 3. This figure shows that precursor incorporation was negligible during mitosis and at a point during mid-interphase, but there were two peaks of RNA synthesis during each cycle. The lower curve of this figure also shows that RNA synthesis is relatively more resistant to actinomycin D inhibition in early interphase. RNA molecules made at different times during the cell cycle have been further characterized by sucrose density gradient analysis of the uridine-labeled RNA (Braun *et al.*, 1966a). The results of sucrose gradient analysis indicate that messenger RNA, transfer RNA and the two size classes of ribosomal RNA are synthesized throughout the cell cycle except during mitosis. RNA patterns in sucrose gradients prepared after 10 minutes of uridine incorporation show that the RNA made at that time is a mixture of ribosomal precursor RNA and messenger RNA while other experiments indicate that most of this pulse-labeled RNA is located in the nucleus. Only later the radioactive RNA is transferred to the cytoplasm (Braun

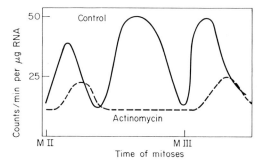

Fig. 3. Uptake of uridine-³H into RNA following a 10-minute pulse treatment of control cultures and cultures pretreated with actinomycin. Results of numerous experiments are combined in two schematic curves. (From Mittermayer *et al.,* 1965.)

et al., 1966b). Even though there were no marked differences in the gradient patterns after 10 minutes of uridine label, Cummins *et al.* (1966b) observed that there were significant differences in the base ratios of RNA molecules synthesized after 10 minutes of exposure to ortho-phosphate ³²P. These results are shown in Fig. 4 as a ratio of adenylic

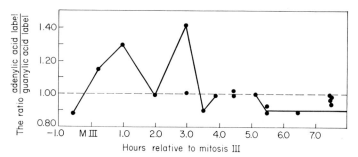

Fig. 4. The ratio of purine label in rapidly labeled RNA following 10-minute pulses with orthophosphate ³²P. *MIII* designates that the third synchronous mitosis had been observed in the cultures. (From Cummins, *et al.,* 1966b.)

acid* to guanylic acid label in RNA. There is a distinct shift in the A to G ratio as the *Physarum* cultures pass through mitosis. Newly synthesized RNA molecules begin to be predominantly adenylic acid-rich rather than guanylic acid-rich as they are at the end of the cycle. The significance of these results will be discussed later in this chapter.

* Abbreviations: A, G, C, T, and U for the nucleotides adenylic acid, guanylic acid, cytidylic acid, thymidylic acid and uridylic acid, respectively.

B. RNA Synthesis *in Vitro*

In vitro studies of nuclei isolated from plasmodia at different stages of the cell cycle by Mittermayer *et al.* (1966a) showed that changes in rate of RNA polymerase activity were similar to changes in the rate of RNA synthesis observed in the intact organism. These results are shown in Fig. 5. This figure gives a measure of radioactive UTP incorporation into RNA following a 10-minute incubation period. The results suggest that the pattern of DNA transcription *in vivo* was retained, to a significant degree, in the isolated nuclei.

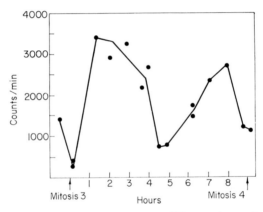

Fig. 5. RNA synthesis in nuclei isolated at different times of the mitotic cycle. Each point represents the rate of incorporation of uridine triphosphate-³H into RNA of nuclei from one stationary culture. (From Mittermayer *et al.*, 1966a.)

Cummins and Rusch (1967) studied the molecular pattern of transcription in isolated nuclei in more detail by employing a nearest neighbor frequency procedure to measure nucleotide sequences in the newly synthesized RNA. This procedure is based on the principle that nucleoside triphosphate precursors of RNA that are labeled in the α-5′-phosphorus position, transfer that label to their nearest neighbor after the polynucleotide is synthesized, isolated, and hydrolyzed to form 2′,3′-mononucleotides. The principles and the procedures for nearest-neighbor frequency analysis were developed by Josse *et al.* (1961) and by Hurwitz *et al.* (1962). The procedures for frequency analysis as employed by Cummins and Rusch (1967) are to isolate nuclei, incubate them with a mixture of the four nucleoside triphosphates, one of which is labeled, and then to isolate and to degrade the product of RNA polymerase action to 2′,3′-nucleotides by hydrolyzing the product in alkali. Finally, the labeled nucleotides are separated by a thin-layer chromatographic procedure, analyzed for radioactivity and the proportion of total incorporated radioactivity bound by each nucleotide is determined. This procedure finally gives a measure of the dinucleotide frequencies in RNA synthesized at different times of the cell cycle, and these dinucleotide frequencies provide fundamental information about the arrangement of nucleotide sequences in newly synthesized RNA.

Cummins and Rusch (1967) have completely analyzed the frequencies of all 16 RNA dinucleotides of polynucleotide synthesized in nuclei isolated at different stages of the cell cycle. Large differences in the

TABLE I

NUCLEIC ACID COMPOSITION[a]

Sample	Adenylic (%)	Guanylic (%)	Cytidylic (%)	Uridylic (T) (%)	GC (%)
Nuclear DNA	28	22	22	28	44
In vitro nuclear RNA[b]					
Early cell cycle	26	24	24	26	48
Late cell cycle	24	27	24	24	51
In vivo nuclear RNA[c]					
Early cell cycle	30	26	23	21	49
Late cell cycle	27	31	23	19	54
Ribosomal RNA	24	32	23	22	55

[a] The nucleotide composition of slime mold DNA and RNA. The composition of nuclear DNA was determined from its density (Braun *et al.*, 1965). The composition of ribosomal RNA and the ^{32}P base ratio analysis of pulse-labeled RNA were completed by Cummins *et al.* (1966b), who employed thin-layer chromatography for the determinations. The composition of RNA synthesized *in vitro* was computed from the nearest-neighbor frequency data given by Cummins and Rusch (1967). The doublet frequencies used for these computations were taken from the points that the regression line passed through the ordinate at "mitosis" and at mitosis plus 10 hours.

[b] Computed from dinucleotide frequencies.

[c] ^{32}P distribution after 10-minute orthophosphate pulses.

nearest-neighbor frequencies were not observed, but the small differences that were observed were significant. The results of these frequency analyses follow a general rule that dinucleotides beginning with G or C increased during the cell cycle and that dinucleotides beginning with A and U tended to decrease in relative frequency as the cell cycle progressed. As the cell cycle progresses, there is therefore, a shift in DNA transcription from AT rich molecular regions to GC rich molecular regions of the nuclear DNA. Table I gives a comparison between the nucleotide composition of nuclear DNA, RNA transcribed *in vitro, in vivo* pulse-labeled RNA, and ribosomal RNA. The composition of RNA synthesized *in vitro* is computed from the nearest-neighbor frequency data.

Table I suggests that the nuclei isolated later in the cell cycle synthesize a greater proportion of ribosomal RNA than those isolated early in the cell cycle. Estimates derived from the base ratio differences indicate that the RNA synthesized in nuclei isolated just after mitosis contains about one-third ribosomal RNA and two-thirds DNA-like RNA

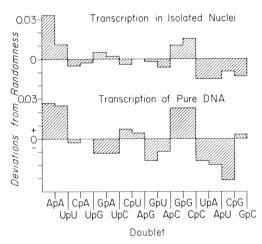

FIG. 6. Nearest-neighbor frequency patterns of RNA transcribed in isolated nuclei and with purified slime mold DNA. These frequencies are normalized to 50% GC content in the DNA templates and expressed as deviations from random expectation in terms of frequency units (0.0625 unit per doublet). This method of treating the data is modified from one described by Subak-Sharpe *et al.* (1966). The frequency data from transcription in isolated nuclei are from Cummins and Rusch (1967). Transcription of pure DNA using bacterial RNA polymerase was as described by Cummins (1967); complete transcription data for purified DNA will be given by Cummins *et al.* (in preparation).

whereas the RNA synthesized in nuclei isolated shortly before mitosis contains two-thirds ribosomal RNA and one-third DNA-like RNA. If relatively GC rich, nonribosomal regions of DNA are preferentially transcribed in these isolated nuclei the estimates over evaluate the proportion of ribosomal RNA.

Figure 6 gives a comparison of the nearest-neighbor (doublet) frequency patterns in nuclear DNA and an average doublet frequency pattern of the RNA synthesized in isolated nuclei. These frequency patterns are corrected for differences in the base ratios of total DNA and newly synthesized RNA. The patterns are quite different indicating that selected molecular regions of total DNA are transcribed in the isolated nuclei.

Another significant aspect of these nearest neighbor frequency studies is that each cell cycle ends while relatively GC-rich DNA molecular regions are being transcribed and begins while AT-rich molecular regions are being transcribed. Apparently mitosis plays a role in organizing transcription during the cell cycle. The next section of this chapter deals in more detail with the proposal that mitosis affects DNA transcription.

C. TRANSCRIPTION AND MITOSIS

The experiments on the role of transcription in regulating mitosis are concerned with the point in the cell cycle in which the last essential messenger RNA for mitosis is synthesized. At present there are no precise estimates of the amounts and kinds of messenger RNA essential for mitosis, nor are there estimates of the duration of synthesis of the quantitatively significant messenger RNA molecules that govern the synthesis of mitotic proteins. The absence of data of this kind hinders complete understanding of the control of cell growth, but estimates of the time during the cell cycle in which the last essential messenger RNA for mitosis is synthesized are cell cycle parameters of no little importance. These parameters constitute the first step toward a complete understanding of the control of mitosis.

Mittermayer *et al.* (1965) made the first attempt to precisely determine a point in the cell cycle when the last essential messenger RNA for mitosis was synthesized. They employed actinomycin D to inhibit messenger RNA synthesis and noted that the last essential messenger RNA for mitosis was formed one to one and one-half cell cycles prior to mitosis. These authors noted that their strain of *Physarum* required relatively high levels of actinomycin D to inhibit mitosis. Sachsenmaier and Becker (1965) repeated these experiments using the same insensitive strain *Physarum*. A strain of *Physarum* somewhat more sensitive to actinomycin D was prevented from undergoing mitosis if the antibiotic was present as early as 2 hours before mitosis. Later Sachsenmaier *et al.* 1967) observed that actinomycin C inhibited RNA synthesis in the insensitive strain more rapidly and completely than did actinomycin D without immediately inhibiting protein synthesis. They also found that actinomycin C prevented mitosis when it was present as early as 1.4 hours prior to telophase. Cummins *et al.* (1965) and Cummins *et al.* (1966a) observed that the last protein essential for mitosis and nuclear reconstruction was synthesized only 14 minutes before metaphase.

In conclusion, it now appears that the last messenger RNA essential for mitosis is completed about 1 hour before metaphase and this messenger RNA continues to be translated until shortly before metaphase. The dramatic structural changes accompanying both mitosis and nuclear reconstruction must be based on changes in existing proteins rather than proteins synthesized during the course of mitosis and these essential proteins must be translated from messenger RNA that was transcribed about 45 minutes earlier than their final translation.

The results on the translation and transcription of the last essential messenger RNA for mitosis in *Physarum* are very similar to the behavior of some mammalian tissue culture cell lines where the last essential messenger RNA molecules for mitosis are transcribed up to 2–3 hours before mitosis (Tobey *et al.*, 1966a, b; Meuller and Kajiwara, 1966b), and these messenger RNA molecules are translated until about 1 hour before mitosis (Tobey *et al.*, 1966a, b; Taylor, 1963).

There is a brief cessation of transcription during mitosis in *Physarum*, as the results of the previous section showed, but other studies indicate that there are no marked changes in the rate of translation or in polysome patterns at that point in the cell cycle (Mittermayer *et al.*, 1966b). There was, however, a shift from polysome to monosomes and a marked reduction in translation about 30 minutes after mitosis, and this phenomenon is probably related to the brief transcription stoppage at metaphase. The results with *Physarum* prove, furthermore, that some messenger RNA molecules can be carried over in the cytoplasm from one cell cycle to the next. These results contrast with observations of Scharff and Robbins (1966), who reported complete disaggregation of polysomes and a cessation of translation during colchicine-arrested metaphase in HeLa cells. It is probable that the basic difference between the results on *Physarum* and HeLa cells is that the nuclear membrane remains intact during mitosis in *Physarum* whereas it disintegrates during mitosis in higher cell types.

Let us now turn to another aspect of the relationship between mitosis and transcription: the effect of mitosis on the cell cycle pattern of RNA synthesis. The discussion indicated earlier that transcription is turned off during mitosis and that a change in the transcription pattern is initiated immediately after mitosis (Cummins and Rusch, 1967). It was argued that mitosis must provide a means of realigning the machinery of transcription at the starting point in the cell cycle. The evidence in support of this hypothesis rests primarily on the observed shift in base ratio and nearest neighbor frequency patterns that occur shortly after mitosis. But since DNA replication begins immediately after mitosis in *Physarum*, it can be argued that the cell cycle transcription pattern is initiated in synchrony with DNA replication rather than by mitosis, as has been suggested by Halvorson (1964) and Tauro *et al.* (Chapter

5) to explain periodic enzyme formation in yeast. There is some preliminary evidence indicating that in *Physarum* transcription is not immediately affected by the presence or the absence of DNA replication. This evidence is from experiments in which DNA replication was blocked immediately after mitosis by the inhibitor fluorodeoxyuridine. During this inhibition there was a sharp increase in rate of RNA synthesis in the inhibited nuclei, and this increase in rate of RNA synthesis was equal in magnitude to the rise in rate of RNA synthesis is uninhibited nuclei (J. E. Cummins, unpublished). Further information concerning the nature of the transcription pattern in the absence of DNA replication will, however, have to be obtained before it will be possible to evaluate fully the role of DNA replication in organizing a cell cycle pattern of RNA synthesis.

Another mode of transcription control has been suggested concerning the regulation of bacterial enzyme synthesis (Cline and Bock, 1966) and bacteriophage multiplication (Kasai and Bautz, 1968). This hypothesis suggests that the regulation of translation may affect the rate of transcription through some sort of feedback control over transcription. It may be possible to relate this type of control mechanism to changes in transcription during the cell cycle of *Physarum* by suggesting that a complete set of messenger RNA molecules, for the entire cycle, are synthesized immediately after mitosis; but that the messenger RNA needed for the later functions—for example, the messenger RNA for synthesis of mitotic proteins cannot be translated, perhaps because an essential transfer RNA molecule is missing or modified, and that this failure of translation may result in feedback inhibition of further mitotic messenger synthesis. Even assuming that this model is reasonable (even appealing because of the two distinct periods of RNA synthesis in *Physarum*), it appears that the chromosome coiling cycle must play a critical role in organizing the transcription pattern either directly upon the genome, or indirectly on translation or replication; because during a series of cell cycles the genome is expressed in exactly the same pattern from one cycle to the next and a unique starting point must be provided for in each cycle. The data presently available are not sufficient for one to decide between the numerous ways in which mitosis might affect transcription. Nevertheless, the data presently available show that mitosis has a role in organizing each cell cycle transcription pattern.

IV. SUMMARY AND CONCLUSIONS

The value of a plasmodial myxomycete, *Physarum polycephalum*, as a model system for studying growth during the cell cycle has been demonstrated. Its essentially perfect mitotic synchrony has placed this organism in a unique position among the subjects of cell cycle investigations and has permitted investigations of the cell processes: DNA replication, DNA transcription, and mitosis.

Studies on nuclear DNA replication yielded the following results:

1. DNA synthesis commences immediately after mitosis and lasts for 2.5–3 hours of the growth cycle of 8–10 hours.

2. DNA molecules which are replicated during a given temporal segment in one S phase are replicated during a similar temporal segment of the following S phase.

3. Mitosis activates replication early in S phase but protein synthesis is necessary for complete nuclear DNA replication.

4. Termination of the early period of DNA replication during the S phase appears to be essential for the synthesis of proteins that activate subsequent DNA replication.

These experiments show that DNA replication is synchronous at the molecular level. The results of these experiments also suggest that protiens necessary for complete nuclear DNA replication are also responsible for maintaining the temporal order in molecular replication. These essential proteins may initiate replication in particular chromosomal regions.

Experiments on DNA transcription yielded the following results:

1. There were two peaks of RNA synthesis during the cell cycle, and RNA synthesis was limited during mitosis and at a point in mid-interphase. Messenger RNA, transfer RNA, and ribosomal RNA were synthesized throughout interphase.

2. There were two peaks of RNA synthesis in nuclei isolated from cultures at different stages of the cell cycle similar to the pattern of RNA synthesis observed *in vivo*.

3. ^{32}P base ratio analysis of RNA synthesized *in vivo* and nearest-neighbor frequency analysis of RNA synthesized *in vitro* in nuclei isolated from cultures at different stages of the cell cycle showed that rapidly labeled RNA made early in the cell cycle was transcribed from AT-rich molecular regions of nuclear DNA while the RNA transcribed later in the cell cycle was richer in GC content.

These results indicate that DNA transcription is synchronous. The abrupt change in transcription pattern after mitosis also suggests that mitosis may be a primary event in organizing the molecular transcription pattern. Besides providing a means of amicably distributing genes between two newborn cells, mitosis appears to provide a mechanism for realigning the transcription machinery at the starting point in its temporal cycle.

ACKNOWLEDGMENT

This article is dedicated to Professor H. P. Rusch.

REFERENCES

Bennett, L. L., Smithers, D., and Ward, C. T. (1964). *Biochim. Biophys. Acta* **87**, 60.

Braun, R., Mittermayer, C., and Rusch, H. P. (1965). *Proc. Natl. Acad. Sci. U.S.* **53**, 924.

Braun, R., Mittermayer, C., and Rusch, H. P. (1966a). *Biochim. Biophys. Acta* **114**, 27.

Braun, R., Mittermayer, C. and Rusch, H. P. (1966b). *Biochim. Biophys. Acta* **114**, 527.

Brewer, E., and Rusch, H. P. (1966). *Biochem. Biophys. Res. Commun.* **25**, 579.

Brewer, E., and Rusch, H. P. (1968). *Exptl. Cell Res.* **49**, 79.

Cline, A., and Bock, R. (1966). *Cold Spring Harbor Symp. Quant. Biol.* **31**, 321.

Cummins, J. E., and Rusch, H. P. (1966). *J. Cell Biol.* **31**, 577.

Cummins, J. E., and Rusch, H. P. (1967). *Biochim. Biophys. Acta* **138**, 124.

Cummins, J., Brewer, E., and Rusch, H. P. (1965). *J. Cell Biol.* **27**, 337.

Cummins, J. E., Blomquist, J., and Rusch, H. P. (1966a). *Science* **154**, 1343.

Cummins, J., Weisfeld, G., and Rusch, H. P. (1966b). *Biochim. Biophys. Acta* **129**, 240.

Cummins, J., Rusch, H., and Evans, T. (1967). *J. Mol. Biol.* **23**, 281.

Cummins, J., Evans, T., and Nygaard, O., manuscript in preparation.

Daniel, J. W. (1966). *In* "Cell Synchrony—Studies in Biosynthetic Regulation" (I. Cameron and G. Padilla, eds.), pp. 117–152. Academic Press, New York.

Daniel, J. W., and Baldwin, H. H. (1964). *In* "Methods in Cell Physiology" (D. Prescott, ed.), Vol. I, pp. 9–42. Academic Press, New York.

Gaulden, M. E. (1956). *Genetics* **41**, 645.

Graham, C., and Morgan, R. (1966). *Develop. Biol.* **14**, 439.

Guttes, E., and Guttes, S. (1963). *Experientia* **19**, 13.

Guttes, E., and Guttes, S. (1964). *In* "Methods in Cell Physiology" (D. M. Prescott, ed.), Vol. I, pp. 43–68. Academic Press, New York.

Guttes, E., Guttes, S., and Rusch, H. P. (1961). *Develop. Biol.* **3**, 588.

Halvorson, H. (1964). *Exptl. Zool.* **157**, 63.

Hinegardner, R., Rao, B., and Feldman, O. (1964). *Exptl. Cell Res.* **36**, 53.

Hurwitz, J., Furth, J., Anders, M., and Evans, E. (1962). *J. Biol. Chem.* **237**, 3752.

Josse, J., Kaiser, A., and Kornberg, A. (1961). *J. Biol. Chem.* **236**, 864.

Kasai, T., and Bautz, E. (1968). *In* "Organizational Biosynthesis" (V. Bryson and H. Vogel, eds.). Academic Press, New York.

Littlefield, J., and Jacobs, P. (1965). *Biochim. Biophys. Acta* **108**, 652.

Mueller, G., and Kajiwara, K. (1966a). *Biochim. Biophys. Acta* **114**, 108.

Mueller, G., and Kajiwara, K. (1966b). *Biochim. Biophys. Acta* **119**, 557.

Mueller, G., Kajiwara, K., Stubblefield, E., and Rueckert, R. (1962). *Cancer Res.* **22**, 1084.

Mittermayer, C., Braun, R., and Rusch, H. P. (1964). *Biochim. Biophys. Acta* **91**, 399.

158   JOSEPH E. CUMMINS

Mittermayer, C., Braun, R., and Rusch, H. P. (1965). *Exptl. Cell Res.* **38**, 33.
Mittermayer, C., Braun, R., and Rusch, H.P. (1966a). *Biochim. Biophys. Acta* **114**, 536.
Mittermayer, C., Braun, R., Chayka, T., and Rusch, H. P. (1966b). *Nature* **210**, 1133.
Nygaard, O., Guttes, S., and Rusch, H. P. (1960). *Biochim. Biophys. Acta* **38**, 298.
Sachsenmaier, W., and Rusch, H. P. (1964). *Exptl. Cell Res.* **36**, 124.
Sachsenmaier, W., and Becker, J. (1965). *Monatsh. Chem.* **96**, 754.
Sachsenmaier, W., Fournier, D., and Gurtler, K. (1967). *Biochem. Biophys. Res. Commun.* **27**, 655.
Scharff, M., and Robbins, E. (1966). *Science* **151**, 992.
Subak-Sharpe, H., Burk, H., Crawford, L., Morrison, J., Hay, J., and Keir, H. (1966). *Cold Spring Harbor Symp. Quant. Biol.* **31**, 737.
Taylor, E. (1963). *J. Cell Biol.* **19**, 1.
Taylor, E. (1965). *Exptl. Cell Res.* **40**, 316.
Tobey, R., Anderson, E. C., and Peterson, D. (1966a). *Proc. Natl. Acad. Sci. U.S.* **56**, 150.
Tobey, R., Peterson, D., Anderson, E. C., and Puck, T. (1966b). *Biophys. J.* **6**, 567.

CHAPTER 8

Control of Enzyme Synthesis during the Cell Cycle of *Chlorella**

Robert R. Schmidt

I. INTRODUCTION

During the past several years, the synchronous culture technique has revealed apparent biochemical diversity in the way different microorganisms control the synthesis of enzymes during their cell cycles. For example, although yeast and bacteria synthesize enzymes in a periodic manner, they have been shown to control enzyme synthesis differently at the transcriptional level. To add further diversity, certain of the unicellular algae (*Chlorella*) have been observed to synthesize individual enzymes continuously at alternating exponential rates throughout cellular development.

Since a primary goal of the cellular biochemist should be to extract unity from apparent biochemical diversity among different types of cells, we have attempted in this chapter to formulate a proposal (1) to explain why different types of transcriptional control might be expected to exist

* The following abbreviations have been used in this chapter: RNA, DNA: ribonucleic acid, deoxyribonucleic acid; ATCase: aspartate transcarbamylase; DHOase: dihydroorotase; GAR: glycinamide ribotide; dCMP, dTMP: deoxycytidine monophosphate, deoxythymidine monophosphate; mRNA: messenger RNA.

in cells with different degrees of cellular compartmentation, and (2) to predict that *Chlorella* and yeast control enzyme synthesis by a similar (if not identical) temporal mechanism at the transcriptional level.

This chapter also includes a description of a new mass-culture technique for synchronized cells which provides nearly constant environmental conditions at high cell densities, thereby resulting in the yield of ample cellular material for most biochemical studies in small harvest volumes. Although the technique was tested using a unicellular green alga, the basic method should be readily adaptable to synchronous cultures of other microorganisms and cells in tissue culture.

II. MASS-CULTURE TECHNIQUE FOR SYNCHRONIZED CELLS

A new mass-culture technique, utilizing continuous dilution with fresh culture medium, recently was developed (Hare and Schmidt, 1968a) to provide cells with nearly constant environmental conditions during synchronous growth. This new technique, when employed to mass-culture synchronized cells of a thermophilic strain (Sorokin and Myers, 1953; Sorokin, 1957) of *Chlorella pyrenoidosa*, yielded approximately three times as much cellular material as a previous nondilution method (Baker and Schmidt, 1963) in less than one-tenth the culture volume during the same time interval.

The technique was developed in the following manner. Synchronized

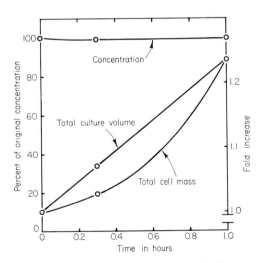

Fig. 1. Calculated increase in culture volume and total cell mass, and the change in concentration of cell mass for exponentially growing (24.12% increase in mass per hour) synchronized cells being diluted at a constant rate (24.12% increase in volume per hour). (From Hare and Schmidt, 1968a.)

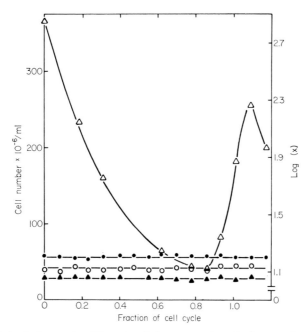

FIG. 2. Actual change in different cellular parameters during synchronous growth of *Chlorella pyrenoidosa* (strain **7-11-05**) in mass culture with continuous dilution. Δ——Δ, Cell number/ml; $x = $ ●——●, mg cellular dry weight/10 ml; ○——○, μg total cellular nitrogen/0.14 ml; ▲——▲, μg total cellular phosphorus/0.6 ml. (From Hare and Schmidt, 1968a.)

cells of the above organism were observed to increase their dry weight (exponentially) by **24.12%** per hour (Schmidt, 1966; Schmidt and Spencer, 1964). Therefore, it seemed reasonable that if culture medium was pumped into a synchronized culture at a constant rate such that the culture volume also was increased by **24.12%** per hour; then, even though the ratio of dry weight to culture volume would not remain constant during the hour interval (i.e., the cells increase their dry weight exponentially while pump rate is constant), this ratio would be the same at the beginning and at the end of the time interval. If that volume in excess of the original culture volume was harvested at the end of the 1-hour interval, it was concluded that the same dilution rate could be continued for the subsequent 1-hour interval. Certain of these relationships were calculated and expressed in graphical form (Fig. 1).

Application of the continuous-dilution technique worked extremely well. Total cellular dry weight, phosphorus, and nitrogen per milliliter of culture were maintained at their initial daughter cell levels throughout synchronous growth (Fig. 2). From a 3-liter culture, it was possible to harvest 3 gm fresh weight of cells per hour in a constant harvest-volume of 583 ml.

In order to calculate the increase in the above growth parameters had there not been a dilution, the actual amount of cellular material (or cell number) per milliliter was multiplied by the following conversion factors: 1.2412 for the first hour, $(1.2412)^2$ for the second hour, etc., or $(1.2412)^n$ for the nth hour (i.e., each harvest contained 24.12% of

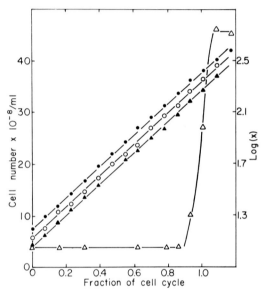

Fɪɢ. 3. Characteristics of synchronous growth of *Chlorella pyrenoidosa* (strain 7-11-05) in mass culture with continuous dilution (after compensation for the dilution effect). △——△, Cell number/ml; $x =$ ●——●, mg cellular dry weight/10 ml; ○——○, μg total cellular nitrogen/0.14 ml; ▲——▲, μg total cellular phosphorus/0.6 ml. (From Hare and Schmidt, 1968a.)

the culture). A convenient method for calculating the conversion factor at the nth hour is: conversion factor $=$ antilog $(n \log 1.2412)$.

After the dilution effect was compensated for, using the above conversion procedure, the growth functions appeared as shown in Fig. 3. Characteristic stepwise count data were obtained, and total cellular phosphorus, nitrogen, and dry weight increased exponentially with essentially identical doubling times as those previously reported (Schmidt, 1966). Although the initial concentration of cells in the new mass-culture system was approximately 30 times higher than in the previous nondilution system (Baker and Schmidt, 1963), it was evident from the exponential nature of the growth curves that light and nutrient saturation was maintained throughout the cell cycle.

In order to express the abscissa of Fig. 3 as fraction of cell cycle, the time (11.70 hours) when one-half the mother cells had divided was defined as one cell cycle. All harvest times were expressed as fraction of that time. The period of nuclear division began at approximately 0.70 of the cell cycle (Curnutt and Schmidt, 1964; Schmidt, 1966).

The synchronization procedure (intermittent illumination), cultural conditions, and growth characteristics of the thermophilic strain (7-11-05) of *C. pyrenoidosa* have been reviewed previously (Schmidt, 1966).

III. CONTROL OF ENZYME SYNTHESIS DURING THE CELL CYCLE

The synthesis of many enzymes has been observed to be periodic during the cell cycles of yeast (Bostock *et al.*, 1966; Gorman *et al.*, 1964; Halvorson *et al.*, 1964; Sylvén *et al.*, 1959) and bacteria (Kuempel *et al.*, 1965; Masters *et al.*, 1964). Although the order of synthesis of individual enzymes in these organisms is related to the position of their respective structural genes on the chromosome (Halvorson *et al.*, 1966; Masters and Pardee, 1965; Tauro and Halvorson, 1966), differences exist in the way each group of organisms regulates gene transcription during the cell cycle. (See also Chapter 5.)

In yeasts the potential for transcription is discontinuous with basal or induced synthesis of different enzymes occurring only at given times in the cell cycle (Tauro and Halvorson, 1966). End-product repression appears to be operative, but only when genes become available for transcription by some ordered or sequential mechanism (Halvorson *et al.*, 1966). However, in bacteria the potential for transcription is continuous (i.e., enzyme synthesis can be induced or derepressed at any time) with end-product repression playing the major role in regulating the time of gene transcription (Donachie, 1965; Kuempel *et al.*, 1965; Masters and Donachie, 1966; see also Chapter 3).

From a comparative biochemical standpoint, one immediately wonders why these two groups of microorganisms should differ in their regulation of a process so basic as gene transcription. However, when the degree of internal cellular organization of these microorganisms is examined, their differences in transcriptional control are not so surprising. For example, since bacteria have minimal compartmentation (i.e., no nuclear membrane or mitochondria, etc.), resulting in rapid exposure of the genome to repressors (cytoplasmic products), it is not likely that there has been selective pressure on these organisms to evolve a higher level of transcriptional control than end-product repression. In contrast, because of the number of membrane barriers (e.g., mitochondrial, nuclear) that end products presumably must pass before reaching the genome in a yeast cell, it seems likely that such an organism would have evolved a mechanism (probably temporal in nature), in addition to end-product repression, to regulate more closely the transcriptional process.

If cellular compartmentation and ordered gene transcription are related from an evolutionary standpoint, it would be reasonable to propose that *Chlorella*, with its high degree of compartmentation (i.e., chloroplast, nucleus, mitochondria, etc.), would also control transcription by

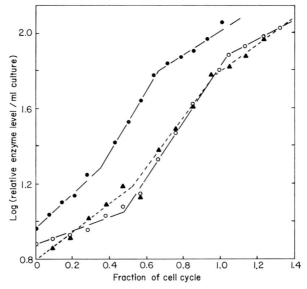

FIG. 4. Relative level of three ribonucleotide biosynthetic enzymes during the cell cycle of *Chlorella pyrenoidosa* (strain 7-11-05). ●——●, Glycinamide ribotide kinosynthetase; ▲---▲, dihydroorotase; ○——○, aspartate transcarbamylase. (GAR kinosythetase, Molloy, 1969; DHOase, Baechtel, 1967; ATCase, Cole and Schmidt, 1964.)

a mechanism similar to that observed in yeasts. The observation (Knutsen, 1965) that the inducibility of nitrite reductase (an inducible enzyme) varies during the cell cycle of *Chlorella* appears to lend support to the above proposal.

To determine whether *Chlorella* also synthesizes noninducible enzymes in a periodic manner, the levels of five different biosynthetic enzymes were measured (Cole and Schmidt, 1964; Baechtel, 1967; Johnson and Schmidt, 1966; Molloy, 1969; Shen and Schmidt, 1966) throughout the cell cycle of a thermophilic species of this organism. Three of the enzymes (Fig. 4) were involved in ribonucleotide biosynthesis (ATCase and DHOase, the first and second enzymes on the pyrimidine nucleotide pathway, respectively; GAR kinosynthetase, the second enzyme on the purine nucleotide pathway), and the other two enzymes (Fig. 5) were on the deoxythymidine triphosphate pathway (dCMP deaminase and dTMP kinase).

Instead of obtaining the anticipated periodic (stepwise) changes in

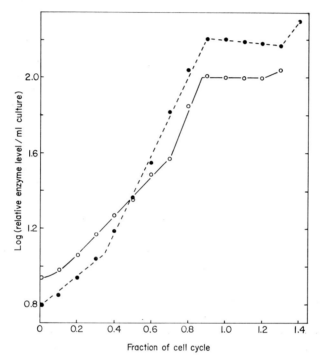

FIG. 5. Relative level of two enzymes on the deoxythymidine triphosphate biosynthetic pathway during the cell cycle of *Chlorella pyrenoidosa* (strain 7-11-05). ○——○, Deoxycytidine monophosphate deaminase; ●---●, deoxythymidine monophosphate kinase. (dCMP deaminase, Shen and Schmidt, 1966; dTMP kinase, Johnson and Schmidt, 1966.)

enzyme level, all five enzymes appeared to be synthesized at approximately alternating exponential rates during most of the cell cycle. The sharp changes in rate of apparent exponential synthesis occurred at characteristic times in the cell cycle for each enzyme.

At first evaluation, these enzyme trends seemed to bear no relationship to those observed in yeast. Furthermore, the sharp changes in exponential rate tended to preclude, from a mathematical standpoint (A. B. Kent, personal communication), solely repressor control of enzyme synthesis by the models proposed for bacteria (Goodwin, 1966; Kuempel *et al.*, 1965). Also, the enzyme trends in *Chlorella* subsequently were shown (Schmidt, 1966) not to be the result of modulation of enzyme activity by changes in intracellular levels of activators or inhibitors (i.e., the enzyme trends reflected changes in actual enzyme level). Thus, the regulatory system for enzyme synthesis in *Chlorella* appeared to have little in common with either the yeast or bacterial systems. Assuming some biochemical unity to exist at least between the *Chlorella* and yeast systems, we rescanned our data for a possible clue to help solve this puzzle.

An examination of the kinetics of synthesis (accumulation) of total cellular DNA revealed the probable reason for exponential enzyme synthesis. Unlike most microorganisms, *Chlorella* usually divides into eight or 16 daughter cells under optimal cultural conditions; therefore, each of the organism's chromosomes (>8, Schmidt, 1967) must replicate three times, i.e., 1, 2, 4, 8, before the end of its cell cycle. Thus, because the chromosomal DNA serves as a template for itself, the amount of DNA associated with each chromosome would be expected to increase in roughly an exponential manner provided the replication time for each step was constant. If each of the chromosomes replicated at different but overlapping periods, the increase in total cellular DNA would approach a smooth exponential function such as that actually observed during the S period of *Chlorella* (Fig. 6). In addition, if each chromosome replicated with more than one replication point (multifork replication), as seen in *Bacillus subtilis* (Oishi *et al.*, 1964; Yoshikawa *et al.*, 1964), this would also significantly contribute to a continuous exponential increase in total cellular DNA. However, in the case of multifork replication, each chromosome must have completed *at least one* full replication prior to initiation of the first of three nuclear divisions (probably amitotic, Wanka and Mulders, 1967) at approximately 0.7 cell cycles (Schmidt, 1966).

The S period for total cellular DNA appears to cover the entire cell cycle (Fig. 6). However, a change in rate of its exponential accumulation occurs between 0.675 and 0.759 cell cycles. Presumably this change in rate could result from an 8-fold increase (exponentially), over a short period of the cell cycle, of a minor DNA component such as that associated with the chloroplast and/or mitochondria. In *Chlamydomonas*, the chloroplast DNA represents approximately 6–14% of the total DNA, and it replicates in the middle of the cell cycle immediately prior to the replication of nuclear DNA (Chiang and Sueoka, 1967; Sager and Ishida, 1963). If in *Chlorella* the chloroplast DNA replicates in a slightly later stage of development than in *Chlamydomonas*, while the nuclear

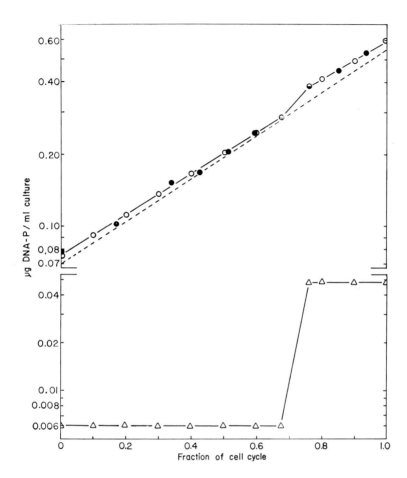

FIG. 6. Actual and simulated levels of total cellular DNA-phosphorus and the simulated levels of nuclear and chloroplast DNA-phosphorus during the cell cycle of *Chlorella pyrenoidosa* (strain **7-11-05**). ●——●, Actual total cellular DNA-P; ○——○, simulated total cellular DNA-P; ----, simulated nuclear DNA-P; △——△, simulated chloroplast DNA-P. Total cellular DNA was assumed to be composed primarily of nuclear and chloroplast DNA. The chloroplast DNA was assumed to be 8% of the daughter cell DNA and to increase 8-fold between 0.675 and 0.759 of the cell cycle. Nuclear DNA (92%) was assumed to increase 8-fold exponentially with the same exponential coefficient (i.e., slope) throughout the cell cycle. Cell number increased 8-fold in this experiment. (Actual DNA-P/ml from Herrmann and Schmidt, 1965.)

DNA replicates over the entire cell cycle, the summation of the S periods of the two species of DNA would result in a total DNA accumulation curve (Fig. 6) almost identical to that actually observed in *Chlorella*. It should be noted that the slope, i.e., exponential coefficient, for the increase in nuclear DNA is also equal to the slope of the straight line between the first and last points on the total cellular DNA trend for one cell cycle.

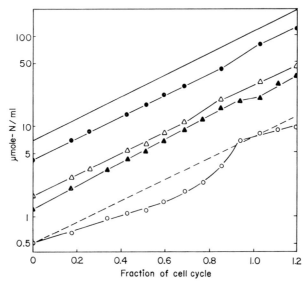

FIG. 7. Levels of cellular-N, protein-N, total nucleic acid-N, RNA-N, and DNA-N during the cell cycle of *Chlorella pyrenoidosa* (strain 7-11-05). ———, Total cellular-N; ●——●, protein-N; Δ——Δ, total nucleic acid-N; ▲——▲, RNA-N; ○——○, DNA-N; ----, slope ($k = 1.20$) nuclear DNA-N. The amount of nitrogen associated with each protein amino acid was measured and then used to calculate protein-N. All parameters increased 16-fold at one cell cycle in this experiment (Hare and Schmidt, 1968b.)

The exponential increase in nuclear DNA predicts that nuclear gene dosage also increases exponentially. If each gene and its progeny are transcribed (with minimum repressor control) for a given period prior to or after each replication, then the synthesis of mRNA ($d\text{mRNA}/dt = \alpha$ gene dosage) and protein (d protein$/dt = \beta\text{mRNA}$) also should increase exponentially provided the rate of translation of mRNA does not obscure the rate of change in gene dosage. The exponential accumulation of total protein and RNA in *Chlorella* (Fig. 7) does indeed support the hypothesis that genes and their progeny are transcribed for a given period prior to or after each replication step.

If the synthesis of nuclear DNA is controlling the rate of synthesis of total cellular RNA and protein, the accumulation curves for these cellular constituents should have the same exponential coefficients (i.e., slopes) as the nuclear DNA. However, for 0.80 of the cell cycle, total protein and RNA have lower (1.14) and higher (1.45) exponential coefficients, respectively, than the one (1.20) calculated for nuclear DNA (Fig. 7).

The small burst in the rate of total protein accumulation near the end of the cell cycle suggests that total protein is composed of a major and minor fraction which change in a manner analogous to that previously described for nuclear and chloroplast DNA (Fig. 6). If a minor protein component (probably chloroplastic) is assumed to represent 8% of the daughter cell protein and is also assumed to increase 16-fold between 0.85 and 1.0 cell cycles, while the major protein component (92%; probably nuclear-directed) is assumed to increase 16-fold exponentially with the same exponential coefficient (1.20) as the nuclear DNA, the plot resulting from the summation of these two components is essentially identical to the one actually observed for total cellular protein (Fig. 8). Thus, the rate of replication of the nuclear genome does appear to be controlling the rate of synthesis of the bulk of the cellular protein.

The exponential coefficient (1.21) for the "overall" rate of accumulation of total RNA (primarily ribosomal RNA) for one cell cycle (i.e., the slope of the straight line between the first and last value on the total RNA curve for one cell cycle; not graphically illustrated) is essentially equal to the coefficient observed for the accumulation of nuclear DNA and nuclear-directed protein. The greater exponential coefficient (1.45), observed during the first 0.84 of the cell cycle, probably means that the nucleus and chloroplast individually are directing ribosomal RNA synthesis during this period of cellular development. Recent evidence indicates that the chloroplast DNA codes for the synthesis of chloroplast ribosomal RNA (Scott and Smillie, 1967). If the nuclear-directed ribosomal RNA accumulates exponentially over the entire cell cycle, as is undoubtedly the case, it can be inferred that the structural genes for nuclear-directed ribosomal RNA are replicated with high frequency and are well scattered throughout the entire nuclear genome.

Once a reasonable hypothesis was developed to explain the exponential nature of protein synthesis, it was possible to evolve a proposal to explain why "individual enzyme activities" increase at essentially alternating exponential rates during most or all of the cell cycle in *Chlorella*. The proposal is comprised of the following assumptions: (1) the genes of each chromosome are replicated in a nonrandom and sequential manner; (2) each gene is replicated three times with a specific doubling time to yield an 8- or 16-fold increase in its original dosage in roughly an exponential manner; (3) gene transcription is also ordered or sequential, genes being available for transcription only for a given period prior to or after their replication; (4) structural genes located at different positions

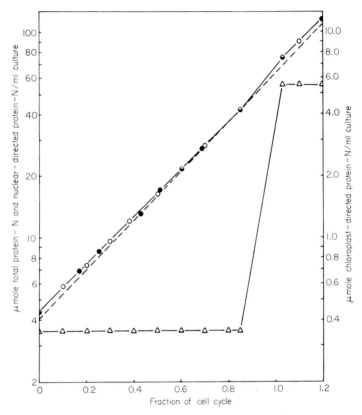

Fig. 8. Actual and simulated levels of total protein-N and the simulated levels of nuclear-directed and chloroplast-directed protein-N during the cell cycle of *Chlorella pyrenoidosa* (strain 7-11-05). ●——●, Actual total protein-N; ○——○, simulated total protein-N; ----, nuclear-directed protein-N; △——△, chloroplast-directed protein-N. Simulated total protein was obtained by summation of simulated nuclear-directed and chloroplast-directed protein. (Actual total protein-N from Fig. 7; Hare and Schmidt, 1968b.)

on the same or different chromosomes code for the synthesis of identical or different proteins with the same enzymatic activity; and (5) certain of these structural genes are replicated and transcribed during specific but overlapping periods of the cell cycle, resulting in a trend of a given "total enzyme activity" which appears to increase at approximately alternating exponential rates.

Graphical simulation of the proposal yielded hypothetical enzyme trends very similar to those actually observed in *Chlorella*. These trends were generated by assuming two proteins with identical enzymatic activities to increase (8-fold) at identical (Fig. 9D) and different (Fig. 9A–C) exponential rates during overlapping periods of the cell cycle. For simplicity, the levels of each enzyme were assumed to be equal and to remain constant in the absence of enzyme synthesis.

Note that certain segments of the hypothetical trends have a slight curvature on the logarithmic plots. This curvature results from their quadratic nature. In each case, the curvature is so slight that it might have gone unnoticed had there been experimental error in the system. As seen in Fig. 9D, only the summation of identical exponential functions gives another exponential function.

A closer examination of the experimental points, comprising the various segments of the actual enzyme trends in *Chlorella,* suggests that these also may have a slight curvature. Because the presence or absence of curvature in portions of the actual trends will play an important role in predicting the rates of synthesis of their possible component enzymes, these trends must be analyzed statistically. For this task, we (A. B. Kent, A. Springall, and R. R. Schmidt) currently are preparing a computer program for curve analysis to determine (1) the exact location of the breakpoints (location of the highest order curvature transitions) in these trends, and (2) whether or not the segments between the break points contain demonstrable finite curvature. Also, we are attempting to resolve the aforementioned enzymes into isozymes, and if present to quantitatively measure their separate levels through the cell cycle. From the number of sharp break points in a logarithmic plot of a given "total enzyme activity," it should be possible to predict the maximum number of (major) multiple forms of the enzyme synthesized during the cell cycle.

Although graphical simulation of the proposal does predict its correctness, the following observations (in total) form a strong experimental basis for the proposal: (1) the synthesis of DNA and the replication of the genome in microorganisms is nonrandom and sequential (Cairns, 1963; Nagata, 1963; Yoshikawa and Sueoka, 1963); (2) the rate of synthesis of an enzyme varies with the dosage of its structural genes (Donachie, 1964; Jacob *et al.*, 1960); (3) the order of synthesis of enzymes in both synchronous yeast and bacteria is related to the relative position of their structural genes on the chromosome (Halvorson *et al.*, 1966; Masters and Pardee, 1965; Tauro and Halvorson, 1966); (4) the

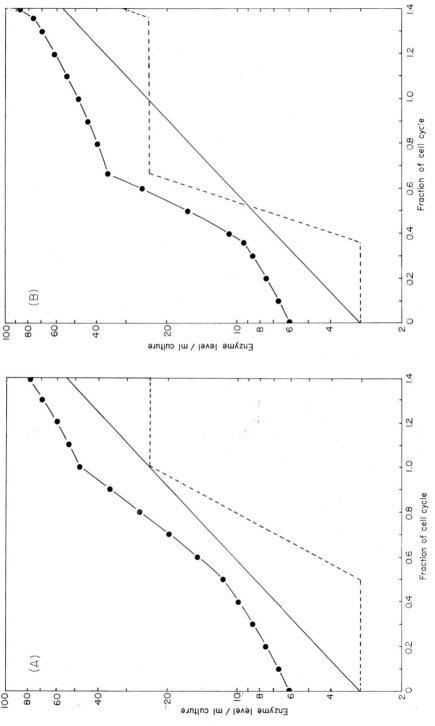

172

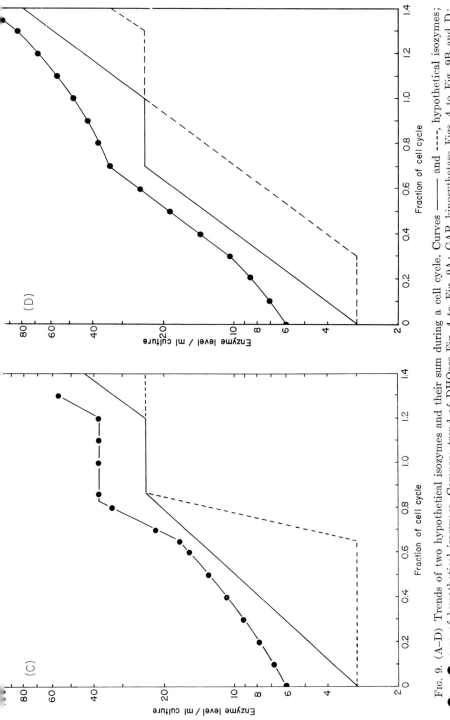

FIG. 9. (A–D) Trends of two hypothetical isozymes and their sum during a cell cycle. Curves ——— and ---, hypothetical isozymes; ●, sum of hypothetical isozymes. Compare trend of DHOase Fig. 4 to Fig. 9A; GAR kinosythetase Figs. 4 to Fig. 9B and D; dCMP deaminase Fig. 5 to Fig. 9C.

inducibility of enzymes in both *Chlorella* and yeasts varies during their cell cycles (Knutsen, 1965; Halvorson *et al.*, 1966; Tauro and Halvorson, 1966); (5) the number of periods of synthesis of a given "enzyme" in yeasts (interspecific hybrids) corresponds to the number of unlinked structural genes of the enzyme (Halvorson *et al.*, 1966; Tauro and Halvorson, 1966); (6) multiple copies of the "same" structural gene appear to be present in certain species of *Chlamydomonas* (Wetherell, 1957; Wetherell and Krauss, 1957); and (7) preliminary evidence indicates that *Chlorella* (F. E. Cole, personal communication) and *Spirogyra* (Frederick, 1967) contain isozymes of a number of their enzymes.

When considered with the above observations, the absence of accumulation of dCMP deaminase or dTMP kinase, observed near the end of the first and the beginning of the second (consecutive) cell cycle (Fig. 5), also indicates that the potential for transcription of enzymes is discontinuous in *Chlorella*. Thus, as originally speculated, the control of transcription in highly compartmented organisms, such as yeast and algae, tentatively appears to be similar. Presumably, the main difference between the yeast and *Chlorella* systems appears to result from the number of times each organism replicates its genome during one cell cycle, i.e., once or three times, respectively.

IV. ISOZYMES AND SUBCELLULAR ORGANELLES

From the number of break points in each of the enzyme trends (Figs. 4 and 5), it would appear that each trend is composed of only two major enzyme components. The question therefore arises whether or not these enzyme components might be isozymes involved in the biosynthesis of nucleotides to be utilized in the replication and transcription of the genome associated with each of the major (mass-wise) subcellular organelles, i.e., chloroplast and nucleus.

The semi-autonomous nature of the chloroplast has recently received considerable attention (Gibor and Granick, 1964; Granick and Gibor, 1967). The chloroplast seems to have particular independence in nucleic acid metabolism. The chloroplast DNA has the potential to replicate itself (Spencer and Whitfeld, 1967), direct the synthesis of its own ribosomal RNA (Scott and Smillie, 1967) and a certain amount of mRNA (Shah and Lyman, 1966; Gnanam and Kahn, 1967), and apparently direct the synthesis of protein (Spencer, 1965). Differences in the time of replication of the nuclear and chloroplast DNA (Chiang and Sueoka, 1967) and the different division times (Gibor and Granick, 1964) of the organelles in the cell cycle also suggest that the organelles might control their nucleic acid metabolism independently.

If the chloroplast synthesizes its own nucleotides by one set of enzymes (located within the chloroplast) while the nucleus utilizes nucleotides synthesized by yet another set located in the cytoplasm and/or nucleus, an interesting question can be asked. Does the genome of each organelle code for most of its own biosynthetic enzymes or does the nuclear genome direct the synthesis of most enzymes that ultimately will function in the different organelles? Because the chloroplast increases so dramatically in size and seemingly replicates its DNA prior to the period of protein synthesis directed by the genome of the chloroplast as suggested earlier (Figs. 6 and 8), it would appear that the nuclear genome directs the synthesis of most of the chloroplast proteins. If so, are these chloroplast proteins synthesized in the cytoplasm and then transferred to the chloroplast to function, or is mRNA transferred from the nucleus to the chloroplast to be translated into proteins by the ribosomes located therein? The synthesis of chloroplast ribosomal RNA in the apparent absence of chloroplast DNA-directed protein synthesis (for 0.80 of the cell cycle) seems to favor the latter possibility. Although the above questions cannot as yet be answered, they serve to illustrate the exciting but difficult challenge that faces those who decide to study the biochemical mechanisms controlling enzyme synthesis in highly compartmented organisms.

V. ENZYMATIC CONTROL OF BIOSYNTHETIC PATHWAYS

Because nuclear-DNA and nuclear-directed RNA increase with essentially the same doubling times during the cell cycle of *Chlorella*, their synthesis is not likely to be rate limited at the precursor level as previously visualized (Schmidt, 1966). In fact, the availability of the DNA template itself seems to be the primary rate-limiting factor in both the replication and transcription processes.

Intuitively it would seem that the order and rate of synthesis of enzymes on the same biosynthetic pathway might play an important role in determining which enzyme might be potentially controlling the rate of accumulation of the end product of the pathway. However, it may prove to be that only the timing and rate of synthesis of the first enzyme on the pathway, with its susceptibility to end-product inhibition, actually controls the rate of synthesis of end product throughout the entire cell cycle.

ACKNOWLEDGMENT

This chapter is the distillate of the efforts of the author's entire research group (Dr. T. A. Hare, Mr. F. S. Baechtel, Mr. W. W. Farrar, Mr. J. B. Flora, Mr. E. C. Herrmann, Mr. G. R. Molloy, Mr. T. O. Sitz, and Mr. L. H. White) and

of Dr. A. B. Kent, whose keen insight into problems of a kinetic nature contributed significantly to the evolution of the proposals described in the present chapter.
Particular thanks are extended to Mr. J. B. Flora for his ideas and critical evaluation of proposals generated by our research group, to Mr. F. S. Baechtel and Mr. T. O. Sitz for their drawing skills, and to Mrs. Lila Eakin for her excellent job of typing and editing the manuscript.

The research discussed in this chapter has been supported in large part by grants from the National Institutes of Health (PHS GM 12042-03) and the National Science Foundation (NSF GB-1960 and GB-4682).

REFERENCES

Baechtel, F. S. (1967). M. S. Thesis, Virginia Polytechnic Institute, Blacksburg, Virginia.
Baker, A. L., and Schmidt, R. R. (1963). *Biochim. Biophys. Acta* **74**, 75.
Bostock, C. J., Donachie, W. D., Masters, M., and Mitchison, J. M. (1966). *Nature* **210**, 808.
Cairns, J. (1963). *J. Mol. Biol.* **6**, 208.
Chiang, K-S, and Sueoka, N. (1967). *Proc. Natl. Acad. Sci. U.S.* **57**, 1506.
Cole, F. E., and Schmidt, R. R. (1964). *Biochim. Biophys. Acta* **90**, 616.
Curnutt, S. G., and Schmidt, R. R. (1964). *Exptl. Cell. Res.* **36**, 102.
Donachie, W. D. (1964). *Biochim. Biophys. Acta* **82**, 293.
Donachie, W. D. (1965). *Nature* **205**, 1084.
Frederick, J. F. (1967). *Phytochemistry* **6**, 1041.
Gibor, A., and Granick, S. (1964). *Science* **145**, 890.
Gnanam, A., and Kahn, J. S. (1967). *Biochim. Biophys. Acta* **142**, 475.
Goodwin, B. C. (1966). *Nature* **209**, 479.
Gorman, J., Tauro, P., LaBerge, M., and Halvorson, H. O. (1964). *Biochem. Biophys. Res. Commun.* **15**, 43.
Granick, S., and Gibor, A. (1967). *Progr. Nucleic Acid Res. Mol. Biol.* **6**, 143.
Halvorson, H. O., Gorman, J., Tauro, P., Epstein, R., and LaBerge, M. (1964). *Federation Proc.* **23**, 1002.
Halvorson, H. O., Bock, R. M., Tauro, P., Epstein, R., and LaBerge, M. (1966). *In* "Cell Synchrony—Studies in Biosynthetic Regulation" (I. L. Cameron and G. M. Padilla, eds.), p. 102. Academic Press, New York.
Hare, T. A., and Schmidt, R. R. (1968a). *Appl. Microbiol.* **16**, 496.
Hare, T. A., and Schmidt, R. R. (1968b). *Arch. Biochem. Biophys.*, submittted.
Herrmann, E. C., and Schmidt, R. R. (1965). *Biochim. Biophys. Acta* **95**, 63.
Jacob, F., Schaeffer, P., and Wollman, E. L. (1960). *10th Symp. Soc. Gen. Microbiol.*, p. 76. Cambridge Univ. Press, London.
Johnson, R. A., and Schmidt, R. R. (1966). *Biochim. Biophys. Acta* **129**, 140.
Knutsen, G. (1965). *Biochim. Biophys. Acta* **103**, 495.
Kuempel, P. L., Masters, M., and Pardee, A. B. (1965). *Biochem. Biophys. Res. Commun.* **18**, 858.
Masters, M., and Donachie, W. D. (1966). *Nature* **209**, 476.
Masters, M., and Pardee, A. B. (1965). *Proc. Natl. Acad. Sci. U.S.* **54**, 64.
Masters, M., Kuempel, P. L., and Pardee, A. B. (1964). *Biochem. Biophys. Res. Commun.* **15**, 38.
Molloy, G. R. (1969). Ph. D. Dissertation, Virginia Polytechnic Institute, Blacksburg, Virginia.
Nagata, T. (1963). *Proc. Natl. Acad. Sci. U.S.* **49**, 551.

Oishi, M., Yoshikawa, H., and Sueoka, N. (1964). *Nature* **204**, 1069.

Sager, R., and Ishida, M. R. (1963). *Proc. Natl. Acad. Sci. U.S.* **50**, 725.

Schmidt, R. R. (1966). *In* "Cell Synchrony—Studies in Biosynthetic Regulation" (I. L. Cameron and G. M. Padilla, eds.), p. 189. Academic Press, New York.

Schmidt, R. R. (1967). Unpublished data.

Schmidt, R. R., and Spencer, H. T. (1964). *J. Cellular Comp. Physiol.* **64**, 249.

Scott, N. S., and Smillie, R. M. (1967). *Biochem. Biophys. Res. Commun.* **28**, 598.

Shah, V. C., and Lyman, H. (1966). *J. Cell Biol.* **29**, 174.

Shen, S. R. C., and Schmidt, R. R. (1966). *Arch. Biochem. Biophys.* **115**, 13.

Sorokin, C. (1957). *Physiol. Plantarum* **10**, 659.

Sorokin, C., and Myers, J. (1953). *Science* **117**, 330.

Spencer, D. (1965). *Arch. Biochem. Biophys.* **111**, 381.

Spencer, D., and Whitfeld, P. R. (1967). *Biochem. Biophys. Res. Commun.* **28**, 538.

Sylvén, B., Tobias, C. A., Malmgren, H., Ottoson, R., and Thorell, B. (1959). *Exptl. Cell Res.* **16**, 77.

Tauro, P., and Halvorson, H. O. (1966). *J. Bacteriol.* **92**, 652.

Wanka, F., and Mulders, P. F. M. (1967). *Arch. Mikrobiol.* **58**, 257.

Wetherell, D. F. (1957). Ph.D. Dissertation, University of Maryland, College Park, Maryland.

Wetherell, D. F., and Krauss, R. W. (1957). *Am. J. Botany* **44**, 609.

Yoshikawa, H., O'Sullivan, A., and Sueoka, N. (1964). *Proc. Natl. Acad. Sci. U.S.* **52**, 980.

Yoshikawa, H., and Sueoka, N. (1963). *Proc. Natl. Acad. Sci. U.S.* **49**, 559

CHAPTER 9

Light-Dependent Formation of Nucleic Acids and Its Relation to the Induction of Synchronous Cell Division in *Chlorella*

Horst Senger and Norman I. Bishop

*Abbreviations: DNA = deoxyribonucleic acid; RNA = ribonucleic acid; DCMU = 3-(3,4-dichlorphenyl)-1,1-dimethylurea.

I. INTRODUCTION

Unicellular algae of the *Chlorella* type remain synchronized during repeated life cycles when exposed to an appropriate light-dark regime, and when combined with regular dilution to a constant cell number after each life cycle (see reviews by Pirson and Lorenzen, 1966, and Tamiya, 1966). Such synchronized cultures have been increasingly used to study the development of the algae and in addition for a variety of physiological experiments in which uniform testing material was required. However, the question concerning the actual mechanism of synchrony itself remains largely unsolved.

In synchronized cultures the burst of cell division is the most apparent evidence for synchrony, and also the light-dependent induction of cell division is directly correlated to the synchronizing light-dark regime. In addition the time between the onset of light and the burst of cell division is endogenously fixed for each strain (Pirson and Lorenzen, 1958; Senger, 1961). All these facts mark the induction of cell division as the most interesting parameter for studying the mechansim of synchrony.

Since photosynthesis is the predominant light reaction in these algae, photosynthesis was expected to be the regulator, if not the mechanism, of synchrony itself. But experiments with mixotrophic cultures (Senger, 1962), inhibitors of pigment synthesis (Metzner and Senger, 1962), pigment mutants (Metzner, *et al.*, 1965), and photosynthesis inhibitor (such as DCMU) (Senger and Schoser, 1966) did not provide evidence that photosynthetic processes are directly involved in the mechanism of synchrony.

As a consequence, our interest was focused on the role of the nucleic acids, especially, since light-dependent nonphotosynthetic nucleic acid formation was found in *Chlorella* (Senger and Bishop, 1966). In the present article, we will present evidence for a potential relationship between light-dependent nucleic acid formation and the light-dependent induction of cell division.

II. ORGANISM AND GROWTH CONDITIONS

The experiments and results to be reported in this article will be restricted to the unicellular green alga *Chlorella pyrenoidosa* Chick strain 211-8b which was obtained from the Pringsheim collection, Göttingen, Germany. Recently this strain has been reclassified as *Chlorella fusca* Shihira and Krauss (Shihira and Krauss, 1965; Soeder, 1965). In experiments where photosynthesis interferred with other light reactions, it was useful to apply an achlorophyllous mutant of *Chlorella*. This mutant,

C-1.1.10.31 from *Chlorella pyrenoidosa* Chick (Emerson strain) was induced by ultraviolet irradiation and described by Bendix and Allen (1962) as: "Achlorophyllous stable yellow in light and darkness. Apparently normal carotenoids."

The algae were grown either autotrophically or mixotrophically in a light thermostat (Kuhl and Lorenzen, 1964) or heterotrophically in a dark incubator at 30°C. The autotrophic culture medium was supplemented with 0.5% glucose for mixotrophic or heterotrophic growth. The cultures were aerated with 3% CO_2 in air. For irradiation white light was provided by a combination of fluorescent lamps (Senger and Bishop, 1966) providing a homogeneous spectrum from about 350 mμ to 750 mμ. Experiments with monochromatic light were carried out in a device similar to that described by Schoser (1966), using Schott or Baird-Atomic interference filters in combination with infrared absorbing filters. Synchronized cultures were obtained by exposing a random culture to a light-dark regime of 16 hours light (9000 lux) and 12 hours darkness. When treated in this way, completely synchronized cultures were obtained after two cycles (Senger, 1961). Such cultures remained synchronized when they were diluted to 1.56×10^6 cells/milliliter (Lorenzen, 1957, 1964) at the end of each dark period.

III. SYNCHRONIZATION

A. REQUIREMENTS FOR IDEAL SYNCHRONIZATION

It is obvious that various organisms require different methods for induction of synchrony. Nevertheless, several methods are used by different investigators for the same organism. As a result, "synchronous cultures" of one algae may differ considerably from one laboratory to another. Therefore, it appears necessary to outline certain standards for an "ideal" synchronous culture.

Certain requirements for good synchronization have already been set forth in several publications. Almost all investigators working with synchronous algae cultures agree that synchrony should be complete, i.e., that all cells should undergo complete division during their synchronous life cycle. Complete synchrony should always be ascertained by microscopic analysis or by distribution curves (see Senger, 1961). Tamiya (1966) requires that all cells of a synchronous culture should "keep as close as possible to their pace of growth and division" and that the number of daughter cells into which a mother cell divides should be the same for all cells. Similar requirements are made by Pirson and Lorenzen (1966) under the expression "homogeneity." Baker and Schmidt (1963) require that at least two criteria be satisfied: First, that after abolishing the synchronization treatment the biochemical and biophysical events recur in succeeding life cycles. This would assure that no single event is a mere reflection of the synchronization method itself, which could be either shock or environmental stress. Second, that possible changes in culture conditions during synchronous growth do not change the growth rate of the cells. Lorenzen and Schleif's (1966) requirements are that (1) cells before and after a synchronous life cycle have the same biochemical and biophysical stage [this corresponds to the first requirement of Baker and Schmidt (1963)]; (2) the shortest possible life cycle is obtaining during the synchronous growth; and (3) synchrony should be continously repeatable under the applied stimulus of the synchronizing conditions.

Summarizing the requirements for an "ideal" synchronized culture, the following criteria should be fulfilled:

1. *Complete synchronization*. The culture should be completely synchronized, i.e., each cell should divide during its synchronous life cycle.

2. *Homogeneity*. Deviation in time and conditions of the developmental stages of all cells in the culture should be as small as possible. For example, the length of time in which the burst of cell division occurs should be minimal and the mother cells should always divide into the same number of daughter cells.

3. *Exponential growth and shortest life cycle*. Culture conditions for synchronous growth should be as close to optimal as possible; growth should be in the exponential phase and the life cycle should be as short as possible.

4. *Nonsusceptibility of the life cycle to the synchronizing procedure*. The synchronizing procedure should not influence the biochemical and biophysical events in the cells, and the cells should be identical before and after a synchronous life cycle.

It is evident that all these requirements can be fulfilled only with an "ideal" synchronous culture. Nevertheless, synchronous cultures should always be compared by these standards.

B. COMPARISON OF ACHIEVEMENTS OF THE PRESENT SYNCHRONIZATION METHOD WITH THE STANDARDS FOR "IDEAL" SYNCHRONIZATION

Chlorella pyrenoidosa strain 211-8b cultured at 30°C and aerated with 3% CO_2 in air becomes completely synchronized when exposed to a light-dark regime of 16 hours light of 9000 lux and 12 hours darkness. The complete synchronization was proved by distribution curves of cell size, which show that the smallest cell before cell division has a greater volume than the biggest daughter cell (Senger, 1961).

The burst of cell division occurs between hours 18 and 25, with a few earlier and later divisions. This is a comparatively short time. The deviation from the average time which the cell needs from the beginning of the light period until separation of autospores seems to represent the deviation of the developmental stages in the culture. In a similar way the number of autospores formed per mother cell varies. The methods of Tamiya and co-workers (1961) and of Schmidt and co-workers (1966) in starting synchronization by using a certain fraction of uniform cells from a random culture would certainly improve the homogeneity of our cultures.

During the light period, growth is exponential and nearly optimal; this ensures that the cells grow with the shortest possible life cycle. Growth can still be improved by raising the light intensity (Ruppel, 1962) or by a slight elevation of temperature (Lorenzen, 1963). However, as a visible result only the number of autospores per mother cell increases whereas the length of the life cycle is not markedly changed. In the dark period, the growth measured as dry weight or total sugar, decreases (Ruppel, 1962). The question arises, however, as to what "growth" means when protein is preferentially formed during this dark period and when synchrony is positively influenced (Ruppel, 1962). Certainly a dark period at the end of the life cycle is more advantageous to the "growth" of *Chlorella* than continuous illumination gained by repeated dilution (Ruppel, 1962).

We know that *Chlorella* cells are extremely light dependent in the initial stages of their life cycle. The dependence on and the necessity for irradiation declines with increasing age. Finally toward the time of cell division dark reactions seem to be dominant; darkness seems to be more suitable to the developmental stage than does light. Therefore, a light-dark regime may be more physiological than continuous light. Continuous illumination itself includes a light-dark gradient for the culture since the average light intensity for the cells drops by self-shadowing of the growing culture to 60% during the life cycle (Senger and Wolf 1964). Growth was inhibited in cultures exposed to equal irradiation when accompanied by repeated dilution (Ruppel, 1962).

In addition to the physiological requirement of the older cells, the dark period is also the actual synchronizing treatment. In darkness, the young autospores are inhibited from beginning their onset of a new life cycle. Each of them has to wait for a uniform start of all cells at the onset of light. By this way the deviation, caused by the slight variation of the lengths of the individual life cycles, is newly corrected in each light-dark regime. This synchronization method does not interfere with the natural life cycle and, in addition, assures the uniformity of cells during repeated life cycles.

Since most of the requirements listed in Section III,A are fulfilled, or at least approximated, our cultures may be considered well synchronized.

C. Advantage of Using Synchronized Mixotrophic Cultures

Up to this point, we have considered only autotrophic cultures. Under autotrophic growth conditions, photosynthetic green algae need a considerable amount of light to provide sufficient photosynthesis to

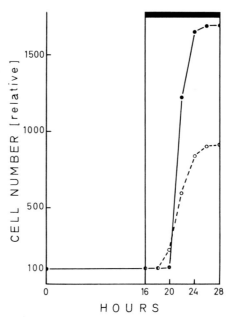

Fig. 1. Cell numbers of an autotrophic ($\bigcirc$---$\bigcirc$) and a mixotrophic ($\bullet$——$\bullet$) synchronous culture of *Chlorella pyrenoidosa*. The cells were grown in a light-dark regime of 16 hours light (9000 lux) and 12 hours darkness, 30°C, aerated with 3% CO_2 in air. For the mixotrophic growth, 0.5% glucose was added. Relative cell number 100 = 1.56 × 10^6 cells/ml.

maintain growth and development necessary for synchrony. It is almost impossible to run experiments under low light intensities or short irradiation periods without disturbing synchrony. For this type of experiment, an additional carbon source is necessary. Since Bergmann (1955) demonstrated that *Chlorella* could grow with glucose in the dark, glucose has been successfully used in synchronized cultures to compensate for photosynthesis (Senger, 1962).

After the addition of glucose (5%) to synchronized autotrophic cultures, cells double their dry weight and increase the number of autospores formed per mother cell from an average of 8 to 16. Cell material production is not restricted to the light period of the mixotrophic cultures. Glucose uptake is the highest in younger cells and decreases toward the time of cell division. The number of autospores not only increases to an average of 16, but also no deviation from this value occurs. In addition, the length of the burst of cell division becomes shorter (Fig. 1). Altogether, this means that homogeneity has been greatly improved in the mixotrophic culture.

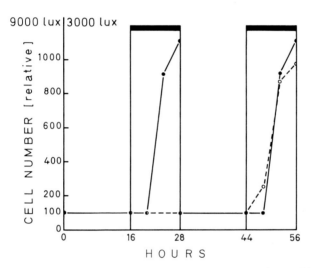

FIG. 2. Cell numbers of an autotrophic (○---○) and a mixotrophic (●——●) synchronous culture of *Chlorella pyrenoidosa* at reduced light intensity. The light intensity was lowered from 9000 to 3000 lux. See Fig. 1 for further experimental details.

If the light intensity in an autotrophic synchronous culture is decreased, the number of autospores is reduced; further reduction of light intensity extends the life cycle (see Section IV,A). These effects can be prevented by using mixotrophic cultures (Fig. 2). The combination of low light intensity and mixotrophic growth may allow for closer examination of the light requirement in the induction of synchrony (see Section IV,C). Obviously, mixotrophic cultures are essential for the determination of the action spectra for nonphotosynthetic processes. To achieve such action spectra for autotrophic cultures would require, in addition to monochromatic irradiation, sufficient light to ensure growth. This became evident when conducting the experiments for the action spectrum of induction of cell division (see Section IVD).

IV. INDUCTION OF CELL DIVISION

A. TIMER (*Zeitgeber*), ENDOGENOUS TIME FACTOR, AND CELL DIVISION

In 1958, Pirson and Lorenzen reported a time relationship between the beginning of the light period and the burst of cell division for synchronized cultures of *Chlorella*. They observed in different light-dark regimes that the time between the onset of light and cell division was constant, and they termed this inherent regulating principle an "endogenous time factor." The change from darkness to light, which marked the beginning of this factor, was postulated to be the *Zeitgeber* (timer) for cell division.

As far as cultures under a synchronizing light-dark regime and normal growth conditions are concerned, there is no doubt about the role of light as a timer, i.e., that light initially starts the induction of cell division. A different timer was discussed by Pirson *et al.* (1963). Their experiments demonstrated inhibition of cell division by high light intensity in synchronized cultures. This inhibition ceased after 6 hours of darkness regardless of the length of the previous light time. Their discussion of the onset of the dark period acting as a timer for cell division led to some confusion (Tamiya, 1966). Formally, one might have called this (onset of the dark period) a timer, but actually this is only a time-dependent recovery process and no "timer" in the above-mentioned sense, i.e., initiating a reaction chain leading to cell division.

The term "endogenous" time factor implies that the time from the onset of illumination until cell division is genetically fixed for a particular strain and that the range of modification under various conditions is restricted. Many investigators have shown the existence of the endogenous time factor and found only little or no deviation of its length (Pirson and Lorenzen, 1958; Lorenzen and Ruppel, 1960; Senger, 1961, 1962; Pirson, 1962; Soeder and Ried, 1962; Pirson *et al.*, 1963). A permanent delay of 2 hours for the release of autospores as a result of aeration with 3% CO_2 in air was reported by Soeder *et al.* (1964) and Soeder and co-workers (1966). Under slightly different growth conditions, this CO_2 effect was not found (Senger, 1962). Soeder and Thiele (1967) noticed a transitory retardation of cell division as a response to calcium deficiency. Certainly other factors could be found to prolong one or more partial reactions during the induction of cell division, but whether or not this manifests an influence on the endogenous time factor might be questioned. It should also be realized that the endogenously controlled reaction chain, beginning with the onset of illumination to cell division, is a very central and essential process of the living cell, and that manipulations to influence it very easily restrict growth and thus inhibit cell division nonspecifically.

How the inhibition of growth influences the induction of cell division can be demonstrated by the following experiment. A synchronized culture of *Chlorella* is switched from 9000 lux to 3000 lux at the beginning of a new light-dark regime. As a result, no cell division occurs during the first light-dark regime under the low light intensity (Fig. 2). Cell division occurs regularly in the second cycle. If in a parallel experiment the lowered photosynthate production is compensated for by the addition of 0.5% glucose, cell division takes place in the usual way and independent of the low light intensity. These experiments demonstrate clearly that a sufficient production of cell material (near optimal growth) is the necessary condition for a successful utilization of light in a specific induction of cell division. For further discussion of this problem see Senger (1962).

B. TEMPERATURE DEPENDENCE OF THE ENDOGENOUS TIME FACTOR

For endogenous rhythmical processes in higher plants, temperature independence is an essential criterion (see Bünning, 1963). In analogy much interest was focused on the action of temperature on the endogenous time factor of *Chlorella* and it was reported to be almost temperature independent in a range from 20°–30°C by several investigators (Pirson and Lorenzen, 1958; Pirson, 1961; Senger, 1961; Soeder, 1966). Figure 3 demonstrates the temperature independence of the endogenous time factor in cultures adapted to 20°, 25°, or 30°C. Only the number of autospores formed per mother cell is influenced. Using different synchronization methods, these results could not be confirmed by Morimura (1959) and Sorokin and Krauss (1965). In discussions of temperature dependence of the endogenous time factor (Tamiya, 1963, 1964, 1966), it was always neglected that investigators claiming temperature independence compared cultures that had been adapted to the temperature for a long time. Conversely, positive temperature dependence was always found in cultures with rapidly changed temperatures. To clarify whether sudden temperature changes would influence synchrony in *Chlorella*, the following experiments were carried out. *Chlorella* cultures were synchronized and the temperature was dropped at the end of one life cycle from 30° to 20°C. The result is shown in Fig. 4a. The first light-dark regime does not induce any cell division, but the burst of cell division occurs at the normal time in the second dark period. To investigate whether this was only a question of inhibited photosynthate production resulting from the lower temperature, a parallel experiment with a mixotrophic culture was treated in the same way. The onset of cell division was the same in both mixotrophic and autotrophic cultures. In another experiment, continuous light was given simultaneously with the lowered temperature (Fig. 4b). In this case cell division starts in both autotrophic and mixotrophic cultures after 46 hours. In the usually synchronized culture, maximal cell division takes place between hours 20 and 24. In the cultures under continuous light and lower temperature, the burst of cell division occurs between the hours 46 and 50. This demonstrates

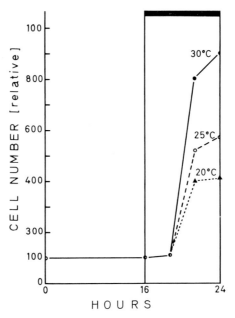

FIG. 3. Cell numbers of autotrophic synchronous cultures of *Chlorella pyrenoidosa* at different temperatures. The cultures were adapted to the different temperatures for 1 week. See Fig. 1 for experimental details. (From Senger, 1961.)

that as a consequence of dropping the temperature 10°C in these synchronized cultures, under intermittent or continuous illumination, an almost exact doubling of the endogenous time factor results. With a different objective, Lorenzen and Schleif (1966) obtained similar results.

Apparently the endogenous time factor of (at least) this strain of *Chlorella pyrenoidosa* shows in well-synchronized cultures (compare Section III,A) temperature independence if the cells are adapted long enough to the respective temperatures. A rapid drop of temperature (30°–20°C) suppresses the cell division for one light-dark-regime, i.e., the endogenous time factor is almost doubled. Endogenous rhythmic processes in higher plants are temperature independent insofar as the period of oscillation in constant temperature is concerned; their amplitude might be lower. A sudden drop of temperature results in a suppression or stretching of one cycle before the oscillation swings in the old period length (see Bünning, 1963). The amplitude of the oscillation in

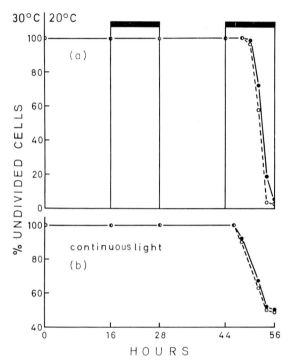

FIG. 4. Induction of cell division in synchronous cultures of *Chlorella pyrenoidosa* after a sudden decrease in temperature. At the beginning of a new light-dark regime (a) or of continuous light (b) the temperature of an autotrophic (○---○) and a mixotrophic culture (●——●) was lowered from 30° to 20°C. See Fig. 1 for culture conditions.

higher plants can be compared with the number of daughter cells formed per mother cell, and this number is certainly lower at lower temperature. If the reaction on temperature should be considered as a criterion for the endogenous nature of processes in algal populations, as in the case of rhythmical processes in higher plants, the time factor between the onset of light and cell division in a well-synchronized *Chlorella* culture is also endogenous.

C. Light Dependence of the Induction of Cell Division

That light serves as the timer for cell division already illustrates the light dependence of induction of cell division. In addition, the experiment demonstrated in Fig. 1 shows that only a certain intensity of light is necessary to induce cell division.

The best way to investigate the light dependence of the induction of cell division is to irradiate a culture, transfer samples of it at subsequent times into darkness, and, after a appropriate time, determine the

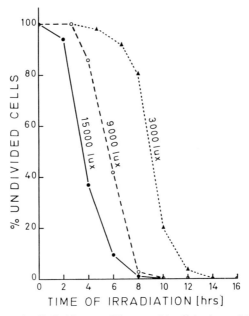

Fig. 5. Induction of cell division at different white light intensities in mixotrophic synchronous cultures of *Chlorella pyrenoidosa*. The percentage of undivided cells was always determined 28 hours after the onset of light. For experimental details see Fig. 1 and text.

number of divided and undivided cells. Experiments of this type have been carried out by Wanka (1959, 1962) and Lorenzen and Ruppel (1960). But these experiments have the disadvantage that autotrophic cultures were used and induction of cell division could not be separated from necessary photosynthesis. This problem was avoided by using mixotrophic cultures (Senger, 1965) (see Section III,C). Synchronized mixotrophic cultures were divided into 3 parallel cultures and irradiated with 3000, 9000, and 15,000 lux of white light. Samples were removed and aerated in the dark. After 28 hours, the percentage of undivided cells, relative to the starting culture, was determined. The results are shown in Fig. 5. The main burst of cell division occurs at 15,000 lux after 2 hours, at 9000 lux after 4 hours, and at 3000 lux after 8 hours. In comparison to the autotrophic cultures, the time of irradiation necessary for induction of cell division was cut down by at least 5 hours.

A plot of the time necessary to induce division of 50% of the cells versus light intensity (Fig. 6) shows the correlation between quantity and intensity of the light. Only a relatively short time of high light intensity is necessary to induce specifically cell division. The observation that more light of low intensity is necessary to induce the same percentage of cell division as observed at high intensity in a short time,

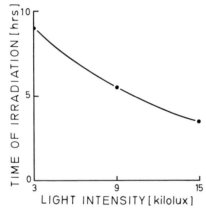

FIG. 6. Relation between time and intensity of irradiation to induce 50% cell division in mixotrophic cultures of *Chlorella pyrenoidosa*. See Fig. 5 for experimental details.

while the endogenous time factor is constant in both cases, is difficult to understand. Two possible explanations may be considered. First, the light reaction fills (fast or slow) a pool which is only slowly depleted by the reaction chain leading to cell division. Second, the observed induction consists of two components: (a) a light-dependent but intensity-independent reaction chain (representing the endogenous time factor) and (b) an intensity-dependent process that provides some compound(s) necessary in the reaction chain or growth of the algae.

D. ACTION SPECTRUM FOR THE INDUCTION OF CELL DIVISION

To learn more about the specific role of light in cell division, an action spectrum for the induction of cell division was determined (see Senger and Schoser, 1966; Senger and Bishop, 1966). For these experiments, it was necessary to use mixotrophic cultures for the reasons mentioned above. Since the energy of monochromatic light was low in comparison to the energy of the white light, cells were exposed to an additional 12 hours of darkness in the presence of 0.5% glucose in order to gain enough substance to render the monochromatic light effective. After 16 hours of exposure to monochromatic light (5000 erg/cm²/sec) and 12 hours of darkness, the cells were counted and the percentage of undivided cells was calculated. Such an action spectrum is shown in Fig. 7. It has its highest peak in the blue region and a lower peak

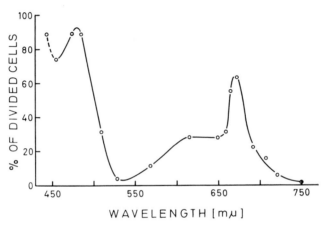

WAVELENGTH [mμ]

Fig. 7. Action spectrum for the light-induced cell division in mixotrophic cultures of *Chlorella pyrenoidosa*. Samples of presynchronized cultures were, after the end of a life cycle, kept in darkness for 12 hours, then irradiated with monochromatic light of 5000 erg/cm²/sec for 16 hours. After an additional 12 hours of darkness, the percentage of divided cells (difference between undivided cells and cell number at the beginning of the irradiation) was estimated. (From Senger and Bishop, 1966.)

in the red. At least one other peak below 450 mμ is probable. Comparison of this action spectrum with those of photosynthesis (Haxo, 1960) shows that the red peak is due to photosynthetic reactions. This seems to be contradictory to the statement in the Introduction, i.e., that photosynthetic processes could not be proved to play an essential role in the mechanism of synchrony. But the cells in the red region show rather abnormal size and seem to divide atypically. If the specific photosynthetic inhibitor DCMU (see Bishop, 1958) is employed, the induction of cell division is inhibited 95% at 674 mμ and only 31% at 485 mμ. This leads to the conclusion that only blue light induces a specific cell division whereas induction in red light is nonspecific (see Senger and Bishop, 1966).

V. NUCLEIC ACID FORMATION IN SYNCHRONIZED
Chlorella CULTURES

A. FORMATION IN LIGHT AND DARK

As demonstrated in Section IV, the synchronous cellular development in *Chlorella* is clearly induced through the action of light. In search for a light process programming cellular development, photosynthesis

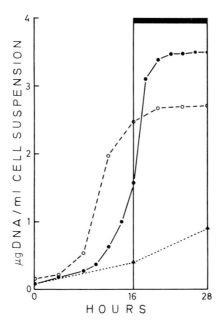

FIG. 8. Course of DNA synthesis during the synchronous life cycle of auto-trophically (○---○) and mixotrophically (●——●) growing *Chlorella pyrenoidosa*. In addition the dark synthesis (▲---▲) of DNA is demonstrated in a heterotrophic culture. At the beginning of the experiment all cultures were started with a cell number of 1.56×10^6 cells/ml. The light intensity was 9000 lux. See Fig. 1 for further experimental details. (Data for the autotrophic culture from Ruppel, 1962.)

was initially considered. But, as pointed out in the Introduction, photo-synthetic processes did not appear to be essential for the mechanism of synchrony. Since the most visible evidence for synchrony, cell division, and the preceding nuclear division, depend upon nucleic acid formation, its possible role in the mechanism of synchrony should be considered.

The formation of nucleic acids during the synchronous growth of *Chlorella* in white light has been followed by several investigators (Iwamura, 1955; Iwamura and Myers, 1959; Iwamura *et al.*, 1963;

Lorenzen and Ruppel, 1960; Ruppel, 1962; Stange *et al.*, 1962; Wanka, 1962; Galling, 1965; Senger and Bishop 1966). Typical curves for DNA formation in autotrophic and mixotrophic synchronous cultures are compared in Fig. 8. In both types of cultures, the initial hours of illumination show very little DNA formation. Subsequently, rapid synthesis occurs followed by a later period of stagnation. It was quite unexpected that in mixotrophic culture, provided with much more substrate and energy,

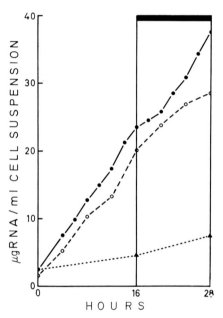

FIG. 9. Course of RNA synthesis during the synchronous life cycle of auto-trophically (○---○) and mixotrophically (●——●) growing *Chlorella pyrenoidosa*. In addition, the dark synthesis (▲---▲) of RNA is demonstrated in a hetero-trophic culture. For experimental details see Fig. 8. (Data for the autotrophic culture from Ruppel, 1962.)

the main DNA is synthesized about 5 hours later than in autotrophic culture. Nevertheless, the burst of cell division in mixotrophic and auto-trophic cultures takes place at almost the same time (Fig. 1). It is also remarkable that the formation of DNA shows less homogeneity than cell division, i.e., synthesis of DNA takes more time than the burst of cell division (compare Figs. 1 and 8). This might be explained by the fact that DNA formation and nuclear division occur in subsequent steps, whereas all autospores of one mother cell can be released at the same time. The facts remain that the main synthesis of DNA takes place at different times in autotrophic and mixotrophic cultures, that induction of cell division is already completed when the main DNA formation starts, and that its homogeneity is less than that of cells in division, and therefore makes it almost impossible to consider DNA formation as a process active in timing cell division.

The increase of RNA (Fig. 9) during the light period is almost linear for both autotrophic and mixotrophic cultures. This synthesis starts concurrently with illumination. In the following dark period, RNA formation in the autotrophic culture slows down while in the mixotrophic culture RNA production continues at the same rate as in the light. In a continuous dark-grown heterotrophic culture, *Chlorella* produces about one-third of the DNA synthesized in an autotrophic culture and only one-fifth of the respective RNA. Whether there is a difference in the RNA formed in light or darkness has not yet been investigated.

For both DNA and RNA, the main synthesis takes place under photosynthetic conditions. Addition of glucose can still stimulate this formation. Obviously, *Chlorella* cells are also able to synthesize both DNA and RNA in darkness without a previous light period.

B. DEPENDENCE OF NUCLEIC ACID FORMATION ON THE INTENSITY OF LIGHT

If nucleic acids are involved in the mechanism of synchrony, their formation should not only be light dependent, but also the rate of nucleic acid synthesis should be intensity dependent like the induction of cell division. An increase of total nucleic acids with higher intensities of white light has been found by Ruppel (1962) and Wanka (1962) for autotrophic cultures. An intensity-dependent formation in mixotrophic cultures has been reported by Senger and Bishop (1966). The influence of different light intensities on the DNA formation in autotrophically and mixotrophically grown cultures is demonstrated in Fig. 10. Both types of cultures show an almost linear increase in DNA. Even at the intensity of 21,000 lux, no saturation level is reached. The absolute difference between the DNA content of the autotrophic and the mixotrophic culture remains almost constant.

Such a parallel increase in autotrophic and mixotrophic cultures is not shown for the intensity-dependent formation of RNA (Fig. 11). The rate of RNA synthesis in mixotrophic cultures increases more markedly with higher light intensities than in autotrophic cultures. Under both types of culture conditions, light saturation of RNA formation occurs at about 15,000 lux.

To provide the necessary precursors and energy for DNA and RNA formation, photosynthesis is the most essential light process. In spite

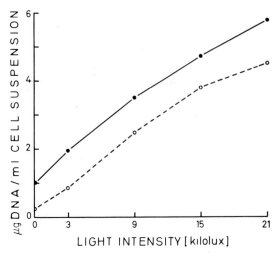

FIG. 10. Amount of DNA formed during the synchronous life cycle of *Chlorella pyrenoidosa* at different intensities of white light in autotrophic (○---○) and mixotrophic (●——●) cultures. All cultures started with a cell number of 1.56×10^6 cells/ml, and the DNA was measured at the end of the 16:12 hours light-dark regime. (From Senger and Bishop, 1966.)

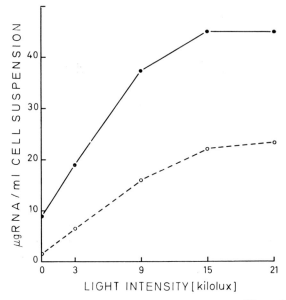

FIG. 11. Amount of RNA formed during the synchronous life cycle of *Chlorella pyrenoidosa* at different intensities of white light in autotrophic (○---○) and mixotrophic (●——●) cultures. See Fig. 10 for experimental details. (From Senger and Bishop, 1966.)

of its action on nucleic acid formation, photosynthesis does not seem to be specifically involved in the mechanism of synchrony. On the other hand, the light-dependent formation of nucleic acids demonstrates an increase upon irradiation parallel to the induction of cell division. Therefore, a light-dependent nucleic acid formation under nonphotosynthetic conditions needs to be demonstrated. The achlorophyllous mutant C-1.1.10.31 of *Chlorella pyrenoidosa* provides an intriguing tool for such experiments. The carotenoids of this mutant and the wild type are obviously the same (see Bendix and Allen, 1962; Senger and Bishop 1966).

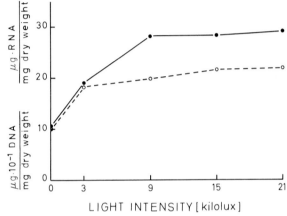

LIGHT INTENSITY [kilolux]

FIG. 12. Ratio of DNA (○---○) and RNA (●——●) to dry weight after irradiation with white light of different intensities in mixotrophic cultures of the achlorophyllous mutant C-1.1.10.31 of *Chlorella pyrenoidosa*. The cultures were grown under a light-dark regime of 10:14 hours. The nucleic acid content was estimated at the end of the dark time. The cultures were started with a number of 3.12×10^6 cells/ml. The temperature was 30°C. (Data from Senger and Bishop, 1966.)

The growth of cells in glucose medium was much slower than for the wild type, and the synchronization under a light-dark regime is certainly not ideal (Metzner *et al.*, 1965). But this was unessential for the experiments concerning light-dependent nucleic acid formation. At the end of a light-dark regime of 10:14 hours, under various light intensities, the nucleic acid contents were measured (Fig. 12). Since synchronization was not complete, the cell number was not reliable for calculation of the DNA and RNA content. Therefore, the ratio of nucleic acids versus dry weight was used. Both DNA and RNA synthesis of the mutant show principally the same course as the mixotrophic cultures of the wild type. The DNA curve has a small but linear increase, and the RNA content reaches saturation at 9000 lux.

These experiments show that a light-dependent formation of nucleic acids independent of photosynthesis (or chlorophyll) occurs in this mutant. The formation of nucleic acids promoted by nonphotosynthetically active light has not yet been demonstrated for the wild type, but it is probable that it occurs. Cells of "glucose-bleached" *Chlorella protothecoides* demonstrate a light-dependent DNA formation under nonphotosynthetic conditions, whereas RNA synthesis was found not to be influenced by light (Sokawa and Hase, 1967).

C. An Action Spectrum for RNA Formation

An action spectrum for RNA formation was surveyed in the following way: Cultures of the mutant C-1.1.10.31 were grown in the dark for

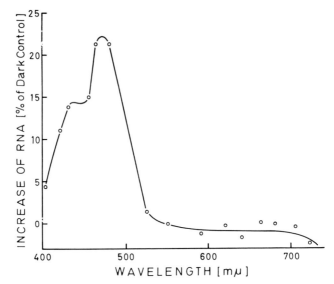

Fig. 13. Action spectrum for the light-induced formation of RNA in the achlorophyllous mutant C-1.1.10.31 of *Chlorella pyrenoidosa*. The increase of RNA was measured after 24 hours of irradiation with monochromatic light $(2.5 \times 10^{-11}$ einstein/cm^2/sec) in a culture that was previously grown in the dark. Each point represents the percentage increase of RNA calculated on the base of the RNA content of the dark control.

2 days. Samples of 50 ml were then exposed for 24 hours to monochromatic light of different wavelengths but of equal incident $(2.5 \times 10^{-10}$ einstein/cm^2/sec). One sample was always maintained in the dark as a control. The RNA content was determined by the difference between the amount of total nucleic acids and DNA. For a detailed reference on modified methods see Senger (1965). Since there is a dark synthesis

of RNA (see Section V,A), the amount of RNA formed at the individual wavelengths is represented as the percentage increase calculated on the base of the dark control (Fig. 13). Irradiation with blue light sharply enhances RNA production. In comparison to the dark control, the increase of RNA under the most effective wavelength is more than 20%. The main peak of the action spectrum for RNA formation in this mutant is around 475 mμ, and a distinct shoulder is present around 430 mμ. The small drop below zero in the region of longer wavelengths appears to be an experimental deviation rather than a photoinhibition.

VI. DISCUSSION OF THE LINKAGE BETWEEN NUCLEIC ACID FORMATION AND INDUCTION OF CELL DIVISION

The two recent review articles by Tamiya (1966) and Pirson and Lorenzen (1966) testify to the wealth of information available from synchronized *Chlorella* cultures. But it remains largely unresolved which biophysical and biochemical processes constitute the actual mechanism of synchrony. Certain facts about this mechanism may be considered as established. These are (1) the light dependence of the induction of cell division with the onset of illumination acting as the timer and (2) the endogenous time factor as manifested by the fixed time between the beginning of illumination and cell division. There is the possibility that this endogenous time factor represents a separate reaction chain parallel to the processes inducing cell division. But since this is not definitely established, we consider the endogenous time factor inherent to the reaction chain leading from the onset of illumination to cell division, and thus representative of the mechanism of synchrony. Three results of our experiments make a role of DNA in the timing mechanism for cell division most improbable: (1) The time between the onset of light and the main synthesis of DNA and the time between DNA synthesis and the burst of cell division vary considerably (Fig. 8). (2) The light-dependent induction of cell division is already completed when the main synthesis of DNA starts (compare Fig. 5 and Fig. 8). Accordingly, it has been reported that DNA synthesis either precedes (Iwamura, 1955), occurs simultaneously with (Lorenzen and Ruppel, 1960), or follows the induction of cell division (Wanka, 1962). (3) The homogeneity in the formation time of DNA is less than in cellular division (see Figs. 1 and 8). Another objection against the direct action of DNA in the mechanism of synchrony comes from Hase *et al.* (1959), who observed that under sulfur and nitrogen deficiencies cell division in *Chlorella* was inhibited but DNA synthesis continued. Nevertheless, there is no doubt about the existence of a light-dependent DNA formation. Also a direct correlation between the amount of DNA formation and the number of autospores formed per mother cell exists (Senger and Bishop 1966; and Wanka and Mulders, 1967).

Like the induction of cell division, RNA synthesis starts with the beginning of the light period and its formation continues throughout the complete life cycle (Fig. 8) (Iwamura, 1955; Ruppel, 1962; Stange *et al.*, 1962; Senger and Bishop, 1966). Another intriguing fact is the coincidence in the blue portion of the action spectra for induction of cell division and for RNA formation (compare Figs. 7 and 13). This suggests RNA synthesis as a possible candidate for the system that controls or regulates the mechanism of synchrony. The discovery of Kowallik (1965) that blue light catalyzes the formation of protein and the finding of Hauschild *et al.* (1965) of a blue light enhancement of amino acid synthesis support the following hypothesis: Blue light enhances formation of RNA, which in turn catalyzes protein synthesis and subsequently enzymes which induce and control the timing mechanism of cell division. The unresolved portion of this hypothesis concerns the question as to which pigment(s) is absorbing the blue light. The three blue light enhanced processes of amino acid, RNA, and protein formation might well be linked in such a way that only one of them is primarily induced by blue light and the others are only secondary reactions. It has been recently reported (Kowallik and Gaffron, 1966; Kowallik, 1967; Pickett and French, 1967) that blue light enhances endogenous respiration. It may, therefore, provide the energy for all the different synthetic processes.

Pigments possibly absorbing in the blue region may be carotenoids (Kowallik, 1965; Senger and Bishop, 1966) and riboflavin (Kowallik, 1967; Pickett and French, 1967). Most action spectra, even ours, are not yet sufficiently refined to warrant a definite statement. But still, carotenoids and/or flavoproteins are the most probable compounds involved in absorption. Although we have pointed out that many experiments do not provide evidence for an active role of photosynthesis in the timing mechanism of synchrony, there is the fact that cell division induced in red light (Fig. 7) can be abolished by DCMU. The possibility of synchronizing *Chlorella* in red light was also reported by Kowallik (1962). Experiments to replace the light-dark regime by a change from a 16-hour photosynthetic period ($+CO_2$) to a 12-hour nonphotosynthetic ($-CO_2$) period under constant illumination did not induce synchronization (Senger, 1961). The different results may be explained by the assumption that in blue or white light only the induction of cell division via RNA occurs and only when no blue light is present will red light induce cell division via photosynthesis. To call the induction in red light "nonspecific" might be justified on the grounds that it represents a substitute reaction. Moreover, the pattern of cell division is different in red light (Senger and Schoser, 1966). Kowallik found a difference also in autospore formation in red and blue light (1962), but after long adaptation of cultures in blue and red light, the same mode of cell division was observed (Kowallik, 1963).

REFERENCES

Baker, A. L., and Schmidt, R. R. (1963). *Biochim. Biophys. Acta* **74**, 75.
Bendix, S., and Allen, M. B. (1962). *Arch. Mikrobiol.* **41**, 115.
Bergmann, L. (1955). *Flora (Jena)* **142**, 493.
Bishop, N. I. (1958). *Biochim. Biophys. Acta* **27**, 205.
Bünning, E. (1963). "Die physiologische Uhr." Springer, Berlin.
Galling, G. (1965). *Flora (Jena)* **155**, 596.
Hase, E., Otsuka, H., Mihara, S., and Tamiya, H. (1959). *Biochim. Biophys. Acta* **35**, 180.
Hauschild, A. H. W., Nelson, C. D. and Krotkov G. (1965). *Naturwissenshaften* **52**, 435.
Haxo, F. T. (1960). *In* "Comparative Biochemistry of Photoreactive Pigments" (M. B. Allen, ed.), pp. 339–360. Academic Press, New York.
Iwamura, T. (1955). *J. Biochem.* **42**, 575.
Iwamura, T., and Myers, J. (1959). *Arch. Biochem. Biophys.* **84**, 267.
Iwamura, T., Kanazawa, T., and Kanazawa, K. (1963). *In* "Studies on Microalgae and Photosynthetic Bacteria," pp. 577–586. Univ. Tokyo Press, Tokyo.
Kowallik, W. (1962). *Planta* **58**, 337.
Kowallik, W. (1963). *Planta* **60**, 100.
Kowallik, W. (1965). *Planta* **64**, 191.
Kowallik, W. (1967). *Plant Physiol.* **42**, 672.
Kowallik, W., and Gaffron, H. (1966). *Planta* **69**, 92.
Kuhl, A., and Lorenzen, H. (1964). *In* "Methods in Cell Physiology" (D. M. Prescott, ed.), Vol. I, pp. 159–187. Academic Press, New York.
Lorenzen, H. (1957). *Flora (Jena)* **144**, 473.
Lorenzen, H. (1963). *Flora (Jena)* **153**, 554.
Lorenzen, H. (1964). *In* "Synchrony in Cell Division and Growth" (E. Zeuthen, ed.), pp. 571–578. Wiley (Interscience), New York.
Lorenzen, H., and Ruppel, H.-G. (1960). *Planta* **54**, 394.
Lorenzen, H., and Schleif, J. (1966). *Flora (Jena)* **156**, 673.
Metzner, H., and Senger, H. (1962). *Vorträge Gesamtgebiet Botan.* **1**, 217.
Metzner, H., Rau, H. J., and Senger, H. (1965). *Planta* **65**, 186.
Morimura, Y. (1959). *Plant Cell Physiol.* **1**, 49.
Pickett, J. M., and French, C. S. (1967). *Proc. Natl. Acad. Sci. U.S.* **57**, 1587.
Pirson, A. (1961). *Colloq. Intern. Centre Natl. Rech. Sci. (Paris)* **103**, 103.
Pirson, A. (1962). *Vorträge Gesamtgebiet Botan.* **1**, 178.
Pirson, A., and Lorenzen, H. (1958). *Naturwissenschaften* **45**, 497.
Pirson, A., and Lorenzen, H. (1966). *Ann. Rev. Plant Physiol.* **17**, 439.
Pirson, A., Lorenzen, H., and Ruppel, H.-G. (1963). *In* "Studies on Microalgae and Photosynthetic Bacteria," pp. 127–139. Univ. Tokyo Press, Tokyo.
Ruppel, H. G. (1962). *Flora (Jena)* **152**, 113.
Schmidt, R. R. (1966). *In* "Cell Synchrony" (I. L. Cameron and G. M. Padilla, eds.), pp. 189–235. Academic Press, New York.

Schoser, G. (1966). *Ber. Deut. Botan. Ges.* **79**, 271.
Senger, H. (1961). *Arch. Mikrobiol.* **40**, 47.
Senger, H. (1962). *Vorträge Gesamtgebiet Botan.* **1**, 205.
Senger, H. (1965). *Arch. Mikrobiol.* **51**, 307.
Senger, H., and Bishop, N. I. (1966). *Plant Cell. Physiol.* **7**, 441.
Senger, H., and Schoser, G. (1966). *Z. Pflanzenphysiol.* **54**, 308.
Senger, H., and Wolf, H. J. (1964). *Arch. Mikrobiol.* **40**, 47.
Shihira, I., and Krauss, R. W. (1965). "*Chlorella*-Physiology and Taxonomy of Forty-One Isolates." Port City Press, Baltimore, Maryland.
Soeder, C. J. (1965). *Arch. Mikrobiol.* **50**, 368.
Soeder, C. J. (1966). *Ber. Deut. Botan. Ges.* **79**, 138.
Soeder, C. J., and Ried, A. (1962). *Arch. Mikrobiol.* **42**, 176.
Soeder, C. J., and Thiele, D. (1967). *Z. Pflanzenphysiol.* **57**, 339.
Soeder, C. J., Ried, A., and Strotmann, H. (1964). *Beitr. Biol. Pflanz.* **40**, 159.
Soeder, C. J., Strotmann, H., and Galloway, R. A. (1966). *J. Phycol.* **2**, 117.
Sokawa, Y., and Hase, E. (1967). *Plant Cell. Physiol.* **8**, 509.
Sorokin, C., and Krauss, R. W. (1965). *Am. J. Botany* **52**, 331.
Stange, L., Kirk, M., Bennett, E. L., and Calvin, M. (1962). *Biochim. Biophys. Acta* **61**, 681.
Tamiya, H. (1963). *Symp. Soc. Exptl. Biol.* **17**, 183–214.
Tamiya, H. (1964). *In* "Synchrony in Cell Division and Growth" (E. Zeuthen, ed.), pp. 247–305. Wiley (Interscience), New York.
Tamiya, H. (1966). *Ann. Rev. Plant Physiol.* **17**, 1.
Tamiya, H., Morimura, Y., Yokota, M., and Kunieda, R. (1961). *Plant Cell Physiol.* **2**, 383.
Wanka, F. (1959). *Arch. Mikrobiol.* **34**, 161.
Wanka, F. (1962). *Ber. Deut. Botan. Ges.* **75**, 457.
Wanka, F., and Mulders, P. F. M (1967). *Arch. Mikrobiol.* **58**, 257.

CHAPTER 10

Effects of High Pressure on Macromolecular Synthesis in Synchronized *Tetrahymena*

Arthur M. Zimmerman

I. INTRODUCTION

Hydrostatic pressure has been used as a tool for studying a variety of cellular activities for nearly four decades. Although most of the early physiological studies were concerned with the direct effects of pressure on form and movement of the cell, during the past 15 years pressure studies have been used to correlate biochemical and physiological activities of the cell.

The use of hydrostatic pressure as a research tool was firmly established following the work of Brown (1934) and Brown and Marsland

(1936) in which a clear relationship between hydrostatic pressure and the structural characteristics of protoplasmic gels was revealed. Protoplasmic gelation reactions are readily affected by both temperature and pressure. Formation of gelated structures within cells represents an endothermic reaction which is accompanied by a volume increase (Marsland and Brown, 1942), conforming to the type II classification of Freundlich (1937). Thus, decreasing temperature or increasing pressure tends to weaken gelational structures within cells by causing a shift in the sol-gel equilibrium toward the sol state. The solating action of pressure is readily reversible; under control temperature-pressure conditions it may be used as an index of gelation reactions. Pressure has been used for analyzing the functional significance of intracellular gelations in various cellular activities such as amoeboid locomotion, cytokinesis, and mitosis, as well as for evaluating the effects of various chemical agents on these intracellular gelations (Marsland, 1956; Zimmerman *et al.*, 1957; Zimmerman, 1964a,b, 1967).

Pressure interference with cytoplasmic cortical gel or with the formation of the mitotic apparatus has been shown to inhibit cell division in marine eggs (Zimmerman and Marsland, 1964). Pressure also affects protoplasmic gelation reactions in *Amoeba* causing a reversible blockade of amoeboid locomotion and associated protoplasmic streaming (Landau *et al.*, 1954). Recently Tilney *et al.* (1966) has shown that high pressure reversibly affects the organization of the microtubules in axopodia of *Actinosphaerium*.

Not only can pressure affect and alter the physiological structure of cells, but pressure can also alter the synthetic activity of the cell. Zimmerman (1963) demonstrated that DNA synthesis continues under 5000 lb of pressure per square inch (psi), although mitosis and cytokinesis are blocked; at greater magnitudes of pressure DNA synthesis is also inhibited (Zimmerman and Silberman, 1967). Murakami (1961), studying RNA synthesis in marine eggs and Landau (1966b), studying nucleic acid and protein synthesis in *Escherichia coli*, reported that high pressure inhibits synthesis. Pressure also inhibits enzymatic activity, as has been recently demonstrated by Pollard and Weller (1966), who studied β-galactosidase from *E. coli*, and by Berger (1958), who studied phenylgalactosidase from *Streptomyces*. Moreover, pressure can alter the amount of adenine nucleotides in cultured cells (Landau and Peabody, 1963) and in marine eggs (Landau, 1966a).

In order to have sufficient material for biochemical analysis in cell division studies, there is need for a synchronous population of cells. Heat-synchronized cultures of *Tetrahymena pyriformis* have been studied for the past decade, since Scherbaum and Zeuthen (1954) reported on a method for synchronization. These cells are excellent models for studies on cell division not only because of the ease in handling and maintaining the cultures, but also because of extensive biochemical and physiological studies (see review, Zeuthen, 1964) that offer the investigator a solid foundation for further study.

The objective of the present paper is to report on our studies of the effects of high pressure on the structural as well as the biochemical aspects of division in *Tetrahymena*.

II. EFFECTS OF PRESSURE ON DIVISION

High pressure has two striking effects on synchronous cultures of *Tetrahymena pyriformis* (GL); it can affect the division schedule, and it can block furrowing activity. If a pulse of pressure is applied to a synchronous population of cells prior to furrow formation, the subsequent division is delayed. However, if comparable pressure is applied at the time of incipient furrow formation (cytokinesis), division of the cell is reversibly blocked.

A. DIVISION DELAYS

The effects of pressure on the division schedule are dependent on the magnitude and duration of the pressure, as well as on the time after the last heat shock (EH) that the pressure is applied. A pressure of 14,000 psi applied at 0 minutes after EH, for a duration of 5 minutes, causes most cells to round up and become distorted, with about 10% surviving 12 hours after decompression. When the duration of exposure is decreased to 2 minutes, many of the distorted cells return to a near normal shape 145 minutes after decompression, and as many as 30% survive and are capable of division. When the intensity of pressure is reduced to 10,000 psi for 2 minutes, the shape of the cells is not markedly altered, although movement is sharply reduced within 30 seconds after the application of pressure. However, if the duration of this pressure is increased to 10 minutes, the cells tend to become bulbous at the posterior end and to exhibit very slow, erratic movement. Nonsynchronized *Tetrahymena* growing in log phase cultures were found to be more susceptible to the rounding effects of pressure than heat-synchronized cultures. A comparison of pressure effects on synchronized cells and those growing in log phase cultures is shown in Fig. 1.

Synchronized *Tetrahymena*, subjected to a 2-minute pulse of pressure (10,000 psi) during the interval following the last heat shock, undergo division delays which are directly related to the time after EH that the cells are pressurized. As shown in Fig. 2, there are progressively increased division delays during the first 40 minutes after EH. Pressures

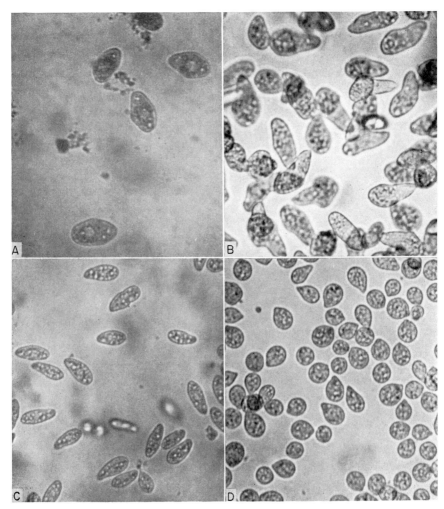

Fig. 1. The effects of pressure on a synchronized culture and a log growth culture of *Tetrahymena pyriformis* GL. The photomicrographs were taken while cells were in the pressure chamber. (A) Control cells photographed in the pressure chamber at atmospheric pressure 10 minutes after EH. Cells are actively swimming. (B) Synchronized cells subjected to 10,000 psi for 10 minutes at 23°C. Movement is markedly decreased. Many cells accumulate on the window. Some cells tend to become spherical, while other cells are bulbous at the posterior. (C) Photomicrograph of cells from a log phase culture at atmospheric pressure. (D) Cells at log phase subjected to 10,000 psi for 10 minutes at 28°C. The cells become round and tear-shaped exhibiting reduced, uncoordinated ciliary movement; however, there was no effective translational movement.

of 10,000 psi applied between 43 and 48 minutes after EH cause a disruption of cell synchrony; however, pressure treatment initiated at 50 and 60 minutes after EH results in no appreciable division delays (Lowe and Zimmerman, 1967).

Comparable experiments conducted at 7500 psi for a duration of 2 minutes exhibit similar results, however, the division delays are markedly reduced and maximum delays occur at an earlier time after EH. At 7500 psi maximum division delay occurs when treatment is initiated

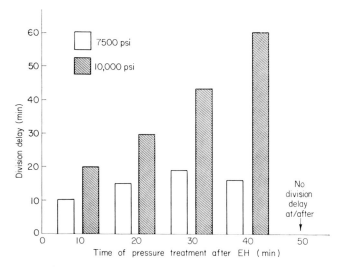

FIG. 2. Effect of pressure on division in synchronized *Tetrahymena*. Synchronized cultures were subjected for 2 minutes to a pressure of 10,000 psi or 7500 psi at 28°C. The division delays at various times after EH are shown. Between 42 and 48 minutes after EH, pressure treatment caused disruption of cell synchrony. At 50 and 60 minutes after EH, there were no appreciable division delays. (Data of Lowe and Zimmerman, 1967.)

30 minutes after EH, whereas at 10,000 psi the maximum delay occurs when treatment is initiated 40 minutes after EH. Nevertheless, at both pressures there is a negligible division delay when treatment is initiated at 50 or 60 minutes after EH.

B. CYTOKINESIS

As previously mentioned, pressure applied at the time of furrowing can prevent the progress of the furrow reaction as well as cause cells with well-developed furrows to abort. Varying magnitudes of pressure were applied to a synchronized culture of *Tetrahymena* just prior to cytokinesis, for a duration of 20 minutes. At a pressure of 7500 psi the furrowing reaction was blocked. This intensity of pressure not only blocked the progression of the furrows, but also resulted in a reversal of furrowing. The cells were distorted and occasionally exhibited a large contractile vacuole; their movement was erratic and markedly reduced. At 1.5 hours after decompression, the cells began to regain their normal shapes, and the following day less than 10% of the cells were distorted.

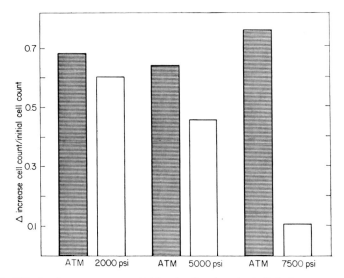

Fig. 3. Effect of pressure on cytokinesis. Synchronized *Tetrahymena* were subjected to varying pressures at 70 minutes after EH for a duration of 20 minutes. The ratio of the Δ increase in cell count to the initial cell at the various pressures are illustrated in the graph. Each experiment is shown with its corresponding atmospheric control. (Data of Lowe, 1968.)

When the cells were exposed to a lower pressure of 2000 psi, furrowing was retarded, but not blocked. After 20 minutes of pressure treatment, the increase in cell count was similar to that in nonpressurized controls. An intermediate pressure value of 5000 psi resulted in a partial blockade of furrowing, and about 30% of the cells were blocked (Fig. 3). Recently, MacDonald (1967a) reported that 250 atmospheres of pressure (3700 psi) immediately arrests cell division in logarithmically growing cultures of *Tetrahymena pyriformis* strain W. It appears from these studies that synchronized *Tetrahymena* are more resistant to pressure inhibiting effects than are logarithmically growing cells.

III. RIBOSOMAL STUDIES

A. RIBOSOME PROFILES FROM SYNCHRONIZED *Tetrahymena*

Isolation and characterization of ribosomes in *Tetrahymena* from both synchronized and nonsynchronized cultures have been carried out in several laboratories (Plesner, 1961, 1963; Lyttleton, 1963; Cameron *et al.*, 1966; Whitson *et al.*, 1966a,b). Our laboratory reinvestigated

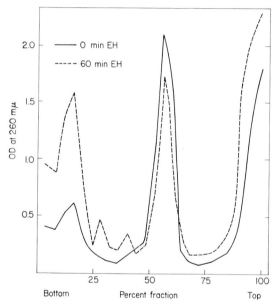

FIG. 4. Optical density profiles of sucrose gradient (5–20% sucrose with a 50% cushion) centrifugations of the 10,000 *g* supernatant fraction from heat-synchronized *Tetrahymena* that were homogenized at 0 and 60 minutes after EH. The material was centrifuged for 3.5 hours at 23,000 rpm in an SW 25.2 rotor at 4°C. Two major OD_{260} peaks are observed. A faster sedimenting peak is recovered at the bottom 20% fraction of the tube. A major fraction was obtained at the 55% fraction. (Data of Hermolin, 1967.)

the ribosomes in heat-synchronized cultures of *Tetrahymena pyriformis* GL in order to look for any possible quantitative changes in the ribosomes and polysomes during the period from the last heat shock to division (the first division cycle). At various times after EH, *Tetrahymena* were washed in inorganic media (Hamburger and Zeuthen, 1957) and homogenized in 0.05 *M* Tris buffer, pH 7.4 with 5×10^{-3} M Mg^{2+}. The 10,000 *g* supernatant fraction from the cell homogenate was subjected to sucrose gradient centrifugation. Analysis of the sucrose density gradient (5–20%

sucrose gradient layered over 50% sucrose cushion and centrifuged for
3.5 hours at 23,000 rpm in a SW 25.2 rotor) indicated the presence
of two major peaks of OD_{260} which were recovered in the 20 and 55%
fractions of the total gradient.

Representative sucrose gradient patterns at 0 and 60 minutes after
EH are shown in Fig. 4. At 60 minutes after EH, the material in the
20% fraction is markedly increased as compared with that found at
EH 0. It is assumed that the material found in this fraction is the
polysomal material. The ribosomal material is found in the 55% fraction.

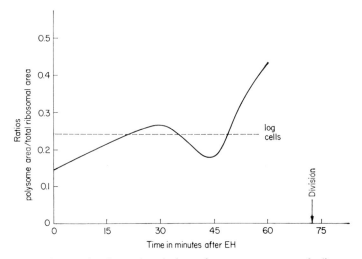

FIG. 5. The changes in the ratio of the polysome area to total ribosome area
are plotted as a function of time after EH. The areas under the polysome peak
and ribosome peak were determined from sucrose gradient profiles. (Data of Hermolin, 1967.)

In order to quantitate the changes in the amount of polysomes during
the period preceding the first synchronous division, the areas under
the polysome peak were measured and compared to the total area (the
sum of the areas under the polysomal and ribosomal peaks). This is
a conservative method for evaluating fluctuations of one of the fractions
with respect to the total amount of material present and will reflect
alterations in the amount of polysomal material. The ratio of the polysome area to total ribosome area was plotted as a function of time
after EH. The results are shown in Fig. 5. This graph indicates a small
peak at 30 minutes after EH and a marked peak at 60 minutes after
EH, the time at which the last measurement was recorded. At 45 minutes
after EH, the ratio is at a minimum and corresponds to that found

at EH 0. Division in these synchronized cultures occurred at 75 minutes after EH. The ratio from log phase cultures was also determined and was found to be intermediate to the 30-minute EH and the 45-minute EH value.

B. RIBOSOME CHARACTERISTICS

If the OD_{260} material in the 20% fraction contains polyribosomes (as seen in Fig. 4), one would expect this material to be sensitive to ribonuclease (RNase) treatment (Warner *et al.*, 1963). If, on the other hand, the OD_{260} material contains ribosomes that are primarily membrane-bound and accumulated at the boundary between the sucrose cushion and the gradient, then one would expect the membranous material to be readily dissolved by treatment with deoxycholate (DOC). In order to investigate these two possibilities, the 10,000 g supernatant fractions of cells at 60 minutes after EH were treated with RNase or with DOC. When the supernatant from the 10,000 g homogenate was treated with RNase (10 μg/ml for 15 minutes at 4°C), the OD_{260} of the rapidly sedimenting material was abolished and there was a corresponding increase in OD_{260} in the ribosomal region. The results from a typical experiment are shown in Fig. 6.

It did not appear that the polysome fraction contained any appreciable amount of membrane-bound ribosomes since, treatment of the 10,000 g supernatant fraction with deoxycholate (0.45% by weight) did not appreciably decrease the OD_{260} of the rapidly sedimenting peak, although there was a slight reduction in this peak.

As has been established in other systems, the integrity of the polysomes in *Tetrahymena* is dependent upon the magnesium concentration. When the magnesium concentration of the homogenizing medium was reduced to 10^{-4} M, the sucrose density gradient profile was markedly changed. Instead of two OD_{260} peaks, as found with the higher concentration of magnesium (5×10^{-3} M), there was only one rather broad, slowly sedimenting peak, which was appreciably lighter than that found with the higher concentrations of magnesium. It appeared, therefore, that the polysomes were disrupted at lowered concentrations of Mg^{2+}, and furthermore that the ribosomes themselves were markedly altered when the concentrations of Mg^{2+} were decreased.

In order to further characterize the ribosomal material in the 10,000 g supernatant fraction, this material was subjected to analytical ultracentrifugation. As shown in Fig. 7, two small peaks are visible. The faster sedimenting peak is representative of the polysomal material, and the slower sedimenting peak is representative of the ribosomal material. This material was obtained from a heat-synchronized culture of *Tetra-*

hymena at 60 minutes after EH. It appears from these patterns that
the relative amount of polysome material is approximately equal to
that of the ribosomal material. The sedimentation coefficient of the
ribosomes were calculated to be **84 S**, and the sedimentation coefficient
of the rapidly sedimenting polysome peak was calculated to be **120 S**.
These values compare favorably with those obtained by Whitson *et*

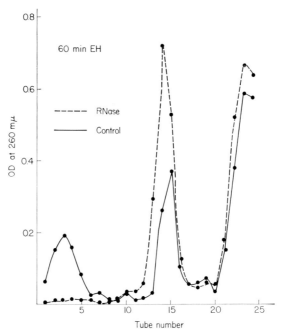

Fɪɢ. 6. Optical density profiles of sucrose gradient centrifugations of a **10,000** *g*
supernatant from an RNase-treated homogenate and control homogenate are
shown. Cells were homogenized at 60 minutes after EH and divided into two
aliquots. One aliquot was treated with 10 μg/ml for 15 minutes at 4°C. The other
aliquot was used as a control. The rapidly sedimenting fraction has been virtually
eliminated, and the ribosome fraction (tube number 12–16) has been markedly
increased. (Data of Hermolin, 1967.)

al. (1966a,b). Whitson and co-workers studied cold synchronized *Tetrahymena pyriformis* strain HSM and found an **82 S** peak and a series
of heavier peaks at **125 S**, **148 S**, and **186 S**. Their electron micrographs
showed that the **82 S** peak consisted of monosomes, the **125 S** of aggregates of two ribosomes, the **148 S** of aggregates of three, and the **186
S** fraction of aggregates of four ribosomes (cf. Cameron *et al.*, 1966).

Different sedimentation values have been obtained by Plesner (1961,

1963) and Lyttleton (1963). Plesner reported the presence of 70, 80, and 100 S particles from heat-synchronized *Tetrahymena pyriformis* strain GL while Lyttleton who worked with nonsynchronized cultures found three different values of 72, 75, and 78 S for the ribosomes, depending upon the method of ribosome preparation. The sedimentation values obtained by various investigators are summarized in Table I.

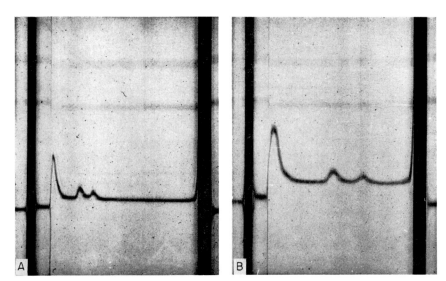

FIG. 7. Analytical ultracentrifugation pattern of a 10,000 g supernatant fraction of cells homogenized at 60 minutes after EH in 0.05 M Tris, pH 7.4 with 5×10^{-3} M Mg^{2+}. (A) The photograph was taken at 5 minutes after a speed of 33,450 rpm was reached, phase angle 60 degrees. (B) Photograph was taken at 13 minutes at a phase angle of 40 degrees. The s_{20} value is 120 for the rapidly sedimenting peak and 84 for the slower sedimenting peak. The protein concentration was 1.9 mg/ml.

C. SUCROSE GRADIENT PROFILES FROM PRESSURIZED CELLS

The mechanism by which ribosomes associate with messenger RNA to form polysomes is not clearly understood. Formation of this complex may involve changes in the spatial configuration of the components. It is conceivable that the polysome complex has a volume that is essentially different from the sum of the volumes of the individual components. If polysome formation is accompanied by a change in configuration of either the mRNA, ribosome or aminoacyl-tRNA and this change is associated with a volume increase, then polysome stability could be affected by high pressure.

It was previously demonstrated that newly formed polysomes are

TABLE I

SEDIMENTATION VALUES

Strain	Synchrony	Sedimentation coefficient			Comments	References
		Subunit	Ribosome	Associated products		
GL	Heat	—	84	120	10,000 g supernatant, 5 mM Mg^{2+}	Present study
GL	Heat	—	70, 80	100	Resuspended pellet, 25 mM Mg^{2+}; at low Mg^{2+} (0.1 mM) 70 S stable only at 50–65 min EH	Plesner (1961)
GL	Heat	50	70	100	5 mM Mg^{2+} at 40 and 60 min EH	Plesner (1963)
		30, 50	—	—	40 min EH, 0.1 mM Mg^{2+}	
		50	70	—	60 min EH, 0.1 mM Mg^{2+}	
GL	Nonsynchronized	—	72	—	Resuspended pellet of cells lysed by indole in 12.5 mM Tris (no Mg^{2+})	Lyttleton (1963)
		—	75	115	Resuspended pellet of cells lysed by indole in H$_2$O (no Mg^{2+})	
		—	78	—	"Ribosomes isolated by a single stage of high speed centrifugation" (no Mg^{2+})	
HSM	Cold	61	82	123, 174	At end of cold period	Whitson et al. (1966a,b)
			83	125, 148, 186	At 1 hr after start of warm period	

cell lysate with 0.5 mM Mg^{2+}

present at 60 minutes after EH. Thus at 60 minutes after EH a synchronized culture was divided into two aliquots. One sample was subjected to 5000 psi of pressure for 2 minutes. Then both the nonpressurized
control sample and the pressurized sample were homogenized and the
10,000 g supernatant fraction was subjected to sucrose gradient analysis.
As shown in Fig. 8, pressure has a marked effect on the polysomes.
The OD_{260} of the rapidly sedimenting material (tube 3–5) from the

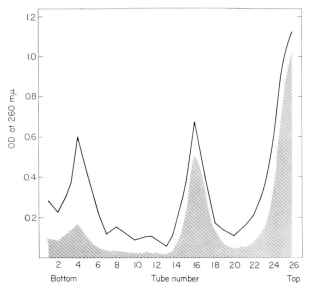

FIG. 8. Optical density profile of the 10,000 g supernatant fraction obtained
from cells subjected to 5000 psi for 2 minutes, at 60 minutes after EH, is compared
to the supernatant from nonpressurized control cells. The solid line illustrates
the OD_{260} of the controls, and the hatched area indicates the OD_{260} of the pressurized cells. The cells were divided into two equal samples prior to pressure
treatment. There was a marked decrease in the OD_{260} of the rapidly sedimenting
peak (tube number 4–5) from the cells which were pressurized.

pressurized sample is much less than that found for the control sample.
At higher pressures (10,000 and 14,000 psi) and at longer pressure pulses
(5 minutes) the patterns were comparable to that shown in Fig. 8. However, when the pressure was reduced to 2000 psi for 2 minutes, the sucrose
gradient profiles were similar to the profiles from the nonpressurized
control cells.

Although it was demonstrated that pressure affects the stability of
the polysomes in *Tetrahymena*, it was not known whether pressure would
also exert an effect on polysome *formation*. In order to investigate this

problem, cells were pressurized at a time, after the last heat shock, which precedes the rapid increase in polysome formation. Thus, at 45 minutes after EH the cells were pulsed for 2 minutes at a pressure of 5000 psi and homogenized at 60 minutes after EH. The 10,000 g supernatant fraction was analyzed by sucrose-gradient studies. The relative amount of the polysomes 13 minutes after decompression (60 minutes after EH) was comparable to the nonpressurized controls, at 60 minutes after EH (see Fig. 9).

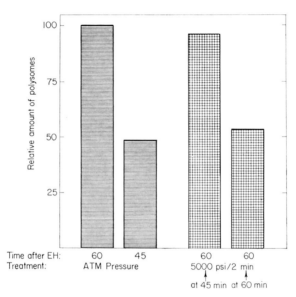

Fig. 9. The relative amount of polysomes from cells at atmospheric pressure is compared with that from cells subjected to 5000 psi of pressure for 2 minutes. The relative amount of polysomes at 60 minutes after EH was arbitrarily given a value of 100. At 45 minutes after EH the relative amount of polysomes has a value of approximately 50. The relative amount of polysomes from cells pressurized at 60 minutes after EH has a value of approximately 50. Cells pressurized at 45 minutes after EH and permitted to recover for 13 minutes at atmospheric pressure have a value comparable to that of control cells at 60 minutes after EH.

The mechanism by which ribosomes bind to messenger RNA is not known. Pressure does not seem to affect the individual components of the polysomes prior to formation, however, it appears that the newly formed polysomes are readily disrupted by pressure. The present studies suggest, but do not prove, that there may be an increase in volume inherent in the newly formed polysomes which is greater than that of the individual components (messenger RNA and ribosomes).

IV. PROTEINS ASSOCIATED WITH DIVISION

Although Zeuthen (1961) and co-workers have proposed the presence of a "division protein" in *Tetrahymena*, the isolation and characterization of such a "division protein" has not been unequivocally established. Recently, however, Watanabe and Ikeda (1965a,b) identified a "division protein" in a water-soluble extract of *Tetrahymena pyriformis* strain W which could be isolated using a DEAE-cellulose column. They report that a distinct peak can be eluted from the DEAE-cellulose column at 45 minutes after EH, whereas at 0 EH, only a small amount of the material

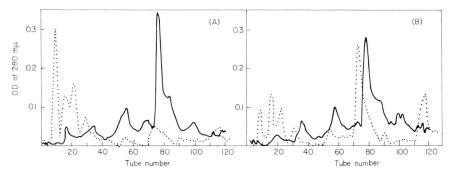

FIG. 10. Effluent patterns obtained by chromatographing water-soluble protein fractions on DEAE-cellulose. (A) OD_{280} of the effluent profile at 0 minutes after EH. (B) Illustrates the OD_{280} of the effluent profile at 45 minutes after EH. The solid line is the OD_{280} profile obtained by L. Lowe using DEAE-cellulose (0.92 meq/gm, Serva). Each tube number represents 4 ml of effluent. The flow rate was 0.88 ml/min carried out at 2°C. The maximum OD_{280} was observed at tube number 78–82. The dotted line is the profile obtained by Watanabe and Ikeda (1965a), which was superimposed on the present data. The "division protein peak" as described by Watanabe and Ikeda (1965a) occurs at tube number 70–75.

can be found. The dotted line in Fig. 10 is the effluent profile from a DEAE-cellulose column redrawn from the study of Watanabe and Ikeda (1965a). These authors propose that the OD_{280} material at tube number 70–75 (Fig. 10B) is the "division protein" material. Using a technique similar to that reported by Watanabe and Ikeda (1965a), L. Lowe, in our laboratory, attempted to isolate this "division protein" and to study the effects of hydrostatic pressure on this material. However, Lowe could not confirm the alterations in the OD_{280} profiles from chromatographed extracts at 0 and 45 minutes after EH. The solid line in Fig. 10 is a representative profile from the study of Lowe. There is a distinct OD_{280} peak which was eluted at tube 78–82. This OD_{280} peak corresponds to that found by Watanabe and Ikeda (1965a) at tube

70–75. However, there is no "quantitative" difference in the amount of material that was eluted in the fractions at 0 and 45 minutes EH.

Although no differences were established at 0 or 45 minutes after EH, this does not deny the existence of "division protein." In view of the extensive work of Zeuthen (1961), Rasmussen and Zeuthen (1962), and Hamburger (1962), which suggested a synthesis of "division protein" during the division schedule in synchronized cultures, it is possible that the "division proteins" exist but could not be resolved by the techniques employed in our laboratory.

When synchronized *Tetrahymena pyriformis* strain W or GL were subjected to varying pressures and durations (10,000 psi for 2 minutes or 14,000 psi for 2 or 5 minutes) at 45 minutes after EH, the protein profiles of aqueous extracts were similar to extracts from nonpressurized controls as shown in Fig. 10 (Lowe and Zimmerman, 1967). Thus, pressure did not cause any changes in the water-soluble cellular proteins that would affect their solubility or their elution patterns from a DEAE-cellulose column. Furthermore, there were no quantitative differences in the OD_{280} elution patterns obtained from *Tetrahymena pyriformis* strain W or GL.

V. PRESSURE EFFECTS ON RNA AND PROTEIN SYNTHESIS

It was previously shown (Section III,A) that in *Tetrahymena* the relative amount of polysomes varies during the period following the last heat shock. Furthermore, it was also demonstrated that pressure affects polysome stability. Thus, studies were conducted to evaluate the effects of pressure on the synthesis of RNA during this period. Aliquots of synchronous cultures of cells were incubated in uridine-^{3}H (10 μc/ml) for 10-minute durations at various times after the last heat shock. The cells were lysed by freezing (in Dry Ice and acetone) and thawing, and the acid insoluble fraction was precipitated with 5% trichloroacetic acid (TCA) that contained 10^{-3} M uridine. The precipitate was washed with additional amounts of the TCA-uridine solution, then with 70% and 100% ethanol, with ether and finally air dried. The precipitate was dissolved in hyamine hydroxide and counted in a Packard spectrometer. Other synchronous cells at 15 and 25 minutes after EH were subjected to 10,000 psi for 2 minutes, and at subsequent 10-minute intervals they were pulsed with uridine-^{3}H. The results of these experiments are shown in Fig. 11B. The curve illustrating incorporation of labeled uridine in nonpressurized control cells is in general agreement with that reported by Moner (1965), who studied incorporation by means of autoradiography. The maximum rate of uridine-^{3}H incorporation occurs at 20–30 minutes after EH and just before division. The effects of pressure on the incorporation of uridine indicate that there is a reduction in the ability of the cells to incorporate the isotope into the cold TCA-extractable material. Furthermore, the curves are displaced to the right, indicating that incorporation is delayed in the pressure-treated cells.

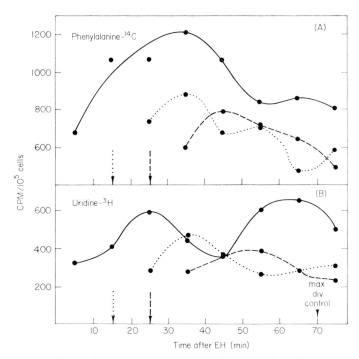

FIG. 11. The effects of hydrostatic pressure on incorporation of phenylalanine-^{11}C and uridine-[3]H. The atmospheric control cells (——) were pulsed for 10 minutes in the labeled isotope. The radioactivity (cpm/10[5] cells) of the acid-insoluble fraction was plotted as a function of the time after EH. In the pressurized series, the cells were pressurized at 15 minutes after EH (......) or at 25 minutes after EH (------) for 2 minutes at 10,000 psi. Subsequently, the cells were incubated in the protein or RNA radioactive precursor for a 10-minute duration and prepared for radioactive counting as above. The cells pressurized at 15 minutes after EH are represented by "......" and at 25 minutes after EH by "------". (From the work of L. Lowe, 1968.)

Similar incorporation studies were conducted using the isotope phenylalanine-^{14}C during the period preceding synchronous division. The cells were incubated in the isotope compound (0.25 μc/ml) for a duration of 10 minutes at various times after EH. Following isotope treatment, the cells were lysed and the acid-insoluble fraction was precipitated with hot 5% TCA. The precipitate was washed with additional TCA with added L-phenylalanine, followed by washes with ethanol and ether, and then counted in a Packard spectrometer. The results are illustrated in Fig. 11A. Maximum incorporation occurred between 30 and 40 minutes after EH. The lowest rate of incorporation occurred immediately after the last heat shock and during the last 20 minutes of the division cycle. When cells were subjected to a pressure pulse of 10,000 psi for 2 minutes at 15 and 25 minutes after EH, the rates of incorporation of phenylalanine-^{14}C were markedly reduced and the curves were shifted toward the right, similar to the results found in the uridine-^{3}H study.

Not only was pressure shown to affect the rate of incorporation of the radioactive precursors, but 10,000 psi of pressure for 2 minutes also affected the accumulative incorporation of uridine-^{3}H and phenylalanine-^{14}C. Cells were placed into the isotopes at 0 EH. At 15 and 25 minutes after EH, they were subjected to a pressure pulse. The accumulative incorporation of both radioactive precursors was markedly reduced following the pressure pulse. These experiments were carried out up to 70 minutes after EH. At that time the incorporation of phenylalanine-^{14}C was reduced by about 30%, whereas the uridine-^{3}H was reduced between 30% and 50%.

VI. DISCUSSION

A. IS THERE A "DIVISION PROTEIN"?

Extensive evidence has been accumulated to suggest that some protein synthesis is necessary, during the 75 minutes preceding division, for division to occur in synchronous cultures of *Tetrahymena* (Zeuthen, 1964). This protein has tentatively been labeled "division protein" by Zeuthen (1961). He proposed that "division proteins" are produced prior to the "physiological transition point" which occurs at about 55 minutes after EH. Although the studies presented in this paper do not resolve the question of the presence or absence of "division protein," they do shed some light on the overall problem. Watanabe and Ikeda (1965a,b) have identified a "division protein" rich in SH groups (Ikeda and Watanabe, 1965). However, Lowe, in our laboratory, was unable to confirm the OD_{280} changes in the protein profiles of aqueous extracts from *Tetrahymena pyriformis* strain GL or W. In her studies, Lowe found that the protein profiles were similar at 0 and 45 minutes after EH. She washed the cells somewhat differently from the method reported by Watanabe and Ikeda (1965a) because it was found that, with their techniques, there was an appreciable amount of cytolysis. In addition, a different grade of DEAE-cellulose was employed by Lowe in her studies. She used two types of DEAE-cellulose, 0.92 meq/gm (Serva) and 1.0 meq/gm (Whatman). Watanabe and Ikeda (1965a), in their original study, employed DEAE-cellulose, 1.0 meq/gm (Serva). They stated that results could be obtained using other types of DEAE-cellulose, but the best results were obtained with 1.0 meq/gm (Serva). As shown in Fig. 10, the OD_{280} peaks from the work of Watanabe and Ikeda (1965a) are "relatively similar" to those of Lowe, if one considers the different types of DEAE-cellulose employed. However, the fact that Lowe consistently obtained a maximum OD_{280} at tube number 78–82 at 45 minutes after EH, as well as at 0 minutes, makes it difficult to designate this peak as "division protein."

If a "division protein" is produced at 45 minutes after EH, it must be concluded from our work that it is relatively resistant to the disrupting effects of high hydrostatic pressure since, no alterations in the protein profiles were found in cells that were subjected to high pressure. Nevertheless, other systems associated with division must be markedly affected since cultures subjected to short pulses of pressure between 42 and 48 minutes after EH exhibit a disruption of synchrony. Moreover, pressure applied during the earlier period of the division cycle (0–40 minutes EH) caused marked division delays. Recently, MacDonald (1967b) suggested from his pressure studies on logarithmically dividing *Tetrahymena* that division delays may be mediated through a protein similar to that proposed by Rasmussen and Zeuthen (1962).

It was demonstrated that pressure affects protein synthesis (see Section V). The effect of pressure on protein synthesis may not be directly involved in furrow formation or in the cleavage process per se, but may be concerned with some other structural aspect of the cell, for example, the oral structure. Simpson (1966) has reported that pressure markedly inhibits the formation of the oral structure. He found that pressure applied at early organizational stages of oral morphogenisis resulted in a loss of organization, and, in general, the oral structure was re-formed prior to division. Furthermore, cells pressurized late in the synchronization period either replaced their "mouths" prior to division or divided and then replaced their oral structures.

B. RELATIONSHIP OF MACROMOLECULAR SYNTHESIS AND CYTOKINESIS

The maximum increase in the relative amount of polysomes during the period from EH 0 to division appears to occur at 60 minutes after EH. At this time, the rate of incorporation of uridine-^{3}H also reaches a maximum, however, the phenylalanine-^{14}C incorporation is relatively low. Originally, it was thought that the increase in polysomes at 60 minutes after EH was preparatory to a synthesis of material which would be necessary for the subsequent cytokinesis. Upon further examination of biochemical synthesis, as well as the effects of pressure at this time, it becomes obvious that this hypothesis is unacceptable, and an alternative will be proposed.

The *Tetrahymena* are relatively unaffected by pressure pulses during the period just preceding division (50–60 minutes after EH), although they lose their synchrony when they are subjected to pressure between 42 and 48 minutes after EH. The most striking division delays occur following a pressure treatment between 30 and 40 minutes after EH. At this time, the rate of incorporation of phenylalanine-^{14}C has reached a peak, which would be compatible with the proposal that an essential protein is being synthesized at this time. Preceding this peak of maximal division delays and protein synthesis, there is an increased synthesis of RNA. The rate of incorporation of uridine-^{3}H reaches a peak between 20 and 30 minutes after EH. This is followed by an increase in the relative amount of polysomes. It is suggested, therefore, that the increased rate of RNA synthesis is required for polysomal formation and these polysomes may be responsible for the increased rate of protein synthesis which follows at 30–40 minutes after EH. Furthermore, the increased rate of phenylalanine-^{14}C may be reflecting the synthesis of a protein for division.

As seen from Fig. 12, the uridine-^{3}H incorporation and the relative amount of polysomes increase markedly at 60 minutes after EH, just prior to cytokinesis. However, the phenylalanine-^{14}C incorporation does not show a corresponding increase at this time. This suggests that the increased RNA synthesis, 60 minutes after EH, may not be related to the subsequent division which follows immediately, but may, indeed, be related to the second division cycle, as proposed by Moner (1965).

When pressure (5000 psi for a 2-minute duration) was applied at 60 minutes after EH, the relative amount of polysomes was markedly decreased to an amount which was comparable to that found at 45 minutes after EH. Division delays at this time were minimal. These data are compatible with the hypothesis that proteins synthesized at this time, 60 minutes after EH, were not essential for the first synchronous division. Further evidence to support this hypothesis is obtained from the pressure pulsing studies at 45 minutes after EH. It was previously reported that there was no reduction in the relative amount of polysomes when cells were pressurized for 2 minutes at 45 minutes after EH and the polysomes were examined at 60 minutes after EH. In view of the fact that pressure did not prevent the formation of the polysomes (at 60 minutes after EH), but caused a loss of division synchrony, it is doubtful that these polysomes were essential for the subsequent division. It is conceivable, although unproven, that the increase

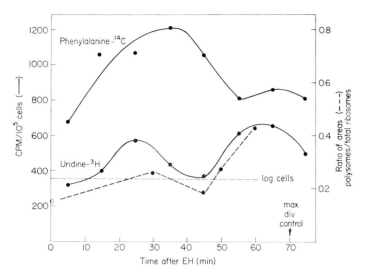

FIG. 12. The rate of incorporation of phenylalanine-^{14}C (0.25 μc/ml) and uridine-^{3}H (10 μc/ml) and the relative amount of polysomes are plotted as a function of the time after EH. RNA synthesis, as reflected by uridine-^{3}H incorporation, and the relative amount of polysomes, reached a peak which precedes the peak of protein synthesis.

in the polysomes at 30 minutes after EH is associated with the production of a protein that may be essential for division.

C. CONCLUDING REMARKS

The previous discussion has demonstrated that hydrostatic pressure may be used to analyze and evaluate synchronous division in *Tetrahymena*. The physiological as well as the biochemical effects of pressure on *Tetrahymena* have been quantitated. The physiological effects of high pressure are manifested in division delays and blockade of cytokinesis. Our studies suggest that the pressure-induced division delays in synchronous cultures of *Tetrahymena* are mediated through interference with specific biochemical systems associated with protein synthesis. This viewpoint is supported by the effects of pressure *in vivo* on polysomes as well as on RNA and protein synthesis. Further confirmation arises from the pressure studies on other biological systems.

ACKNOWLEDGMENTS

Experiments from the author's laboratory have been carried out with the support of research grants from the National Research Council of Canada and the National Cancer Institute of Canada. Their assistance is gratefully acknowledged. The author is grateful to Mrs. Helen Laurence for her invaluable technical assistance, and to Dr. Selma Zimmerman for her suggestions and help in the preparation of this manuscript.

REFERENCES

Berger, L. R. (1958). *Biochim. Biophys. Acta* **30**, 522.
Brown, D. E. S. (1934). *J. Cell. Comp. Physiol.* **5**, 335.
Brown, D. E. S., and Marsland, D. A. (1936). *J. Cell. Comp. Physiol.* **8**, 159.
Cameron, I. L., Cline, G. B., Padilla, G. M., Miller, O. L., Jr. and Van Dreal P. A. (1966). *J. Natl. Cancer Inst. Monog.* **21**, 361.
Freundlich, H. (1937). *J. Phys. Chem.* **41**, 901.
Hamburger, K. (1962). *Compt. Rend. Trav. Lab. Carlsberg* **32**, 359.
Hamburger, K., and Zeuthen, E. (1957). *Exptl. Cell Res.* **13**, 443.
Hermolin, J. (1967). M.Sc. Thesis, University of Toronto, Toronto, Canada.
Ikeda, M., and Watanabe, Y. (1965). *Exptl. Cell Res.* **39**, 584.
Landau, J. V. (1966a). *J. Cell Biol.* **28**, 408.
Landau, J. V. (1966b). *Science* **153**, 1273.
Landau, J. V., and Peabody, R. A. (1963). *Exptl. Cell Res.* **29**, 54.
Landau, J. V., Zimmerman, A. M., and Marsland, D. A. (1954). *J. Cell. Comp. Physiol.* **44**, 211.
Lowe, L. (1968). Ph.D. Thesis. University of Toronto, Toronto, Canada.
Lowe, L., and Zimmerman, A. M. (1967). *J. Protozool. Suppl.* **14**, 9.
Lyttleton, J. W. (1963). *Exptl. Cell. Res.* **31**, 385.
MacDonald, A. G. (1967a). *Exptl. Cell Res.* **47**, 569.
MacDonald, A. G. (1967b). *J. Cell Physiol.* **70**, 127.
Marsland, D. (1956). *Intern. Rev. Cytol.* **5**, 199.
Marsland, D., and Brown, D. E. S. (1942). *J. Cell. Comp. Physiol.* **20**, 295.
Moner, J. G. (1965). *J. Protozool.* **12**, 505.
Murakami, T. H. (1961). *Symp. Soc. Cell. Chem.* **11**, 223 (in Japanese).
Plesner, P. (1961). *Cold Springs Harbor Symp. Quant. Biol.* **26**, 159.
Plesner, P. (1963). *In* "Cell Growth and Cell Division" (R. J. C. Harris, ed.), pp. 77–91. Academic Press, New York.
Pollard, E. C., and Weller, P. K. (1966). *Biochem. Biophys. Acta* **112**, 573.
Rasmussen, L., and Zeuthen, E. (1962). *Compt. Rend. Trav. Lab. Carlsberg* **32**, 333.
Scherbaum, O., and Zeuthen, E. (1954). *Exptl. Cell Res.* **6**, 221.
Simpson, R. E. (1966). Ph.D. Thesis. University of Iowa, Iowa City, Iowa.
Tilney, L. G., Hiramoto, Y., and Marsland, D. (1966). *J. Cell Biol.* **29**, 77.
Warner, J. R., Knopf, P. M., and Rich, A. (1963). *Proc. Natl. Acad. Sci. U.S.* **49**, 122.
Watanabe, Y., and Ikeda, M. (1965a). *Exptl. Cell Res.* **39**, 443.
Watanabe, Y., and Ikeda, M. (1965b). *Exptl. Cell Res.* **39**, 464.
Whitson, G. L., Padilla, G. M., and Fisher, W. D. (1966a). *Exptl. Cell Res.* **42**, 438.
Whitson, G. L., Padilla, G. M., and Fisher, W. D. (1966b). *In* "Cell Synchrony—Studies in Biosynthetic Regulation" (I. L. Cameron and G. M. Padilla, eds.), pp. 289–306. Academic Press, New York.
Zeuthen, E. (1961). *In* "Biological Structure and Function II" (T. W. Goodwin and O. Lindberg, eds.), pp. 537–548. Academic Press, New York.

Zeuthen, E. (1964). *In* "Synchrony in Cell Division and Growth" (E. Zeuthen, ed.), pp. 99–158. Wiley (Interscience), New York.

Zimmerman, A. M. (1963). *Exptl. Cell Res.* **31**, 39.

Zimmerman, A. M. (1964a). *Biol. Bull.* **127**, 345.

Zimmerman, A. M. (1964b). *Biol. Bull.* **127**, 538.

Zimmerman, A. M. (1967). *J. Protozool.* **14**, 451.

Zimmerman, A. M., and Marsland, D. (1964). *Exptl. Cell Res.* **35**, 293.

Zimmerman, A. M., and Silberman, L. (1967). *Exptl. Cell Res.* **46**, 469.

Zimmerman, A. M., Landau, J. V., and Marsland, D. (1957). *J. Cell. Comp. Physiol.* **49**, 395.

CHAPTER 11

The Role of Microtubules in the Cell Cycle

John R. Kennedy, Jr.

I. INTRODUCTION

Although some of the earliest descriptions of microtubules were of microtubules in the cytoplasm of nerve cells (Schmitt and Geren, 1950; Palay, 1956), these structures were thought until recently to be a common component of only certain specialized (ciliated and flagellated) cells and protozoa (Pitelka, 1963). In the past few years, reference to microtubules in a variety of cell types has considerably increased (for recent reviews see Slautterback, 1963; Sandborn *et al.*, 1965; Behnke and Forer, 1967). Thus, many investigators now consider microtubules to be a ubiquitous cellular organelle. In view of this vast accumulation of literature, it is logical to consider the possible roles of microtubules during the cell cycle and to determine whether there exists any constant functional pattern.

The known functions and distribution of microtubules will be briefly reviewed, and their basic structure will be considered. Recent evidence concerning specific division-associated functions will be discussed in relation to their possible role in division-oriented cell synchrony of metazoan and protozoan cell populations.

II. FUNCTION AND DISTRIBUTION OF MICROTUBULES

Several functions have been ascribed to cellular microtubules. These may be grouped into three major categories: motility, form, and metabolism.

A. MICROTUBULES FOR MOTILITY

Organelle motility is the primary function of spindle fibers and ciliary filaments. These elements represent one group of microtubules derived from two homologous organelles—centrioles and basal bodies. The inclusion of spindle fibers into the microtubule class has occurred only recently. Roth and Daniels (1962) clearly demonstrated the microtubular nature of spindle fibers in dividing *Pelomyxa*. Although centrioles are lacking in this organism, the organization of these fibers is comparable to that of mitotic spindles in other forms. Observations on sea urchin eggs (Kane, 1962; Kiefer *et al.*, 1966) and embryos (Harris, 1962), avian and mammalian tumor cells (de-Thé, 1964), newt heart tissue (Barnicot, 1966), human fibroblasts (Barnicot, 1966; Krishan and Buck, 1965), and spermatid of the earthworm (Anderson *et al.*, 1966) have further established the existence of a common microtubular organization to spindle fibers. The mechanism of anaphase chromosome movement and the precise role of spindle microtubules is still unclear; however, its importance at that time in the cell cycle is undisputed since disruption of spindle formation by various means is a common mechanism for division inhibition (Biesele, 1958; Kihlman, 1966).

The basic $9 + 2$ pattern of microtubules in cilia (flagella) is well known, and several possible mechanisms of contraction have been proposed. This literature has recently been extensively reviewed by Sleigh (1962) and Rivera (1962). The similarity of basal bodies and their cilia to centrioles and their spindle fibers has been cited only recently with any frequency (Sleigh, 1962; Ledbetter and Porter, 1963; Pease, 1963; Randall and Disbrey, 1965; Krishan and Buck, 1965; Barnicot, 1966; Kiefer *et al.*, 1966). Such a similarity should be emphasized in view of the capacity of centrioles to develop cilia under certain conditions (Sotelo and Trujillo-Cenóz, 1958; Sorokin, 1962; Renaud and Swift, 1964). For example, Stubblefield and Brinkley (1966) found that, upon removal of Colcemid inhibition from Chinese hamster fibroblasts, centrioles become associated with vesicles. These vesicles, which may arise from the Golgi complex, flatten across the end of the centriole to form the ciliary sheath. The centriole expands into the vesicle, forms the ciliary bud, and undergoes rapid elongation. Vesicular elements are added, enlarging the ciliary shaft and forming a mature cilium. Thus, it might be expected that factors which affect microtubules derived from centrioles would also affect those microtubules arising from basal bodies. An active role of cilia (flagella) in cell morphogenesis and division seems to be confined primarily to the protozoa. Much work in this area has been performed on the oral morphogenesis in *Tetrahymena* (Frankel, 1962, 1964, 1965; Gavin and Frankel, 1966).

B. MICROTUBULES FOR MAINTENANCE OF FORM

The majority of microtubules, aside from those of cilia (flagella) and mitotic spindles, probably function in the maintenance of cellular form. Some recent evidence suggests that microtubules may also be responsible for establishing cellular form in differentiating embryonic tissue.

One of the most extensively studied group of this class of microtubules is in the protozoa (reviewed completely by Pitelka, 1963). In the ciliates, these tubules appear to be derived from basal bodies and, as such, constitute a major portion of the oral and cortical architecture. In *Blepharisma undulans*, for example, the peripheral ciliary microtubules of the adoral zone of membranelles extend proximally to form the walls of the basal bodies. Extensions of these same microtubules pass inward to the cytoplasm and arch over to form a microtubular basket continuous with the undulating membrane on the right margin of the mouth (Kennedy, 1965). A similar set of microtubules comprises the oral area of *Tetrahymena* but is not as extensive (Nilsson and Williams, 1966; Williams and Zeuthen, 1966). Although numerous functions have been proposed for these microtubules (see Jahn and Bovee, 1967 for a detailed review), the major roles are probably support, maintenance of form (Allen, 1967; Kennedy, 1965), and preservation of cortical ciliary patterns from organism to organism and through the cell cycle.

Additional evidence for the supportive function of microtubules comes from studies of microtubules in *Actinosphaerium*. Tilney and Porter (1965) described an extensive microtubular network in the radiating arms of this heliozoan. When the organism was subjected to high pressure, the arms lost their rigidity. Examination of the cell revealed a degradation of the microtubules into fine filaments (Tilney *et al.*, 1966). A similar function has been accorded to cytoplasmic microtubules in the cells of tumors from the kidney, liver, and ovaries of birds (de-Thé, 1964). Fawcett and Witebsky (1964) suggest that circumferential microtubular bands in nucleated erythrocytes of *Amphiuma* may represent an internal cytoskeleton capable of preserving normal cell shape under various conditions of stress.

Microtubular components of other cells such as neurons (Schmitt and Geren, 1950; Gray, 1959; Palay, 1956, 1960; Sotelo and Palay, 1968) and numerous other vertebrate (Behnke, 1964) and invertebrate (Anderson et al., 1966; Slautterback, 1963) cells not in division may also function in form stabilization. Although this type of role is difficult to establish, evidence from embryonic tissue studies seems to support such an interpretation. Recent reports indicate the occurrence of microtubules in embryonic cells of the mouse (Herman and Kauffman, 1966) and rat (Behnke, 1964). Moreover, at the time of cell elongation in chick developing lens rudiments, microtubules 250 Å in diameter appear. These tubules are oriented parallel to the direction of cell elongation (Byers and Porter, 1964) and may establish the form to be taken by the differentiated cells. Pickett-Heaps (1967) observed that colchicine treatment of wheat xylem cells caused a disappearance of most wall microtubules resulting in a loss of organization in the pattern of wall deposition. As suggested by Green's (1965) observations on growth in Nitella, the microtubules may be "directly involved in the stability of the orientation of wall synthesis."

Microtubular function in form stabilization of differentiated cells may appear first at the time of differentiation and persist throughout the life of the cell. This might best be exemplified by the ciliated protozoa mentioned above. Since at each cell division the entire cortical ciliary network must be duplicated, cortical microtubules in addition to stabilizing interphase cells could form the pattern by which new cilia, basal bodies, and microtubules would develop for the daughter cells of subsequent generations.

C. MICROTUBULES IN METABOLISM

The wide distribution of cytoplasmic microtubules has resulted on a purely speculative basis in their being attributed with metabolic functions. Among these have been the movement of various small molecules and ions or water in the cell (Slautterback, 1963) and the elaboration of various secretory products (Slautterback, 1963; Herman and Kauffman, 1966). If microtubules are capable of such cellular activities, they must play a continuous role in the cell cycle.

III. STRUCTURE OF MICROTUBULES

Since numerous functions can be assigned to cellular microtubules, one might expect dissimilarity between microtubules from various sources. Available evidence indicates quite the contrary. The general pattern of microtubule structure within all three major categories (Section II,A–C) has persisted from organism to organism. When examined in longitudinal section, the microtubule appears as two dark parallel lines with a central area slightly more dense than the surrounding cytoplasm (Figs. 1 and 4). This increased central density is a result of most of the microtubule lying within the thickness of the section. When viewed in cross section, they have a dense-staining circular outer wall and an apparently empty central area (Figs. 2, 3, and 5) giving the appearance of a tubular structure. The general absence of internal structure has been consistently observed in sectioned material. Absence of internal structure has been further substantiated in negatively stained whole-mount preparations of microtubules in which the internal portion of the tubule accumulates stain forming a dark central core (Pease, 1963; André and Thiéry, 1963; Gall, 1966; Barnicot, 1966; Grimstone and Klug, 1966; Behnke and Forer, 1967). Probably the greatest variation has been in tubular size: measurements range from 120 to 270 Å. It has been suggested that the 270 Å tubules represent a functionally elastic form while the 120–200 Å tubules are active in synthesis or metabolism and mitosis (Slautterback, 1963). This type of classification seems quite arbitrary since both ciliary (270 Å group) and spindle (120–200 Å group) microtubules can be synthesized from centrioles. It is more likely that size variation represents structural modification of a basic subunit to correspond to a specific function.

The "unit membrane" structure so characteristic of most cellular membranes is not evident in microtubules (Ledbetter and Porter, 1963; Slautterback, 1963). However, substructure organization in microtubules has been suggested by Ledbetter and Porter (1963) who observed 13 circular filamentous subunits in the walls of sectioned plant cell microtubules. Similar substructure has been reported in microtubules from a variety of tumor cells (de-Thé, 1964), rat tissue (Sandborn et al., 1965; Anderson et al., 1966), and some invertebrate tissues (Anderson et al., 1966).

Through utilization of negative straining techniques and whole-mount preparations, it has been possible to further resolve the basic macromolecular structure in microtubules from a variety of organisms. Thus far, studies have been conducted on spindle fibers from sea urchin eggs (Kiefer et al., 1966) and newt and human (Barnicot, 1966) cells, newt nucleated erythrocytes (Gall, 1966), vertebrate and invertebrate sperm fibrils (Pease, 1963; André and Thiéry, 1963; Behnke and Forer, 1967), flagellar fibrils of protozoa (Grimstone and Klug, 1966; Ringo, 1967), and cilia from rat trachea (Behnke and Forer, 1967). They have revealed the microtubules to be composed of a series of longitudinally arranged subfilaments. Each subfilament consists of 30–40 Å longitudinally con-

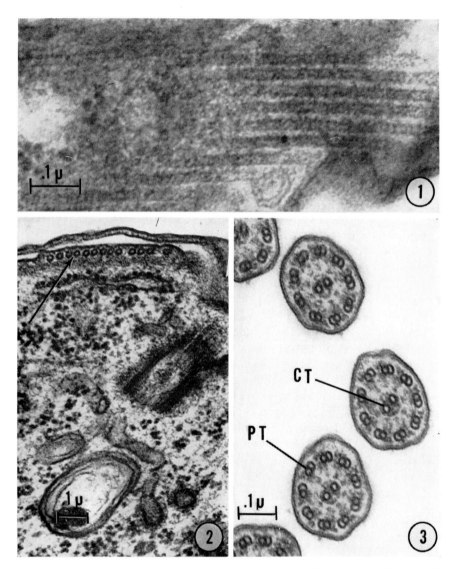

FIG. 1. A section through the cortex of *Tetrahymena pyriformis* shows the parallel arranged cortical microtubules. ×134,000.

FIG. 2. A cross section through the microtubules (arrow) seen in Fig. 1 reveals their characteristically dark staining outer wall and their apparently empty inner area. ×72,000.

FIG. 3. The paired peripheral (*PT*) and single central (*CT*) microtubules are evident in this section of *Tetrahymena pyriformis* cilia. ×97,000.

nected granules separated by a space. A center-to-center periodicity between granules of the same filament ranges from 50 Å in spindle (Barnicot, 1966) and erythrocyte (Gall, 1966) to 80–88 Å in sperm (André and Thiéry, 1963; Pease, 1963). Variation has also been reported in the number of subfilaments per tubule with 9–12 (Barnicot, 1966) or 13 (Kiefer et al., 1966) in spindle fibers, 12–14 in nucleated erythrocytes (Gall, 1966), and 10 (Pease, 1963) or 10–11 (André and Thiéry, 1963) in spermatozoan microtubules. As mentioned above, Ledbetter and Porter (1963) estimated that there were 13 circular subunits in sectioned plant microtubules. Phillips (1966) utilized sectioned material and Markham's rotation method to estimate the most probable number of tubules in *Sciara* spermatid at thirteen. Behnke and Forer (1967) found no significant difference in basic subfilament structure from that of earlier investigators; however, they did indicate that numerical variation in subfilaments occurred depending on which ciliary microtubule was counted. They estimated that central microtubules contain 12–14 subfilaments, A tubules contain 10–12, and B tubules 6–8. Cytoplasmic microtubules which are generally unpaired contain 12–14 subfilaments, as do the unpaired central filaments. These observations support those of Ringo (1967), who suggests that ciliary doublets are probably composed of microtubules which share 3–4 subunits at their line of fusion. Variation in subfilament number may represent a genetic or functional difference, or it may represent variation in counting by individual investigators of a microtubule population containing paired and single filaments. Spindle fiber counts would probably be most consistent in this regard; but even here, Barnicot (1966) placed the number between 9 and 12 subunits in vertebrate tissue, and Kiefer et al. (1966) at 13 subunits in sea urchin eggs.

Behnke and Forer (1967) found no significant difference between cytoplasmic, accessory and ciliary microtubule structure following negative staining. Nonetheless, they did observe a differential degradative effect of such agents as colchicine, cold (0°C), heat (50°C), and brief and prolonged exposure to pepsin on various portions of different microtubules. They concluded that at least part of the microtubules were proteinaceous. They do not believe that all microtubules are the same as suggested by many of the above-mentioned investigators but rather should be grouped on the basis of their response to the above-mentioned treatments, into four major categories: (1) A tubules, (2) B tubules, (3) central and accessory tubules, (4) cytoplasmic and spindle tubules.

Recent studies concerning the effects of colchicine on microtubules suggest that this classification may not be valid. Shelanski and Taylor

(1967) found that colchicine-^{3}H was selectively bound to a protein subunit with a sedimentation constant of 6 S by zone centrifugation. This protein subunit, extracted from sea urchin sperm tails, comprised the central ciliary filaments. They concluded that the 6 S colchicine-binding protein is a subunit of microtubules. Borisy and Taylor (1967b) also found that a 6 S component of isolated mitotic apparatus of sea urchin eggs selectively bound colchicine-^{3}H. However, no binding could be detected in the 27 S protein which makes up the bulk of the mitotic spindle. Thus, a similarity exists between the effects of colchicine on central ciliary filaments and on some spindle fibers, members of different microtubule groups according to Behnke and Forer's classification.

Behnke and Forer (1967) did observe that in normally dividing crane fly spermatocytes spindle fibers responded differentially to colchicine depending on whether they were associated with chromosomes, passed through them, or were associated with mitochondria. Earlier Ris (1949) had found that he could separate anaphase chromosome movement from spindle elongation in spermatocytes of the grasshopper by varying the concentration of chloral hydrate. At concentrations above 0.1%, the spindle became shorter and narrower and finally disappeared. At a concentration of 0.08%, the chromosomes moved to the poles but spindle elongation was inhibited. These observations support those of Borisy and Taylor (1967b) described above and suggest that not all spindle fibers respond the same to mitotic inhibitors but vary depending upon their specific function in mitosis.

Thus, another possibility should be considered. The basic unit of all microtubules may be a structural protein, possibly the 30–40 Å granules which comprise the individual filaments. By addition and modification of certain biochemical and morphological components necessary for specific tubular function, as the presence of peripheral arms containing ATPase in cilia and flagella (Gibbons, 1963, 1965), or the sharing of common subfilaments by peripheral ciliary microtubules (Ringo, 1967), variation in stability of a basic microtubule substructure might be obtained. This might account for the differential binding of colchicine to potential active sites of unpaired central ciliary filaments (Shelanski and Taylor, 1967) or certain (6 S) spindle filaments (Borisy and Taylor, 1967b). Comparable active sites of other microtubules may already be bound with other biochemical components or adjacent microtubules. This might also explain Behnke and Forer's (1967) observations of a differential effect of the various degradative agents (colchicine, cold, heat, and pepsin) on different parts of the microtubule population. Nonetheless, the basic substructural unit, possibly a structural protein, at this time appears to be a common feature of cellular microtubules.

IV. MICROTUBULES IN THE CELL CYCLE

In an analysis of cell division, James (1966) discusses division-oriented versus cycle-oriented synchrony. He indicates that the main concern in division-oriented synchrony is to make a population of cells divide without regard to the interphase activities of the cell. Cycle-oriented induced synchrony, on the other hand, is concerned with fitting the synchrony technique with the cell cycle. Centriole (basal body)-derived microtubules are most probably responsible for division-related functions. In most cells the formation of the mitotic spindle is the primary function, form maintenance being secondary. In the case of ciliated protozoa, basal body derived microtubules are functional in one of the most sensitive stages of division, stomatogenesis, as well as in the earlier events of cortical morphogenesis. By interrupting microtubule formation at these times, division-oriented synchrony has been obtained.

Several investigators have failed to observe significant structural difference between mitotic and cytoplasmic microtubules (Ledbetter and Porter, 1963; de-Thé, 1964; Anderson et al., 1966). Thus, they have suggested that cytoplasmic microtubules found during interphase may be spindle microtubule remnants, a synthesis or storage form of these fibers (Herman and Kauffman, 1966), or spindle fibers modified for metabolic functions. Biosynthetic capabilities of these microtubules and their ability to transport substances within the cell are presently speculative. They are based strictly on morphological evidence such as a relationship between secretory structures and the cell membrane or simply their abundance in many cells. While microtubules may function in these capacities, it should be recalled that little or no physiological basis exists for many of these proposals. If cytoplasmic microtubules are capable of such activity, they must continually function during the cell cycle.

A. SPINDLE INHIBITION AND DIVISION SYNCHRONY

Although a wide variety of both physical and chemical agents are capable of inhibiting cell division (Kihlman, 1966), only a few have been studied with regard to their effects on microtubules. Most prominent is colchicine and its derivative Colcemid (Pickett-Heaps, 1967; Stubblefield and Brinkley, 1966; Brinkley et al., 1967). Recently in studies on human cell cultures, Taylor (1965) demonstrated that concentrations of colchicine above 2×10^{-7} M blocked metaphase cells. Lower concentrations blocked only prophase cells. This suggests that higher concentrations are capable of disrupting existing spindles (microtubules) while lower colchicine concentrations only inhibit formation of spindle microtubules as reported by Eigsti and Dustin (1955). This conclusion is supported by the fine structure studies discussed below. Taylor (1965) also observed that colchicine uptake is rapid reaching equilibrium in less than 15 minutes. Radioactive tracer studies show that it is bound to the cell but at a concentration of 10^{-7} M has no measurable effect on synthesis of DNA, RNA, or protein. These and other factors led to the hypothesis that the cell contains sites capable of binding colchicine, in turn, preventing normal mitotic spindle formation. Further, these sites may be a structural protein. In conjunction with the studies of Borisy and Taylor (1967b) and Shelanski and Taylor (1967) mentioned above (Section III), such a structural protein must be a component of cellular microtubules.

The degradative effect of high colchicine concentrations (Eigsti and Dustin, 1955) has been observed in dividing plant cells (Pickett-Heaps, 1967). At concentrations of 0.1–0.3%, microtubules disappear from the spindle no matter what the stage of mitosis, and cytoplasmic microtubules are also destroyed. Microtubular elements of human polynuclear leukocytes are also destroyed by high (2.5×10^{-4} to 2.5×10^{-5} M) colchicine concentrations (Malawista and Bensch, 1967), as are the supportive tubular elements in the axopods of *Actinosphaerium* (Tilney, 1965).

As mentioned above, lower concentrations of colchicine are apparently capable of only inhibiting microtubule formation (Eigsti and Dustin, 1955). Thus, in Chinese hamster fibroblasts (Brinkley *et al.*, 1967; Stubblefield and Brinkley, 1966), Colcemid inhibited formation of the continuous spindle fibers, blocking the cells in metaphase. Other microtubules were present, but whether they existed prior to Colcemid treatments is not known. Upon removal of the inhibitor, the continuous microtubules re-form; and the cells complete division. Stubblefield and Klevecz (1965) first utilized this system to obtain large populations of synchronously dividing cells for biochemical studies. Such synchronized cultures recover adequately (from Colcemid treatment) and show "no significant deviations from control cultures in their mitotic interval, generation time, DNA synthesis kinetics, or proliferative capacity" (Stubblefield *et al.*, 1967). Thus, although Colcemid has provided a division-oriented synchronous population, the postmitotic growth more approximates normal cell growth thereby avoiding the imbalanced growth of other systems.

Other mitotic spindle inhibitors are now known to have a degradative effect on microtubules. Cold which disrupts spindle birefringence and inhibits chromosome movement (Inoué, 1964) prevents microtubule formation in the mitotic apparatus of *Chaos carolinensis* (Roth, 1967). A similar antimitotic action by pressure (Marsland, 1966) probably also results in spindle microtubule inhibition and degradation since pressure is capable of microtubule disruption in *Actinosphaerium* (Tilney *et al.*, 1966) and *Tetrahymena* (Kennedy and Zimmerman, 1968).

It is apparent that microtubules play an important role in cell elongation and chromosome movement during the division interval. Since blockage of this system with Colcemid for a short interval induces synchronous division in some cells, other spindle inhibitors such as low temperature and pressure might also accomplish similar results. Utilizing low temperature becomes problematic since total cell metabolism would be reduced. However, by alternating periods of low temperature with periods of normal growth at optimal temperature, it might be possible to block mitotic spindle formation and accumulate cells in metaphase. Such a system has been utilized to induce synchronous division in *Tetrahymena* (Padilla and Cameron, 1964) and may represent inhibition of microtubule formation.

B. ORAL MICROTUBULE INHIBITION AND DIVISION SYNCHRONY

Little attention has been given to the functions of microtubules in the cell cycle of ciliated protozoa; however, the extent and complexity of organization of these structures in ciliates is evident when the recent reviews by Pitelka (1963, 1968) are examined. In *Tetrahymena*, one of the most thoroughly studied ciliates, the cortical infraciliature is a network of microtubules (Allen, 1967) probably originating from basal bodies. In Fig. 4, cortical microtubules arise from the tubules of the basal body wall and extend up under the cell membrane. In cross section (Fig. 5) their tubular nature is evident. These tubules must be duplicated at some time during the cell cycle in order for cell division to occur.

Since the successful induction of synchronous cell division in *Tetrahymena* (Scherbaum and Zeuthen, 1954), the stages of oral development and factors controlling its inhibition have been extensively studied. In addition to ciliary membranelles, the oral region contains an elaborate microtubular network (Nilsson and Williams, 1966). A comparable microtubular architecture must develop in the area of the anarchic field (Williams and Zeuthen, 1966) apparently as the final event before cytokinesis. Figure 6 shows some of these microtubules forming between the basal bodies in a developing oral area. The sequential events of oral development can be blocked or delayed if certain metabolic inhibitors or heat shocks are applied to the culture prior to stage four or five in the staging employed by Frankel (1962). The point in the cell cycle beyond which the cell is insensitive to these agents, the "physiological transition point" or "stabilization point," corresponds closely with the interval at which oral subcortical fiber formation is nearing completion (Williams and Zeuthen, 1966). It is interesting to speculate about

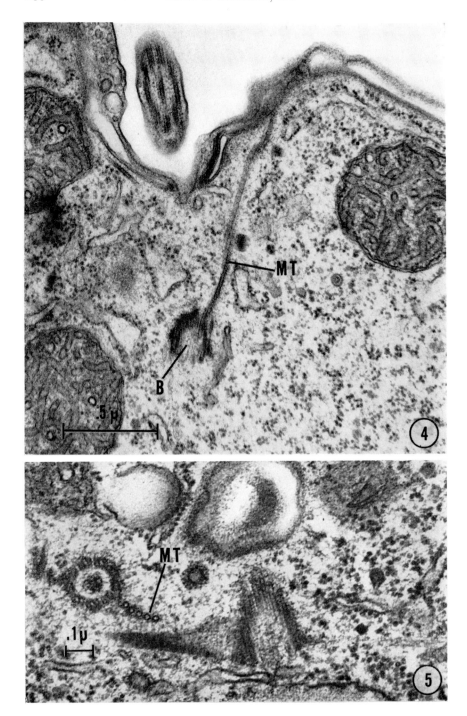

the function of oral microtubules in this process. If the final event prior to cytokinesis in *Tetrahymena*, or any ciliate, is formation of the oral apparatus, inhibition of oral morphogenesis (specifically microtubule formation) may result in inhibition or delay of division. Williams and Zeuthen (1966) consider this possibility in discussing the relationship of oral fiber formation to oral morphogenesis in *Tetrahymena*. They suggest that temperature shocks and various metabolic inhibitors which take effect prior to the "stabilization point" may block the synthesis of oral fiber protein as well as remove previously synthesized oral fiber protein whether present as an assembled structure or a precursor.

Although the various metabolic inhibitors may be preventing synthesis of oral microtubule precursors, heat shocks may be acting later in the cycle to block polymerization of the precursors into microtubules since synthesis of other molecular components is not completely inhibited. For example, Christensson (1959) found that although protein synthesis is substantially greater at lower temperatures during early heat shocks, there is progressive loss of temperature inhibition in the later part of synchrony treatment. Other data suggest that DNA synthesis also continues during heat shock periods (Scherbaum, 1964).

We were interested in determining what effect an inhibitor of microtubule formation would have on oral development and cytokinesis. Thus, colchicine, which is selectively bound to a variety of cellular microtubules including ciliary microtubules in *T. pyriformis* (Borisy and Taylor, 1967a), was used. Heat-synchronized populations of *T. pyriformis* (GL) (modification of the technique of Scherbaum and Zeuthen, 1954) were exposed to 0.005 M colchicine at the end of the last heat shock (EH). While division inhibition did not occur, a delay in division of about 45 minutes beyond the control culture was obtained. Lower concentrations of colchicine resulted in less delay, and higher concentrations extended the delay interval (Fig. 7) suggesting a dose-dependent action by colchicine. Addition of colchicine at EH + 45 minutes caused oral resorption and extended the delay period for an additional 45 minutes giving a total delay of 90 minutes beyond the control culture. Examination thus far of the anarchic field of colchicine delayed cells (Fig. 8) shows a general absence of the well-developed microtubules present in normal cells at the same interval (Fig. 6) and a retarded ciliary growth.

FIG. 4. Microtubules (*MT*) arising from the wall of a basal body (*B*) can be seen in this micrograph. These microtubules extend upward to just beneath the cortex of *Tetrahymena*. ×50,000.

FIG. 5. This cross section through microtubules (*MT*) arising from basal bodies reveals their tubular nature better than similar microtubules seen in Fig. 4. ×75,000.

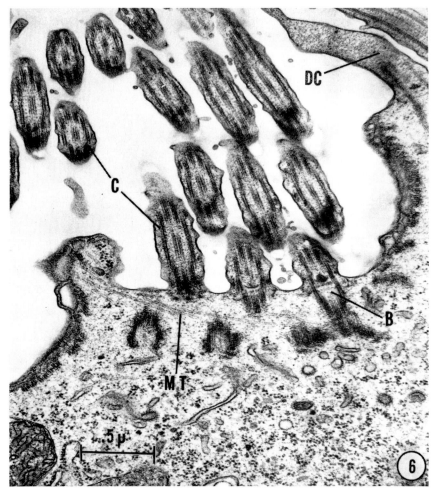

FIG. 6. This developing mouth contains mature basal bodies (*B*) with well-developed cilia (*C*). A cilium in early development is also visible (*DC*), as are the microtubular connections (*MT*) between basal bodies. ×38,000.

Similar observations on the inhibitory effect of colchicine on ciliary growth have recently been reported for *T. pyriformis* (Rosenbaum and Child, 1967).

Unfortunately, electron microscopic study of oral development and oral resorption has not been thoroughly carried out on *T. pyriformis*. We know that oral development in colchicine-delayed organisms is behind that of control cells as indicated by the presence of incompletely developed basal bodies (Fig. 8) as well as the absence of well developed cilia and oral microtubules. However, we are unable to relate the events of colchicine delay with those of oral resorption since only light microscope studies have been conducted on the latter. The question remains as to whether all oral organelles (i.e., basal bodies and microtubules) are broken down or only cilia are resorbed and membranelle organization inhibited. That is, do the events of oral resorption actually represent complete degradation of oral structures or an inhibition and recycling of the orderly sequence of membranelle organization?

Microtubules are present also during micronuclear (as spindle fibers) and macronuclear division (Fig. 9) in *T. pyriformis*. Holz *et al.* (1957) reported that at the end of successive heat shocks the number of cells in preparation for division increased as characterized by the presence

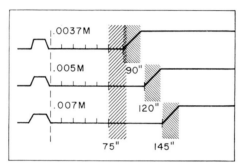

Fig. 7. Colchicine-induced dose-dependent delay in division synchrony. Dotted line represents end of heat treatment (EH). Control division occurred at EH + 75 minutes (heavy diagonal lines). Light diagonal lines represent division delay at the three concentrations employed.

of the "anarchic field" stage of stomatogenesis and micronuclear anaphase (strain WH-6). We found that after colchicine-induced delay macronuclear microtubules are absent in the amicronucleate strain (GL) employed. Thus, although micronuclear spindle inhibition is not a factor in division delay of strain GL, both spindle and macronuclear microtubule inhibition may be involved in division delay in other strains of *T. pyriformis*. This may be another factor in division delay but will require further analysis. However, the absence of microtubules from the developing oral area and the macronucleus following colchicine treatment of strain GL suggests that microtubule polymerization may at least be necessary for completion of division in *T. pyriformis*. This is consistent with the observations of Zimmerman (Chapter 10, Section II,A) that a 2-minute pulse of pressure (10,000 psi) following the end of heat treatment delays or inhibits division in synchronized *Tetra-*

hymena if it is assumed that degradation of existing or inhibition of developing microtubules occurs. This time and pressure is comparable to that employed by Tilney *et al.* (1966) to break down microtubules in *Actinosphaerium.*

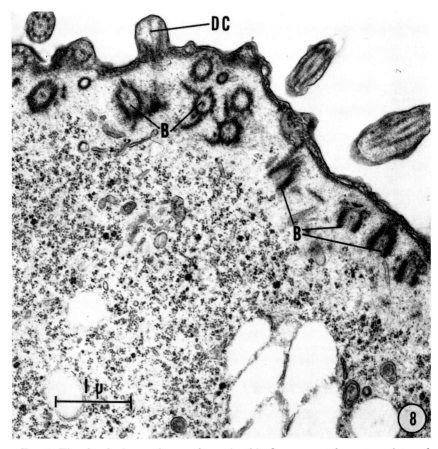

FIG. 8. The developing oral area shown in this figure was taken at an interval comparable to that of Fig. 6. However, the *Tetrahymena* were inhibited from dividing with 0.005 *M* colchicine. No distinct microtubules are visible. Basal bodies (*B*) are incompletely developed, as are the few cilia (*DC*) present. ×21,000.

Colchicine also affects division in *Blepharisma undulans* (Hirshfield and Pecora, 1955). Lower concentrations (10^{-4} *M*) inhibit micronuclear division, but cytokinesis continues. If the level of colchicine is raised (10^{-3} *M*), cytokinesis can also be blocked. We have found that extended exposure of *Blepharisma* to 0.005 *M* colchicine causes rounding of the

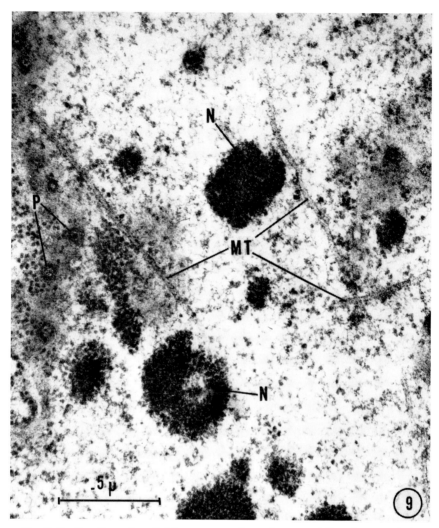

FIG. 9. Section through a dividing macronucleus of a normal *Tetrahymena*. Microtubules (*MT*) are visible between nucleoli (*N*). Nuclear pores (*P*) are also evident. ×54,000.

cells. By placing these cells in fresh media free of colchicine, they will regain their normal shape after a few hours. Preliminary examination of rounded cells indicates a breakdown of the normal cortical microtubule system. The above suggests that micronuclear spindle tubules (Jenkins, 1967) are more sensitive to colchicine than are those of the cortex and may be blocked by lower concentrations; but by raising the concentration to a high enough level, existing microtubules may also be broken down. These interpretations correspond with observations of other investigators as described above (Section IV,A). It would be interesting to know if a similar sensitivity differential exists between oral and nuclear microtubules in *T. pyriformis*.

C. CORTICAL MICROTUBULE INHIBITION AND DIVISION SYNCHRONY

In view of the ability of colchicine to delay division in synchronized *Tetrahymena*, the possibility of obtaining synchronized populations from

TABLE I

THE EFFECT OF COLCHICINE (0.005 M FOR 45 MINUTES) ON ORAL DEVELOPMENT IN LOGARITHMICALLY GROWING *Tetrahymena pyriformis*

Time	Stages of oral development							
	0	R[a]	1	2	3	4	5	6
0 (control)	73	—	8	5	5	3	1	5
EC[b]	73	10	11	—	—	—	3	3
After EC (minutes)								
20	81	2	16	1	—	—	—	—
40	81	4	11	3	—	1	—	—
50	71	6	16	7	—	—	—	—
60	65	—	16	14	5	—	—	—
75	49	1	19	11	12	8	—	—
90	54	1	13	5	4	14	4	5
105	41	—	15	9	8	10	5	12
120	38	—	18	8	10	13	5	8
135	56	—	5	7	5	12	6	9
150	58	—	3	5	5	12	5	12
165	75	—	1	3	3	6	7	5
185	82	—	7	3	1	3	1	4

[a] R = Oral membranelles in the process of resorption.
[b] EC = End of colchicine treatment.

logarithmic cultures by exposure to this agent was examined. The number of cells in oral development could be increased by as much as 30% over control cultures (see Table I) by single exposure to colchicine, but it was not possible to obtain the degree of synchronous division realized with the heat shock method. However, it was possible to bring a greater number of cells into the same phase of growth. As indicated by cell counts (Table I) several groups of cells moved through the stages of oral development together with a total of 62% of the cells in oral development at 2 hours after exposure to colchicine. Longer exposure to colchicine did not appreciably increase the number of cells in oral development. Failure to obtain a high degree of synchrony is not surprising if the state of the cortex in logarithmic and heat-synchronized cells is examined. According to the estimates of Williams and Scherbaum (1959), heat-synchronized populations have duplicated their somatic ciliature and therefore their cortical microtubule content by the end of heat treatment. Most logarithmic cells, however, are in the process of cortical microtubule formation. Thus, such agents as colchicine and possibly heat and metabolic inhibitors would only be affecting the developing oral area in heat-synchronized populations. In logarithmic populations, on the other hand, the cortical microtubule system may also be inhibited from developing. Examination of these questions in continuing in our laboratory.

Rosenbaum et al. (1966) attributed cortical and nuclear alterations in T. pyriformis (variety 1, mating type II) to heat damage of microtubules. They observed "extreme variation in size and shape, fractured and displaced kineties and abnormalities of karyokinesis" in logarithmic cultures grown at supraoptimal temperatures (39°–40°C). In order to obtain growth at these temperatures, culture media were supplemented with synthetic or natural phospholipids. Thus, damage was attributed to the effects of high temperature on structures either known or thought to contain phospholipids. Primarily, they associated the observed changes to "alterations in lipid form and function at microtubule interfaces" and possibly to cellular membranes.

The characteristic lipoprotein organization of cell membranes is not present in microtubules (Section III). Also, Behnke and Forer (1967) and Taylor and his co-workers (Taylor, 1965; Borisy and Taylor, 1967a, b; Shelanski and Taylor, 1967) suggest that the basic unit of microtubules may be a structural protein. The existence of lipid in microtubules has thus far not been established. In the absence of such evidence, it seems premature to assume that heat is affecting microtubule lipid.

Since the silver impregnation stain employed by Rosenbaum et al. (1966) reveals cell surface configurations, these investigators may have been observing distortion of cortical lipoprotein membranes which resulted in alterations of kineties. It is generally accepted that the integrity of both the micro- and macronuclear membranes are maintained during division. Disruption of these membranes may account for impairment of micronuclear division and the production of amicronucleates and abnormal macronuclei. In view of the variation in temperature sensitivity of different strains of T. pyriformis, examination of these questions at the ultrastructure level to determine whether the alterations observed are due to effects on cell membranes, microtubules, or both is certainly justifiable.

V. SUMMARY

It is apparent that microtubules have two basic roles in any cell, those sustaining normal cellular activity and those associated with the division process. While microtubules may be active in cell metabolism and synthesis, these "interphase" microtubules are more likely responsible for form maintenance.

Those cells which are involved in continuous growth and division must synthesize microtubules prior to each division. By inhibiting spindle microtubule formation it is possible to accumulate metaphase stages in a cell population. Upon removal of inhibition, the cells are often able to complete division in some cases yielding a division-oriented synchronous population. Thus, the major role of microtubules seems to be to carry the earlier steps of biosynthesis to completion, allowing the cell to distribute its component DNA, RNA, protein, and organelles to the two daughter cells. Cytoplasmic ("interphase") microtubules may represent remnants of spindle tubules from previous divisions which have been adapted to various functions for maintaining interphase activities.

Microtubular function in the ciliates is more complex in view of the extensive cortical network of tubules and cilia. This entire system responsible for form and motility must be duplicated sometime prior to cell division. However, the key factor in control of cell division seems to be the developing oral microtubule system. Colchicine which specifically inhibits microtubule formation induces oral resorption and delays division. Thus, the oral anlage in ciliates might be compared with the mitotic spindle of other cells as the key to cytokinesis. Nevertheless, microtubules are definitely of great significance during the division interval of the cell cycle, their regulation being a prime factor in obtaining division-oriented cell synchrony. [See Note Added in Proof, p. 248.]

ACKNOWLEDGMENTS

The author wishes to thank Miss Ellen Brittingham for expert technical assistance during the course of these studies and Dr. S. H. Richardson for suggestions in preparation of the manuscript.

Unpublished research discussed in this chapter was supported by a grant from the National Institute of Health (GM 13386).

REFERENCES

Allen, R. D. (1967). *J. Protozool.* **14**, 553.
Anderson, W. A., Weissman, A., and Ellis, R. A. (1966). *Z. Zellforsch.* **71**, 1.
André, J., and Thiéry, J. P. (1963). *J. Microscop.* **2**, 71.
Barnicot, N. A. (1966). *J. Cell Sci.* **1**, 217.

Behnke, O. (1964). *J. Ultrastruct. Res.* **11**, 139.
Behnke, O., and Forer, A. (1967). *J. Cell Sci.* **2**, 169.
Biesele, J. J. (1958). "Mitotic Poisons and the Cancer Problem." Elsevier, Amsterdam.
Borisy, G. G., and Taylor, E. W. (1967a). *J. Cell Biol.* **34**, 525.
Borisy, G. G., and Taylor, E. W. (1967b). *J. Cell Biol.* **34**, 535.
Brinkley, B. R., Stubblefield, E., and Hsu, T. C. (1967). *J. Ultrastruct. Res.* **19**, 1.
Byers, B., and Porter, K. R. (1964). *Proc. Natl. Acad. Sci. U.S.* **52**, 1091.
Christensson, E. (1959). *Acta Physiol. Scand.* **45**, 339.
de-Thé, G. (1964). *J. Cell Biol.* **23**, 265.
Eigsti, O. J., and Dustin, P. (1955). "Colchicine." Iowa State Univ. Press, Ames, Iowa.
Fawcett, D. W., and Witebsky, F. (1964). *Z. Zellforsch.* **62**, 785.
Frankel, J. (1962). *Compt. Rend. Trav. Lab. Carlsberg* **33**, 1.
Frankel, J. (1964). *J. Exptl. Zool.* **155**, 403.
Frankel, J. (1965). *J. Exptl. Zool.* **159**, 113.
Gall, J. (1966). *J. Cell Biol.* **31**, 639.
Gavin, R. H., and Frankel, J. (1966). *J. Exptl. Zool.* **161**, 63.
Gibbons, I. R. (1963). *Proc. Natl. Acad. Sci. U.S.* **50**, 1002.
Gibbons, I. R. (1965). *Arch. Biol. (Liège)* **76**, 317.
Gray, E. G. (1959). *J. Anat.* **93**, 420.
Green, P. B. (1965). *Excerpta. Med. Intern. Congr. Ser.* **77**, 21.
Grimstone, A. V., and Klug, A. (1966). *J. Cell Sci.* **1**, 351.
Harris, P. (1962). *J. Cell Biol.* **14**, 475.
Herman, L., and Kauffman, S. L. (1966). *Develop. Biol.* **13**, 145.
Hirshfield, H. I., and Pecora, P. (1955). *Exptl. Cell Res.* **9**, 414.
Holz, G. G., Scherbaum, O. H., and Williams, N. E. (1957). *Exptl. Cell Res.* **13**, 618.
Inoué, S. (1964). *In* "Primitive Motile Systems in Cell Biology" (R. D. Allen and N. Kamiya, eds.), pp. 549–594. Academic Press, New York.
Jahn, T. L., and Bovee, E. C. (1967). *In* "Research in Protozoology" (T. T. Chen, ed.), Vol. I, pp. 41–200. Macmillan (Pergamon), New York.
James, T. W. (1966). *In* "Cell Synchrony–Studies in Biosynthetic Regulation" (I. L. Cameron and G. M. Padilla, eds.), pp. 1–13. Academic Press, New York.
Jenkins, R. A. (1967). *J. Cell Biol.* **34**, 463.
Kane, R. E. (1962). *J. Cell Biol.* **15**, 279.
Kennedy, J. R. (1965). *J. Protozool.* **12**, 542.
Kennedy, J. R., and Zimmerman, A. (1968). Unpublished data.
Kiefer, B., Sakai, H., Solari, A. J., and Mazia, D. (1966). *J. Mol. Biol.* **20**, 75.
Kihlman, B. A. (1966). "Actions of Chemicals on Dividing Cells." Prentice Hall, Englewood Cliffs, New Jersey.
Krishan, A., and Buck, R. C. (1965). *J. Cell Biol.* **24**, 433.
Ledbetter, M. C., and Porter, K. R. (1963). *J. Cell Biol.* **19**, 239.
Malawista, S. E., and Bensch, K. G. (1967), *Science* **156**, 521.
Marsland, D. (1966). *J. Cell Physiol.* **67**, 333.
Nilsson, J. R., and Williams, N. E. (1966). *Compt. Rend. Trav. Lab. Carlsberg* **35**, 119.
Padilla, G. M., and Cameron, I. L. (1964). *J. Cell. Comp. Physiol.* **64**, 303.
Palay, S. L. (1956). *J. Biophys. Biochem. Cytol. Suppl.* **2**, 193.
Palay, S. L. (1960). *Anat. Record* **138**, 417.
Pease, D. C. (1963). *J. Cell Biol.* **18**, 313.

Phillips, D. M. (1966). *J. Cell Biol.* **31**, 635.
Pickett-Heaps, J. D. (1967). *Develop. Biol.* **15**, 206.
Pitelka, D. R. (1963). "Electron-Microscopic Structure of Protozoa." Macmillan (Pergamon), New York.
Pitelka, D. R. (1968). *In* "Progress in Protozoology Research" (T. T. Chen, ed.), Vol. 3, in press. Macmillan (Pergamon), New York.
Randall, J., and Disbrey, C. (1965). *Proc. Roy. Soc.* **B162**, 473.
Renaud, F. L., and Swift, H. (1964). *J. Cell Biol.* **23**, 339.
Renaud, F. L., Rowe, A. J., and Gibbons, I. R. (1968). *J. Cell Biol.* **36**, 79.
Ringo, D. L. (1967). *J. Ultrastruct. Res.* **17**, 266.
Ris, H. (1949). *Biol. Bull.* **96**, 90.
Rivera, J. A. (1962). "Cilia, Ciliary Activity and Ciliated Epithelium." Macmillan (Pergamon), New York.
Rosenbaum, J. L., and Child, F. M. (1967). *J. Cell Biol.* **35**, 117A.
Rosenbaum, N., Erwin, J., Beach, D., and Holz, G. G. (1966). *J. Protozool.* **13**, 535.
Roth, L. E. (1967). *J. Cell Biol.* **34**, 47.
Roth, L. E., and Daniels, E. W. (1962). *J. Biophys. Biochem. Cytol.* **12**, 57.
Sandborn, E., Szeberenyl, A., Messier, P., and Bois, P. (1965). *Rev. Canad. Biol.* **24**, 243.
Scherbaum, O. H. (1964). *In* "Synchrony in Cell Division and Growth" (E. Zeuthen, ed.), pp. 177–196. Wiley (Interscience), New York.
Scherbaum, O. H., and Zeuthen, E. (1954). *Exptl. Cell Res.* **6**, 221.
Schmitt, F. O., and Geren, B. B. (1950). *J. Exptl. Med.* **91**, 499.
Shelanski, M. L., and Taylor, E. W. (1967). *J. Cell Biol.* **34**, 549.
Shelanski, M., and Taylor, E. W. (1968). *J. Cell Biol.* **38**, 304.
Slautterback, D. B. (1963). *J. Cell Biol.* **18**, 367.
Sleigh, M. A. (1962). "The Biology of Cilia and Flagella." Macmillan, New York.
Sorokin, S. (1962). *J. Cell Biol.* **15**, 363.
Sotelo, C., and Palay, S. L. (1968). *J. Cell Biol.* **36**, 151.
Sotelo, J. R., and Trujillo-Cenóz, O. (1958). *Z. Zellforsch.* **49**, 1.
Stubblefield, E., and Brinkley, B. R. (1966). *J. Cell Biol.* **30**, 645.
Stubblefield, E., and Klevecz, R. R. (1965). *Exptl. Cell Res.* **40**, 660.
Stubblefield, E., Klevecz, R., and Deaven, L. (1967). *J. Cell Physiol.* **69**, 345.
Taylor, E. W. (1965). *J. Cell Biol.* **25**, 145.
Tilney, L. G. (1965). *Anat. Record* **151**, 426.
Tilney, L. G., and Porter, K. R. (1965). *Protoplasma* **60**, 317.
Tilney, L. G., Hiramoto, Y., and Marsland, D. (1966). *J. Cell Biol.* **29**, 77.
Williams, N. E., and Scherbaum, O. H. (1959). *J. Embryol. Exptl. Morphol.* **7**, 241.
Williams, N. E., and Zeuthen, E. (1966). *Compt. Rend. Trav. Lab. Carlsberg* **35**, 101.

NOTE ADDED IN PROOF

Since the preparation of this manuscript two important papers have appeared relating to the structure of microtubules (Section III). Renaud, *et al.* (1968) report that outer ciliary fiber protein of *Tetrahymena*, composed of a 55,000 M. W. subunit is similar to muscle actin. In addition, in a recent report on the composition of sea urchin sperm tail microtubules Shelanski and Taylor (1968) state that comparison of isolated protein properties with electron microscopic evidence of microtubule structure suggests that the chemical subunit (M. W. = 120,000) is composed of two 40 Å morphological subunits.

CHAPTER 12

Biochemical Aspects of Morphogenesis in the Brine Shrimp, *Artemia salina*

F. J. Finamore and J. S. Clegg

I. INTRODUCTION

The extensive and highly organized division of cells constitutes an indispensable part of the overall pattern of normal embryonic development in multicellular organisms. It is of interest, therefore, that a significant part of embryonic development in the brine shrimp *Artemia* *salina* takes place without the occurrence of any cell division. This observation was made by Nakanishi *et al.* (1962), who clearly demonstrated that the encysted gastrula,* after a period of dormancy, differentiates into a partially formed larva ("prenauplius") in the absence of cell division.

Nakanishi *et al.* (1962) also noted that cell division is resumed in the prenauplius after it has emerged from the cyst, and that the frequency of cell division increases markedly during the formation of a complete nauplius larva.

Because the general nature of brine shrimp development is not familiar to many, and since knowledge of it is required for an understanding of the research to be presented here, we will briefly summarize the relevant aspects of its sexual reproduction and some salient features of laboratory manipulations in which this organism is used. Before doing so, we should point out that this species exists in several varieties or races found throughout the world; these may be bisexual diploids or tetraploids (reviewed by Barigozzi, 1957), or parthenogenetic forms having diploid, triploid, tetraploid, pentaploid, and even octaploid chromosome numbers (reviewed by Goldschmidt, 1952). This article will be restricted to the bisexual races from the Great Salt Lake, Utah, and from salterns near San Francisco, California.

A. The Reproductive Cycle

This account is based chiefly on the work of Fautrez-Firlefyn (1951) and Bowen (1962). The relevant features are summarized in Scheme 1:

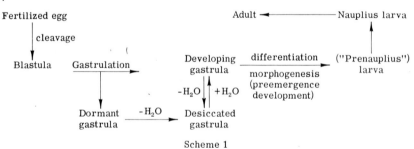

Scheme 1

After oogenesis and vitellogenesis, the mature oocytes, all synchronized at metaphase of the first meiotic division, move in discrete clutches

---

* Nakanishi *et al.* (1962) consider this to be a blastula. However, the work of Fautrez-Firlefyn (1951) and Dutrieu (1960) indicates that gastrulation has occurred. The term gastrula will be used here.

from the lateral sacs (oviducts) into the ovisac (uterus) of the female, where fertilization occurs. Meiosis is then completed, and cleavage results in a blastula which then undergoes gastrulation. From this point, one of two developmental routes can be followed: the gastrula either continues its development and gives rise to a nauplius larva, which is released from the female ovisac, or the gastrula encysts, enters a dormant state, and is released into the environment. The fate of these encysted gastrulae is desiccation, which occurs either osmotically (their environment is usually a concentrated brine) or after being washed onto shore and exposed to air. By means yet unknown, desiccation "activates" the gastrula and terminates the obligate dormancy (see Dutrieu, 1960). These dried gastrulae, now dormant only because they lack water, will rapidly resume metabolic activity and development when suitably rehydrated (Muramatsu, 1960; Emerson, 1963; Clegg, 1964). Still retained, however, is their ability to tolerate desiccation. As indicated in the diagram, removal of water from the gastrula, even after it has experienced several hours of metabolic activity, once again results in a reversible cessation of metabolism (Clegg, 1967). Clearly, this is an unusual ability for an embryo presumably undergoing differentiation.

If its metabolism is not interrupted, the gastrula will continue to develop and will give rise to a partially formed nauplius larva (prenauplius) which emerges from the cyst. The total time required for this "preemergence development" is between 8 and 15 hours under optimal conditions (Clegg, 1964). The histological pictures of the gastrula and newly emerged prenauplius are shown in Fig. 1. It is clear that the processes of morphogenesis are well under way in the prenauplius and, although not evident at this magnification owing to the large amount of yolk present, considerable cellular differentiation has taken place, particularly in the anterior regions, appendages, and gut.

It should be recalled that these morphogenetic events of preemergence development take place in the complete absence of cell division (Nakanishi *et al.*, 1962). These authors counted the total number of nuclei in gastrulae and in embryos during preemergence development, and found this number (about 4000) to be statistically constant. The possibility that cell death and cell division were occurring simultaneously was ruled out, since no mitotic figures were observed throughout preemergence development. We have confirmed their results in full although we have observed different numbers of nuclei in batches of gastrulae from different sources. The fact that differentiation and morphogenesis occurring during preemergence development do not involve cell division has some ramifications that we will consider later.

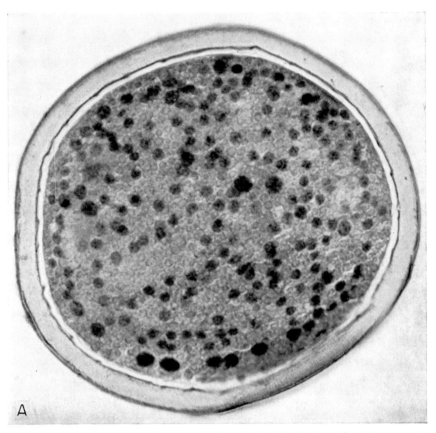

Fig. 1. (A) Section of dormant encysted gastrula. (B) Section of newly emerged prenauplius (opposite page).

Following emergence, cell division resumes (Nakanishi *et al.*, 1962), and the prenauplius completes its development into a swimming nauplius larva. These larvae are packed with yolk and can survive for about 4 days without feeding. Early larval development has been described in detail by Anderson (1967). If conditions are suitable, the nauplii undergo about 14 molts to reach the adult stage. The "egg-to-adult" period can be as short as 2 weeks.

Although relatively little is known about the detailed biochemical events associated with this unusual development (see reviews by Green, 1965; Dutrieu, 1960; and Urbani, 1962, for earlier work), recent studies have indicated these embryos to be worthy of further attention. Thus, Finamore and Warner (1963) discovered the presence of a unique class of guanine-containing nucleotides in the encysted gastrulae (cysts) and

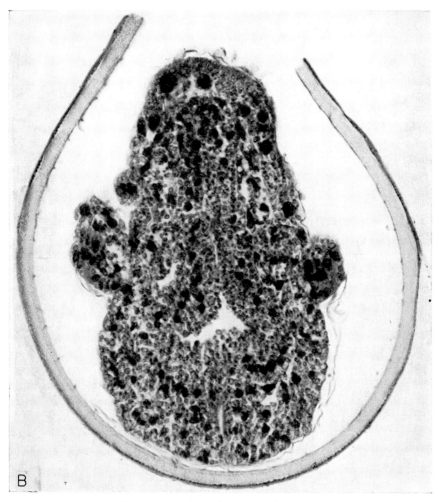

FIG. 1B. For legend see opposite page.

have carried out a series of investigations dealing with these compounds (Finamore and Warner, 1963; Warner and Finamore, 1965a,b, 1967; Clegg *et al.*, 1967). In the present article, one of us (F.J. Finamore) will consider some studies on one of these nucleotides, namely, P^1,P^4-diguanosine 5′-tetraphosphate and its potential role as a regulator of DNA synthesis accompanying the onset of cell division in nauplii. The fact that embryonic differentiation occurs to some extent without cell division has implications also with respect to the synthesis of proteins, and recent research on this subject will be presented (Section III),

B. Source, Preparation, and Treatment of
   Artemia Embryos and Nauplii

The dried encysted gastrulae ("brine shrimp eggs or cysts") can be obtained commercially in large amounts (Sanders Brine Shrimp Co., Ogden, Utah). They exhibit spectacular resistance to environmental extremes (reviewed by Dutrieu, 1960, also see Littlepage and McGinley, 1965) and, as a result, are readily stored for long periods of time at very low temperatures or under vacuum. These populations are, however, heavily contaminated with debris and cryptobiotic microorganisms, some of which are closely associated with the shell of the *Artemia* embryos and cannot be removed by washing with saline solutions. Treatment of the embryos with 7% antiformin (Nakanishi *et al.*, 1962) for 15–20 minutes removes these contaminants, and the embryos can then be redried over $CaCl_2$ or any other suitable desiccant to produce an essentially aseptic preparation. Merthiolate can also be used with success (Provasoli and Shiraishi, 1959). The results of any study on these encysted embryos which involves the use of radioactive isotopes but does not utilize an appropriate sterilization method for the embryos should be viewed with reservation.

In some cases it is desirable to utilize gastrulae that have been rehydrated but have not yet broken dormancy. This can be achieved by placing the encysted gastrulae in seawater or dilute NaCl solutions maintained at 0°C for 3 hours. Under such conditions, hydration is essentially complete but metabolism does not occur to any measurable extent (Iwasaki, 1964). Thus, one can begin an experiment with essentially aseptic preparations of either dried or prehydrated, but still dormant, gastrulae. Because the dried gastrula is ametabolic (Clegg, 1967), one can also define in detail any measurable parameter of a given population of embryos (enzyme levels, metabolic concentrations, etc.) prior to beginning a series of experiments. The experimental advantages of this are obvious.

Disruption of the encysted embryos for *in vitro* studies is not easy because of the tough shell. For studies on polyribosomes this consideration is critical, and for that reason we have evaluated a variety of disruption methods. The gentlest and most effective of these methods involves the use of Dounce tissue grinders equipped with large clearance pestles (Kontes Glass Co.) and embryos that have been treated with 5% sodium hypochlorite for about 15 minutes. This treatment removes the granular parts of the shell without damaging the embryo, and produces a preparation that can be gently disrupted.

Newly hatched nauplii may be collected in large quantities by pouring the entire contents of the hatching vessels into large separatory funnels. After a few minutes, the nauplii will collect at the stopcock, permitting them to be drained off and separated from empty shells. They may then be allowed to mature in crystallizing dishes containing artificial seawater fortified with penicillin (100 units/ml) and streptomycin (0.1 mg/ml).

II. NUCLEOTIDES AND NUCLEIC ACID METABOLISM IN ENCYSTED EMBRYOS AND NAUPLII

A. DISCOVERY OF P^1,P^4-DIGUANOSINE 5'-TETRAPHOSPHATE

Finamore and Warner (1963) reported that encysted gastrulae of *Artemia salina* contain an unusual symmetrical pyrophosphate anhydride, P^1,P^4-diguanosine 5'-tetraphosphate, whose structure is:

$$\text{Gu} \quad \text{CH}_2\text{OP}-\text{O}-\text{P}-\text{O}-\text{P}-\text{O}-\text{P}-\text{OCH}_2 \quad \text{Gu}$$

The compound can be considered as two guanosine 5'-diphosphate molecules linked by a pyrophosphate bond, and it represents the first of its kind to be found in the nucleotide pool of living organisms. Approximately 2.8% of the dry weight of *Artemia* encysted gastrulae is composed of diguanosine tetraphosphate. These authors suggested at the time of its discovery that the compound represents a primary source of energy for further development and that the unusual structure of the molecule probably protects vital high-energy bonds from degradation during the dormant period.

Two years later, Warner and Finamore (1965b) found the acid-soluble nucleotide pool of *Artemia* encysted gastrulae to contain a compound considered to be a homolog of diguanosine tetraphosphate, namely, P^1,P^3-diguanosine 5'-triphosphate:

$$\text{Gu} \quad \text{CH}_2\text{OP}-\text{O}-\text{P}-\text{O}-\text{P}-\text{OCH}_2 \quad \text{Gu}$$

This compound is not as prevalent in the nucleotide pool as diguanosine tetraphosphate, since it constitutes only approximately 0.04% of the dry weight of the cysts.

The structures of both of these compounds have been verified by chemical syntheses performed by Adam and Moffatt (1966).

That these pyrophosphate anhydrides are not merely biochemical eccentricities of the brine shrimp is shown by the work of Oikawa and Smith (1966), who found that encysted embryos of *Daphnia magna* also contain large quantities of both diguanosine tetraphosphate and diguanosine triphosphate. In this organism diguanosine triphosphate is found in higher concentrations than diguanosine tetraphosphate. These results suggest that encysted dormant embryos of other crustacean species, including fresh water forms, might also contain diguanosine polyphosphate anhydrides and that their presence may be related to encystment and dormancy.

B. NATURE OF THE NUCLEOTIDE POOL

Six guanine-containing compounds are found in large amounts in the nucleotide pool at all developmental stages through hatching of the embryos (Warner and Finamore, 1967) (Table I). In the dormant cyst, approximately 90% of all nucleotides contain guanine, 50% of which is diguanosine tetraphosphate. No free guanine is found, and virtually no ATP is present (i.e., less than 1% of the total nucleotides). Preemergence development (0–5 hours) is characterized by a 14% drop in diguanosine tetraphosphate, but during emergence (5–11 hours) there is no significant change in this compound. After hatching (11.5–36 hours), however, there is a rapid decline in diguanosine tetraphosphate. This table also shows that during early development there is a rapid increase in ATP, which remains constant until after hatching, when its production resumes again.

It is clear from this study that diguanosine tetraphosphate undergoes marked changes in concentration only after hatching, a fact that suggests the compound does not play a major role in early embryonic development but is held in reserve for some specified process(es) occurring in the nauplii. As a result, we devoted our attention to changes occurring in the acid-soluble nucleotide pool in the larvae (Clegg *et al.*, 1967). Table II shows changes in concentration of the major guanine-containing compounds of the nucleotide pool in newly hatched, 1-day-old, and 3-day-old nauplii.

Diguanosine tetraphosphate in the 3-day-old nauplii shows a 4-fold decrease; GDP decreases to a barely detectable level; GMP decreases 5-fold, GTP 2-fold, diguanosine triphosphate 10-fold, and guanosine

TABLE I

GUANOSINE AND NUCLEOTIDE CONCENTRATIONS[a] AT VARIOUS TIMES (HOURS) OF DEVELOPMENT

Compound	Hours of development							
	0	2	5.5	9	12	22	27	36
Guo	4.30 ± 0.50	5.32 ± 0.36	5.78 ± 0.26	5.59 ± 0.43	5.30 ± 0.28	4.78 —[b]	4.72 ± 0.26	4.45 ± 0.45
CMP	0.42 ± 0.05	0.25 ± 0.05	0.25 ± 0.03	0.23 —	0.15 —	0.32 ± 0.01	0.34 ± 0.09	0.26 ± 0.03
AMP	0.76 —	0.38 —	0.32 —	0.29 —	0.26 —	0.23 —	0.19 —	0.15 —
UMP	0.51 —	0.30 —	0.22 —	0.27 —	0.22 —	0.30 —	0.33 —	0.56 —
UDP-X	0.86 —	0.84 —	0.91 —	1.01 —	1.21 —	1.43 —	1.52 —	1.70 —
GMP	4.14 ± 0.17	2.70 ± 0.18	2.02 ± 0.25	1.79 ± 0.34	1.64 ± 0.26	1.45 ± 0.30	1.12 ± 0.22	1.15 ± 0.19
ADP	1.00 ± 0.05	0.51 ± 0.04	0.42 ± 0.06	0.42 ± 0.10	0.43 ± 0.10	0.51 ± 0.08	0.49 ± 0.10	0.60 ± 0.19
GDP	3.50 ± 0.08	2.80 ± 0.26	2.54 ± 0.18	2.39 ± 0.10	2.30 ± 0.13	2.01 ± 0.14	1.54 ± 0.21	1.22 ± 0.23
ATP	0.41 ± 0.14	1.73 ± 0.14	2.08 ± 0.20	2.20 ± 0.17	2.21 ± 0.16	2.59 ± 0.26	2.81 ± 0.06	2.93 ± 0.19
GTP	3.67 ± 0.60	4.22 ± 0.30	4.30 ± 0.24	4.22 ± 0.42	3.91 ± 0.32	3.79 ± 0.14	3.47 ± 0.29	3.23 ± 0.19
Diguanosine triphosphate	1.75 ± 0.05	1.65 ± 0.05	1.74 ± 0.11	1.75 ± 0.05	1.69 ± 0.07	1.63 ± 0.25	1.51 ± 0.21	1.41 ± 0.13
Diguanosine tetraphosphate	16.95 ± 0.44	15.65 ± 0.44	14.72 ± 0.43	14.60 ± 0.48	13.80 ± 0.66	11.23 ± 0.84	9.88 ± 0.70	9.53 ± 0.40
Totals	38.27	36.35	35.30	34.76	33.12	30.47	27.92	27.19

[a] Expressed as micromoles per 200,000 embryos ± standard deviation of the mean with $n - 1$ degrees of freedom. Acid-soluble fractions from embryos and nauplii were deacidified with Alamine 336s and applied to DEAE-cellulose columns (1 × 20 cm). Data from Warner and Finamore, 1967.

[b] —, Indicates standard deviation not determined.

about 2-fold. During this time, the total acid-soluble guanine changes from about 55 μmoles of guanine per gram of protein to about 14 μmoles of guanine per gram of protein, principally because of the decline in diguanosine tetraphosphate.

Because there is no increase in GMP, GTP, or any other guanine-containing compound in the acid-soluble pool, the decrease in diguanosine tetraphosphate cannot be accounted for by hydrolysis to GMP and GTP by virtue of the action of its asymmetrical pyrophosphohydrolase (Warner and Finamore, 1965a). Furthermore, the loss of guanine from the pool is not the result of excretion into the medium, nor can it be quantitatively accounted for solely by an increase in nucleic acid guanine

TABLE II

CONCENTRATION OF ACID-SOLUBLE GUANINE COMPOUNDS IN NAUPLII[a]

Age of nauplii	Concentration (μmoles/gram protein)					
	Guo	GMP	GDP	GTP	Diguanosine triphosphate[b]	Diguanosine tetraphosphate[b]
Newly hatched	8.5	2.6	1.4	4.1	2.2	17.4
1-day-old	5.5	0.9	0.6	2.7	0.8	8.6
3-day-old	3.7	0.5	0.008	2.2	0.2	3.7

[a] Acid-soluble fractions were prepared from nauplii at times indicated. After charcoal adsorption, elution, and subsequent concentration, samples were applied to and eluted from DEAE-cellulose columns (1 × 20 cm). Data from Clegg et al., 1967.

[b] Each mole of diguanosine tetraphosphate and diguanosine triphosphate contains 2 moles of guanine.

(Clegg et al., 1967). Thus, our inability to balance the remarkable overall loss of guanine compounds from the acid-soluble pool with an increase in some other fraction forced us to consider the possibility that guanine compounds are being converted by nauplii to non-guanine-containing compounds, perhaps to other purines.

Figure 2 shows the quantitative changes occurring in total adenine, total guanine, and the total purine level in the nauplii. During this time the total purine level is relatively constant but total guanine decreases and total adenine increases. Furthermore, we have shown (Clegg et al., 1967) that the nauplii are incapable of synthesizing purines de novo during this time, so that the organism must convert guanine to adenine as it is required for further development.

Although it is evident from the foregoing discussion that nauplii of Artemia salina maintain a constant purine level in the absence of de *novo* synthesis and convert guanine to adenine, the question remains as to where diguanosine tetraphosphate fits into this metabolic pattern. A clue to its role in purine metabolism was suggested by the observation that, of all the acid-soluble nucleotides in the nauplii, only ATP shows an increase concomitant with the decrease in diguanosine tetraphosphate.

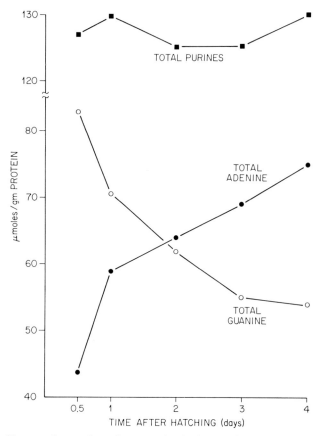

FIG. 2. Changes in total purines, total adenine, and total guanine in nauplii. (Data taken from Clegg *et al.*, 1967.)

The increase in ATP is variable, depending on culture conditions, but is always present. This observation suggests that diguanosine tetraphosphate is converted to ATP which, in turn, may be utilized for a variety of energy-requiring reactions as well as nucleic acid synthesis. Indeed we found that the decrease in diguanosine tetraphosphate and the theoretical yield of adenine resulting therefrom agree closely with the actual increment of total adenine in the nauplii (Clegg *et al.*, 1967).

C. Incorporation of Purines into Nucleic Acids

In an effort to gain some insight into the mechanism by which guanine derivatives are converted to adenine compounds in the nauplii, we exposed the larvae to ^{14}C-labeled purine mononucleotides for 4 hours at 25°C and then transferred them to unlabeled media. At various times thereafter, we determined the specific activities of the adenine and guanine in both the acid-soluble nucleotide pool and the total nucleic acids. Table III shows that, when GMP-^{14}C is used, it is converted to adenine nucleotides as well as guanine nucleotides but, more significantly, the labeled guanine of the nucleotide pool serves as a precursor of nucleic acid guanine, whereas the adenine nucleotides derived from labeled guanine are not efficient precursors of nucleic acid adenine. This observation suggests that another unlabeled (or only slightly labeled) source of adenine is the major precursor of nucleic acid adenine. To check this point, in separate experiments AMP-^{14}C and IMP-^{14}C were administered to nauplii under identical conditions (Table III). Here, too, the results are indicative of an unlabeled source of adenine being the primary precursor of nucleic acid adenine. This conclusion implies that the proposed unlabeled (or slightly labeled) precursor of nucleic acid adenine does *not* mix with the acid-soluble pool of adenine to any great extent but is somehow compartmentalized. Thus, if diguanosine tetraphosphate serves as a source of adenine, as our previous results indicate, then it probably functions as the principal source of nucleic acid adenine and only secondarily as a source of acid-soluble adenine nucleotides. Furthermore, the entire conversion process must be intimately linked to nucleic acid synthesis.

D. Diguanosine Tetraphosphate and DNA Synthesis

As mentioned previously, the embryos of *Artemia salina* undergo considerable differentiation without mitosis or cell multiplication of any sort (Nakanishi *et al.*, 1962). Only in the prenauplius, and primarily in the nauplius, is there significant mitotic activity, and its onset corresponds precisely to the time when diguanosine tetraphosphate starts to decline. Temporally, then, these two events appear to be related.

To test the relationship, if any, between DNA synthesis and the metabolism of diguanosine tetraphosphate, the following experiment was performed. Immediately after hatching, the nauplii were collected and divided into two groups. One group was cultured under sterile conditions in the presence of AMP-^{14}C, uridine, and 5-fluorodeoxyuridine to inhibit DNA synthesis (Volkin and Ruffilli, 1962), and the control group was cultured under identical conditions but without 5-fluorodeoxyuridine.

TABLE III

SPECIFIC ACTIVITIES OF GUANINE AND ADENINE[a,b]

Time (hours)	GMP-C14				AMP-C14				IMP-14C			
	Acid-soluble		Nucleic acid		Acid-soluble		Nucleic acid		Acid-soluble		Nucleic acid	
	Guanine	Adenine	Guanine	Adenine	Guanine	Adenine	Guanine	Adenine	Guanine	Adenine	Guanine	Adenine
0	212,000	40,400	42,800	1,880	25,700	298,000	24,000	26,100	—	—	—	—
10	33,700	35,300	27,400	1,190	13,100	159,000	37,800	45,200	—	—	—	—
24	29,100	32,000	42,300	3,360	11,100	101,000	60,700	58,000	—	—	—	—
44	20,000	45,700	37,900	4,070	9,300	86,700	37,300	45,000	—	—	—	—
0	24,700	13,100	13,600	1,360	—	—	—	—	5,940	47,000	14,900	8,600
8	9,700	12,600	11,400	1,750	—	—	—	—	6,670	70,500	13,400	12,300
20	5,370	12,310	11,200	1,730	—	—	—	—	1,500	29,400	7,620	7,000
32	1,440	8,250	6,820	1,300	—	—	—	—	1,110	17,800	3,830	3,250

[a] Specific activities of guanine and adenine after exposure to labeled precursors. Nauplii were incubated with either ^{14}C-labeled GMP (specific activity, 42 mc/mmole), ^{14}C-labeled AMP (specific activity 34.6 mc/mmole), or ^{14}C-labeled IMP (specific activity 33 mc/mmole) in a final concentration of 0.5 μC per milliliter of incubation medium. After incubation at 25° for 4 hours, nauplii were removed, washed, and placed in sterile unlabeled medium. At various times thereafter, acid hydrolyzates of acid-soluble and nucleic acid fractions were subjected to Dowex 50-H+ column chromatography (Cohn, 1955). Zero hour times refer to samples taken at 4 hours of exposure to isotopes.

[b] Values are expressed as counts per minute per micromole.

After 4 hours of incubation at 25°C, both groups were washed free of labeled precursor and placed in unlabeled medium containing only 5-fluorodeoxyuridine and uridine in the case of the experimental group and only uridine in the control group. At various times thereafter, nauplii

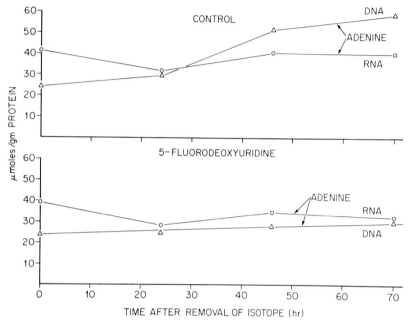

Fig. 3. Concentrations of nucleic acid adenine in control nauplii and in nauplii treated with 5-fluorodeoxyuridine. As soon after hatching as possible, nauplii were collected and divided into two groups: (1) control larvae incubated in the presence of ^{14}C-labeled AMP (specific activity 42 mc/mmole) in a final concentration of 0.5 μc per milliliter of incubation medium) and in the presence of uridine (20 μg per milliliter of incubation medium), and (2) larvae incubated in mixture of ^{14}C-labeled AMP and uridine as above plus 5-fluorodeoxyuridine (10 μg per milliliter of incubation medium). After incubation at 25°C for 4 hours, nauplii from control group were placed in unlabeled medium containing only uridine, whereas nauplii from second group were placed in unlabeled medium containing uridine and 5-fluorodeoxyuridine. At various times thereafter, nucleic acids were extracted by a hot NaCl procedure (Tyner et al., 1953), separated by acid precipitation after alkaline hydrolysis, and acid hydrolyzates were subjected to Dowex 50-H[+] column chromatography (Cohn, 1955). Zero hour times refer to samples taken at 4 hours of exposure to isotope.

were collected from both cultures, the acid-soluble fractions were prepared and subjected to DEAE-cellulose column chromatography (Warner and Finamore, 1967), and the nucleic acids were extracted by a hot NaCl procedure (Tyner et al., 1953). DNA was separated from RNA by alkaline hydrolysis of the RNA and subsequent acid precipitation of the DNA. The nucleic acid purines were obtained by acid hydrolysis and were separated by column chromatography using Dowex 50-H⁺ (Cohn, 1955).

Figure 3 shows the effectiveness of 5-fluorodeoxyuridine as an inhibitor of DNA synthesis. Note that in the control animals RNA adenine remains relatively constant but there is a marked net increase in DNA adenine; on the other hand, nauplii treated with 5-fluorodeoxyuridine show no net increase in DNA adenine. Figure 4 indicates that 5-fluorode-

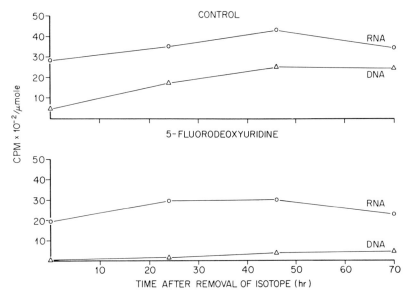

Fig. 4. Specific activities of nucleic acid adenine in control nauplii and nauplii exposed to 5-fluorodeoxyuridine. Incubation conditions, extraction procedure, and column chromatography same as described in Fig. 3.

oxyuridine strongly inhibits incorporation of AMP-¹⁴C into DNA while its incorporation into RNA is affected only slightly, if at all. These results indicate that RNA synthesis and/or turnover is not prevented in the presence of the drug but that DNA synthesis virtually ceases. Of particular interest is the fact that between 20 and 70 hours after hatching, the concentration of DNA adenine increases markedly and actually surpasses that of RNA adenine. However, the specific activity of DNA adenine, although increasing, always remains considerably less than that of RNA adenine. This observation clearly points to two separate pools of nucleic acid adenine: the more heavily labeled pool being used for RNA synthesis and the relatively less labeled pool being used for DNA synthesis. These results are consistent with the data presented earlier in Table III, but the conclusion must now be modified to state that the unlabeled (or slightly labeled) source of adenine is the major precursor of *DNA* adenine.

It is of interest to point out that nauplii that are exposed constantly to 5-fluorodeoxyuridine for several days continue to elongate and differentiate even though DNA synthesis ceases. For the first 2 days of treatment with the drug, there is no noticeable morphological difference between control and experimental nauplii. Only on the third day of exposure is the effect of 5-fluorodeoxyuridine observable. At this time their motility is decreased and they tend to move in an erratic fashion. However, their gross morphological characteristics are virtually indistinguishable from untreated nauplii. Thus, between gastrulation and the formation of the prenauplius, DNA synthesis does not occur (F. J. Finamore,

TABLE IV

CONCENTRATIONS OF DIGUANOSINE TETRAPHOSPHATE AND ATP[a]

Time (hours)	Diguanosine tetraphosphate (μmoles/gram protein)		ATP (μmoles/gram protein)	
	Control	FUdR[b]	Control	FUdR[b]
0	23.7	22.2	15.3	13.7
24	16.5	21.2	13.5	12.3
46	5.8	20.0	20.7	14.8
70	5.3	14.8	22.0	12.1

[a] Concentrations of diguanosine tetraphosphate and ATP at various times after hatching. Nauplii were collected as soon after hatching as possible and incubated in the presence of ^{14}C-labeled AMP as described in Table III. Acid-soluble fractions were separated by DEAE-HCO$_3$ column chromatography as described by Warner and Finamore (1967).

[b] 5-Fluorodeoxyuridine.

unpublished); and now it appears that after hatching, further differentiation may occur in the absence of DNA synthesis.

An investigation of changes occurring in the nucleotide pool of control nauplii and nauplii treated with 5-fluorodeoxyuridine reveals that inhibition of DNA synthesis prevents utilization of diguanosine tetraphosphate and the net increase in ATP (Table IV). However, the *turnover* of adenine in ATP is unaffected by 5-fluorodeoxyuridine (Table V). This observation suggests, once again, that diguanosine tetraphosphate serves only as a secondary source of acid-soluble adenine and that its principal function is to supply adenine directly for DNA synthesis (not for RNA

TABLE V

SPECIFIC ACTIVITIES OF DIGUANOSINE TETRAPHOSPHATE AND ATP[a]

Time (hours)	Diguanosine tetraphosphate (cpm/μmole)		ATP (cpm/μmole)	
	Control	FUdR[b]	Control	FUdR[b]
0	152	149	56,200	50,000
24	1,000	1,420	10,250	16,000
46	1,380	1,800	6,260	7,630
70	1,330	2,000	4,800	5,200

[a] Nauplii were treated as described in Tables III and IV.
[b] 5-Fluorodeoxyuridine.

synthesis). Furthermore, it should be noted that adenine from AMP-^{14}C can serve as a source of guanine in diguanosine tetraphosphate in both the control nauplii and nauplii treated with 5-fluorodeoxyuridine. The fact that the *concentration* of diguanosine tetraphosphate is decreasing rapidly in the control animals while it is remaining essentially constant under 5-fluorodeoxyuridine treatment, plus the fact that the specific activity of diguanosine tetraphosphate is higher when DNA synthesis is in-

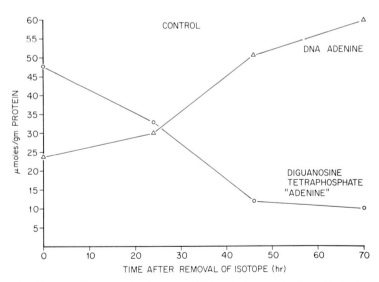

FIG. 5. Changes in concentration of DNA adenine and theoretical change in diguanosine tetraphosphate "adenine." Results of DEAE column chromatography of acid-soluble fraction compared with results of Dowex 50-H$^+$ column chromatography of DNA hydrolyzates as described in Fig. 3.

hibited, indicate that the *rate* of labeling of diguanosine tetraphosphate is several times higher in the presence than in the absence of 5-fluorodeoxyuridine.

The final piece of evidence that diguanosine tetraphosphate serves as a direct precursor of DNA adenine comes from the quantitative relationship between the theoretical yield of adenine from diguanosine tetraphosphate (where 1 mole of the diguanosine compound is equivalent to 2 moles of DNA adenine) and the actual increment in DNA adenine as shown in Fig. 5. This figure depicts the loss of "adenine" from diguanosine tetraphosphate to be quantitatively similar to the increase in adenine of DNA. When DNA synthesis is inhibited by 5-fluorodeoxyuridine, no such relationship exists.

Thus the experimental data are consistent with the proposition that diguanosine tetraphosphate functions as the principal source of DNA adenine, not RNA adenine, and that the conversion process is intimately linked to DNA synthesis.

E. Proposed Mechanism for Participation of Diguanosine Tetraphosphate in DNA Synthesis

On the basis of available evidence, we speculate that the role of diguanosine tetraphosphate in DNA synthesis is as shown in Scheme 2.

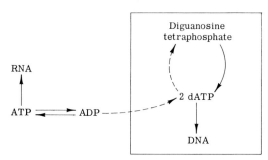

Scheme 2

By a series of enzymatic steps including ribose reduction, conversion of guanine to adenine, cleavage of the pyrophosphate bonds and possibly phosphorylation, diguanosine tetraphosphate is converted to dATP, which, in turn, is incorporated into DNA. Very little dATP escapes to mix with the nucleotide pool, and only minor amounts of ATP and/or ADP are used normally for synthesis of dATP. Upon inhibition of DNA synthesis, conversion of diguanosine tetraphosphate to dATP is shut down, and now only the ATP/ADP conversion system contributes to the dATP pool. The resulting dATP is channeled back into diguanosine tetraphosphate.

This scheme would account for the following observations:

1. Inhibition of DNA synthesis by 5-fluorodeoxyuridine increases the rate of labeling of diguanosine tetraphosphate when AMP-^{14}C is used.

2. When AMP-^{14}C is supplied, the specific activity of DNA adenine never surpasses that of RNA adenine, although the concentration of DNA adenine increases about 2-fold and exceeds that of RNA adenine between 20 and 70 hours after hatching.

3. ATP turnover is unaffected by 5-fluorodeoxyuridine, which inhibits DNA synthesis and diguanosine tetraphosphate utilization.

Thus, our evidence strongly suggests that diguanosine tetraphosphate performs a key function in DNA synthesis in *Artemia salina*. Of perhaps more importance, however, is the following question: *Is DNA synthesis in the nauplii of Artemia intimately linked to and controlled by the utilization of diguanosine tetraphosphate?* The answer of this question must await further experimentation.

III. STUDIES ON PROTEIN SYNTHESIS DURING PREEMERGENCE DEVELOPMENT

Since embryonic differentiation is completely divorced from cell division throughout the preemergence development of *Artemia*, this system presents a unique one for examining the relationship between protein synthesis, its regulation, and the processes of differentiation in a metazoan embryo. Accordingly, such a study was undertaken (Clegg, 1966; Golub and Clegg, 1968). Some of the more recent results are presented in this section.

A. ON THE STATUS OF POLYRIBOSOMES IN DORMANT GASTRULAE

Since it is now generally accepted that protein synthesis occurs on polyribosomes, the question whether or not the dormant gastrula contains polyribosomes is of great interest. This question was examined initially by sucrose density gradient centrifugation. Figure 6 illustrates the profiles obtained for 14,000 g supernatants prepared by extracting the gastrulae directly from the dried state (Fig. 6A) and from the hydrated but still dormant condition (Fig. 6B). The only appreciable difference between these two profiles is the lack of resolution of the 36 S peak and 61 S shoulder in preparations from dried gastrulae. (All sedimentation coefficients reported here are preliminary.) Both preparations contain large and approximately equivalent amounts of 81 S ribosomes that we will refer to hereafter as monosomes. Although polyribosome peaks are not present, some ultraviolet-absorbing material does sediment below the monosome peak. Moreover, most of this material is sensitive to ribonuclease treatment (5 μg/ml for 30 minutes at 0°C) as shown in Fig. 6C.

Because of the sensitivity of polyribosomes to endogenous ribonuclease activity, ionic composition of the homogenizing buffer, and method of disruption, these parameters were evaluated in the *Artemia* system (Golub and Clegg, 1968). The use of deoxycholate was also examined. The results indicate that the lack of discrete classes of polyribosomes

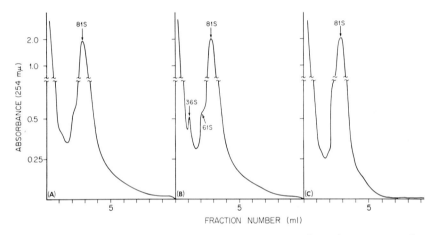

FIG. 6. Sucrose gradient profiles of 14,000 g supernatants from dormant gastrulae. The embryos were homogenized with TMK buffer (0.05 M Tris-HCl pH 7.8, 0.01 M MgCl$_2$, and 0.1 M KCl), and the homogenate was centrifuged for 30 minutes at 14,000 g. The supernatant (about 12 OD$_{260}$ units) was layered onto 15–50% w/v sucrose gradients, and these were centrifuged for 80 minutes at 39,000 rpm in a Spinco SW 41 rotor. Gradients were analyzed with an Isco gradient-fractionating system with continuous recording of absorbancy at 254 mμ. (A) Gastrulae extracted from the dried condition; (B) gastrulae extracted from the hydrated condition; (C) same as (B) except for treatment of the supernatant for 30 minutes at 0°C with 5 μg of ribonuclease per milliliter.

in dormant gastrula preparations does not appear to be the result of any of these variables.

Further study of the question of polyribosomes in dormant gastrulae was made using a cell-free amino acid incorporating system. The data given in Table VI show that preparations from both dried and prehydrated dormant gastrulae do incorporate leucine-^{14}C into protein (hot trichloroacetic acid precipitable material). Although the extent of incorporation is quite low, it requires an energy source and GTP for maximal activity, is dependent upon the presence of the 105,000 g pellet, and is inhibited by the presence of ribonuclease and other inhibitors of protein synthesis, but not by deoxyribonuclease. It has been determined that this incorporation is not due to contamination by microorganisms, so we are confident that dormant gastrula preparations are capable of carrying out amino acid incorporation *in vitro* and thus may contain functional polyribosomes.

It was next determined whether or not this incorporation actually was occurring on polyribosomes by analyzing the distribution of incorporated radioactivity on sucrose gradients. Preparations from dormant gastrulae were incubated *in vitro* as described in Table VI, except that increased amounts of the ribosome preparation (0.1 ml) and leucine-^{14}C (0.5 µc) were used. After 10 minutes and 25 minutes of incubation *in vitro*, aliquots of the reaction mixture were analyzed on sucrose gradients (15–50% w/v) prepared as described by Iverson and Cohen (see Chapter

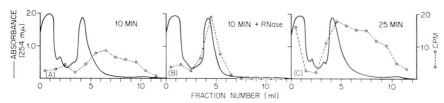

FIG. 7. Analysis of the cell-free amino acid-incorporating system from dormant gastrulae on sucrose gradients. Incubation was carried out as described in Table VI. Aliquots of one reaction mixture were applied to 15–50% (w/v) sucrose gradients after (A) 10 minutes, (B) 10 minutes followed by ribonuclease treatment (1 µg/ml for 30 minutes at 0°C), and (C) 25 minutes. The gradients were analyzed for absorbance at 254 mµ, and 1-ml fractions were collected and plated for determination of radioactivity as described in Fig. 6 and Table VI.

14). The results (Fig. 7A,C) clearly show that most of the incorporated radioactivity is associated with the polyribosome region of these gradients. Moreover, this radioactivity is displaced to the monosome region by ribonuclease treatment of the reaction mixture (1 µg/ml for 30 minutes at 0°C) as shown in Fig. 7B. These results strongly suggest that small amounts of functional polyribosomes are present in preparations from dormant gastrulae.

But it is also apparent that ribonuclease treatment does not result in much of a decrease in the absorbance associated with the polyribosome region (Fig. 7A,B). We have evidence that this ribonuclease-resistant material is glycogen and does not represent stable polyribosomes. The concentration of glycogen resulting from high-speed centrifugation accounts for the increase in amount of material sedimenting below the monosomes in gradients of pelleted preparations (Fig. 7) as compared with gradients of 14,000 *g* supernatants (Fig. 6).

The results shown in Fig. 7 have not been consistently obtained from experiments of this type. In some cases the incorporation of leucine-^{14}C is negligible throughout the gradients. Such negative results might be accounted for simply on the basis of insufficient incorporation since preparations from dormant gastrulae are, at best, not very active (Table VI, Fig. 7). However, these negative results might have some significance.

TABLE VI

LEUCINE-^{14}C INCORPORATION IN CELL-FREE PREPARATIONS FROM DRIED
AND HYDRATED DORMANT GASTRULAE[a]

| | Incorporation (cpm/mg protein/hour) | | | |
| | Dried gastrulae | | Hydrated gastrulae | |
Incubation mixture	1	2	1	2
Complete	226	202	249	187
Omit ATP	32	41	56	20
Omit GTP	101	111	127	85
Omit 105,000 g pellet	9	13	13	12
Complete	233	219	230	241
Add 1μg RNase	18	12	6	22
Add 20 μg DNase	241	210	222	252
Add 0.4 μmole cyclohexamide	46	51	67	58
Add 0.15 μmole puromycin	15	14	18	22

[a] The complete reaction mixture contained the following: 7 μmoles MgCl$_2$; 50 μmoles KCl; 48 μmoles Tris-HCl pH 7.8; 3 μmoles 2-mercaptoethanol; 1 μmole GTP; 1.5 μmole ATP; 8 μmoles creatine phosphate; 50 μg creatine kinase; 0.02 ml 105,000 g supernatant (about 60 μg protein); 0.05 ml 105,000 g pellet (about 300 μg protein and 190 μg RNA); and 0.08 μC leucine-^{14}C (uniformly labeled, 204 mc/mmole) in a final volume of 0.7 ml. After incubation the reaction was stopped by the addition of 1 ml of 10% trichloroacetic acid, the tubes were heated for 30 minutes at 95°C, and the contents were plated on glass pads and counted with a thin-window counter (background < 2 cpm). The 105,000 g pellet from 6 ml of 14,000 g supernatant (200 mg wet weight of embryos per milliliter of TMK buffer) was rinsed, resuspended with 1.5 ml of TMK buffer, and used without further treatment.

For example, it is possible that polyribosome assembly, or activation, might occur to a greater or lesser degree during the process of disruption, depending on the circumstances of that particular experiment.

Of relevance in this regard is the fact that dormant gastrula preparations are able to incorporate phenylalanine-^{14}C at high rates in the presence of polyuridylic acid (Table VII). Sucrose gradient analyses of these reaction mixtures show that polyuridylic acid-directed phenylalanine incorporation is restricted to the 81 S monosome region. Therefore, the monosomes of dormant gastrulae are competent to support protein synthesis as far as artificial messenger is concerned.

Although further study is needed before a definite conclusion is warranted, the weight of present evidence favors the view that dormant gastrulae preparations do contain small amounts of polyribosomes that

TABLE VII

INCORPORATION OF PHENYLALANINE-^{14}C BY PREPARATIONS FROM DRIED
AND HYDRATED DORMANT GASTRULAE WITH AND WITHOUT
ADDED POLYURIDYLIC ACID

| | Incorporation (cpm/mg protein/hour) | | | |
| | Dried gastrulae | | Hydrated gastrulae | |
Sample	1	2	1	2
Without poly U	57	46	21	62
Plus 50 μg poly U	4,280	5,600	4,740	6,050
Plus 100 μg poly U	7,650	11,100	9,100	11,550

[a] The conditions of these experiments are the same as those described in Table I except for the use of phenylalanine-^{14}C (0.075 μc) and polyuridylic acid. The values above were corrected for incorporation not due to the 105,000 g pellet, which varied between 16 and 37 cpm.

are functional in *in vitro* protein synthesis. Whether or not they are physiologically significant is another matter.

B. POLYRIBOSOME FORMATION IN THE METABOLIZING GASTRULA

Of considerable interest is the rate of polyribosome formation when gastrulae are permitted to resume metabolism. Figure 8 illustrates the formation of polyribosomes in prehydrated dormant gastrulae when placed at 30°C and incubated in artificial seawater for periods up to 4 hours. Typical polyribosome peaks are detected after only 5 minutes of incubation, the earliest period we have analyzed thus far. The polyribosomes formed initially seem to be relatively small ones which appear chiefly in the region immediately below the monosomes.

It is significant that this same region of the gradient contains polyribosome-like material present in dormant gastrula preparations (Figs. 6 and 7). As incubation proceeds, the amount of polysomes increases, and it is interesting to note that most of this increase apparently results from the formation of heavier polyribosomes. One interpretation of this is that polyribosome formation in these gastrulae occurs by the sequential attachment of ribosomes to a strand of messenger RNA. Such a sequential attachment has been proposed to occur in *Escherichia coli* following recovery from starvation (Dresden and Hoagland, 1967). Of course, it is also possible that the light polyribosomes detected initially represent either the synthesis of proteins of small molecular weight or the breakdown products of heavy polyribosomes. It is felt that the latter is unlikely since our methods are adequate for the demonstration of

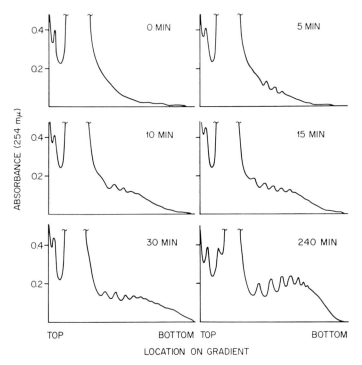

FIG. 8. Polyribosome formation following dormancy in hydrated gastrulae. The conditions are the same as those described in Fig. 6 except that the gastrulae were incubated at 30°C in seawater for the times indicated prior to homogenization. The tops of the 81 S monosome peaks were not measured in this experiment, since the 0.5 scale on the ultraviolet analyzer was used to obtain maximum sensitivity in the polyribosome region.

heavy polyribosomes, as illustrated in Fig. 8. Nonetheless, this possibility cannot be excluded at the present time.

The experiments described above were performed by using prehydrated embryos. Polyribosome formation has also been examined in dried gastrulae incubated at 30°C in seawater without prior hydration at 0°C. Under these conditions characteristic polyribosome peaks have been detected within the first 25 minutes of incubation. At this time, the embryos have hydrated to only about 25% of their maximum water content (Iwasaki, 1964; J. S. Clegg, unpublished results).

By experiments like those described for the dormant gastrula (Section IIIA), it has been shown that the polyribosomes formed during the resumption of embryonic metabolism are functional with regard to amino acid incorporation (Golub and Clegg, 1968). This is illustrated in Fig. 9, which shows the results obtained by using embryos that had been incubated for 4 hours in seawater at 30°C prior to analysis. The use

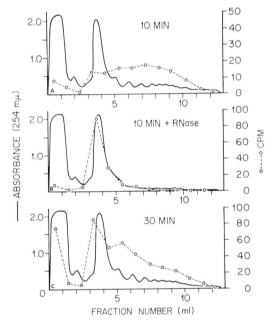

FIG. 9. Incorporation of leucine-^{14}C *in vitro* by preparations from gastrulae that had been incubated for 4 hours at 30°C in seawater prior to homogenization. Note that the scale for radioactivity in profile (A) differs from that for (B) and (C). (See Fig. 7 for further details.) (Data taken from Golub and Clegg, 1968.)

of embryos incubated for greater or lesser periods of time *in vivo* yields qualitatively similar results. Such studies, in addition to demonstrating the functionality of polyribosomes, have also indicated that the incorporation rate of leucine-^{14}C *in vitro* is directly related to the polyribosome content of that particular preparation. This relationship, summarized in Fig. 10, indicates that all the polyribosomes, or a constant percentage of them, are functional during the periods of development we have studied.

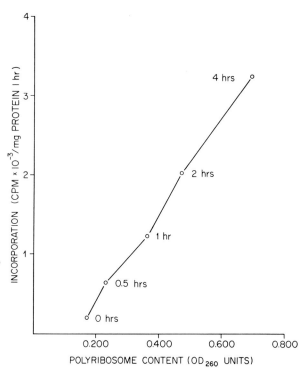

FIG. 10. Relationship between polyribosome content and rate of leucine-^{14}C incorporation *in vitro* by preparations from dormant and developing embryos. The polyribosome content was estimated from the ribonuclease-sensitive absorbance at 260 mμ of the polyribosome region of the gradients. These values represent, approximately, the number of absorbance units of polyribosomes per 0.3 ml of 14,000 g supernatant. The numbers in parentheses near each point represent the duration of incubation at 30°C of hydrated dormant gastrulae which were used in those experiments.

C. THE FATE OF POLYRIBOSOMES IN REDESICCATED EMBRYOS

Another remarkable feature of these embryos is that they can be redesiccated at any time during preemergence development without affecting their ability to complete normal development upon rehydration (Clegg, 1967, and unpublished results). It has been of interest to determine the means whereby these embryos regulate the biosynthesis of proteins and the components of the protein-synthesizing system when undergoing the transition from the hydrated metabolizing state to the desiccated ametabolic state. For example, what happens to the polyribosomes when developing embryos are redesiccated? The answer that has been obtained is given in Fig. 11. When encysted gastrulae are incubated for 4 hours at 30°C, extensive polyribosome formation occurs (Fig. 11A).

The polyribosomes are not retained upon redesiccation of these 4-hour embryos (Fig. 11B) whether the embryos are desiccated at 0°C, or at a temperature (26°C) where metabolism can continue as desiccation occurs (Clegg, 1967). The profile shown in Fig. 11B was obtained by extracting the embryos from the desiccated condition. It has been found, however, that rehydrating the embryos at 0°C prior to extraction produces a similar profile with regard to polyribosomes. Consequently, the apparent absence of discrete classes of polyribosomes in redesiccated embryos is not the result of extraction from the dried condition. If embryos which had been incubated for 4 hours and then redesiccated are

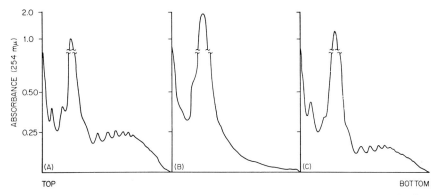

FIG. 11. The fate of polyribosomes in embryos which were redesiccated during preemergence development. Hydrated dormant gastrulae were incubated for 4 hours at 30°C in seawater. They were then either analyzed immediately (A) or redesiccated over CaCl₂ for 12 hours and then analyzed (B). The profile shown in part (C) is from redesiccated 4-hour embryos which had been hydrated for 3 hours at 0°C and then allowed to incubate for an additional 2 hours at 30°C in seawater.

once again hydrated and incubated at 30°C, polyribosome formation likewise reoccurs as shown in Fig. 11C. A detailed comparison of the formation of polyribosomes in these embryos with that occurring in gastrulae which had not experienced previous incubation and redesiccation has not yet been made (Fig. 8). One can imagine several provocative possibilities regarding the differences and similarities, both qualitative and quantitative, of polyribosome formation and function under these circumstances. For example, it will be interesting to compare the proteins being synthesized in developing embryos at the time of redesiccation with those synthesized upon the resumption of metabolism and development when these same embryos are rehydrated and once again incubated at 30°C. By these experiments, and variations of them, it might be possible to determine which proteins are associated most directly with differentiation and which are not.

Another point of interest concerning the profiles from incubated and redesiccated embryos (Fig. 11B) is their comparison with those from unincubated dried gastrulae (Fig. 6A). Both profiles show about the same amount and distribution of material sedimenting below the monosomes, and it should be recalled that this material may contain functional polyribosomes (Fig. 8, Table VI). Whether this material is simply a useless remnant of polyribosomes present in the embryos prior to desiccation, or represents stored polyribosomes which play a physiologically significant role during the transition from dormancy to active metabolism and vice versa, remains to be determined.

A rather large number of investigators have considered the presence of polyribosomes in other dormant systems. Polyribosomes apparently are not retained in the spores of several species of bacteria (Halvorson, et al., 1966; see also Slepecky, Chapter 4) but some dormant plant seeds do appear to contain a small complement of polyribosomes (Marcus and Feeley, 1964, 1965; Allende and Bravo, 1966). The dormant (unembryonated) eggs of the nematode *Ascaris lumbricoides* appear to contain very large quantities of polyribosomes, only some of which are functional (Kaulenas and Fairbairn, 1966). In unfertilized sea urchin eggs, all components of the protein-synthesizing system are apparently present, but whether or not polyribosomes are present has not been fully resolved (see reviews by Monroy, 1965; Spirin, 1966; Nemer and Infante, 1967; see also Chapters 14 and 15, this volume). When comparing these systems, however, it should be remembered that they exhibit varying degrees of dormancy. Thus the significance of polyribosomes in desiccated *Artemia* gastrulae, which are essentially ametabolic, may well be different from that in systems such as unfertilized sea urchin eggs or unembryonated *Ascaris* eggs which, although dormant, are unquestionably metabolizing at a detectable rate (see also Chapter 13).

IV. CONCLUDING REMARKS: ADAPTATION AND THE DEVELOPMENT OF ENCYSTED EMBRYOS

In concluding this paper on *Artemia*, it is appropriate to consider the high degree of adaptation exhibited by these embryros (literature on this subject is contained in the bibliography by Littlepage and McGinley, 1965). The production of the encysted gastrula is clearly essential to the survival of *Artemia* in nature under the peculiar circumstances of its environment. Not only can this gastrula tolerate essentially complete desiccation, exposure to tremendous variations in temperature, and a variety of other hazards that rapidly destroy the vast majority of living organisms, but the embryos must retain these abilities when metabolism is resumed and as the complexities of morphogenesis occur during preemergence development. One can expect, therefore, the presence of adaptations at all levels of organization which comprise the basis of such abilities (Clegg, 1962, 1964, 1965, 1967). One of these adaptations appears to be the ability of the gastrula to undergo extensive development without the occurrence of cell division and, as a result, without the lability that this activity would impose when the embryos were frozen, desiccated, and so forth, during preemergence development. This appears to be accomplished by synchronizing the onset of cell division with the emergence of the embryo from its protective shell. The evidence presented here at least suggests that the metabolism of diguanosine tetraphosphate could be a major regulator of cell division through its participation in DNA synthesis. By synchronizing the utilization of diguanosine tetraphosphate with emergence, the premature onset of cell division in the unemerged embryo may be prevented.

Since the synthesis of specific proteins is generally held to be a necessary condition for successful embryonic development, it is likely that such activity would be extremely sensitive to natural selection and that appropriate mechanisms for maintaining the structural integrity of the protein-synthesizing system during the dormant state would be acquired. The same argument can be made for the acquisition of suitable regulatory mechanisms involving gene transcription and translation. Consequently, the study of these embryos should not only provide insight into the control of nucleic acid and protein synthesis during abrupt metabolic transitions associated with dormancy, but might also be useful in further study of the relationship between nucleic acids, proteins, and embryonic differentiation.

ADDENDUM

After the writing of this article, Drs. T. Hultin and J. Morris were kind enough to supply me (J. Clegg) with a copy of their manuscript prior to its publication [Hultin, T. and Morris, J. E. (1968). *Develop. Biol.* **17**, 143–164]. These authors have carried out an electron microscopical and biochemical study on the ribosomes of *Artemia* embryos during cryptobiosis and after the resumption of development. Although most of our biochemical results are in general agreement with theirs, major differences in techniques and interpretation make comparison difficult at this time. The reader should consult their paper for further details.

ACKNOWLEDGMENT

This research was jointly sponsored by the National Institutes of Health (Grant H. D. 03478) and by the U.S. Atomic Energy Commission under contract with the Union Carbide Corporation.

REFERENCES

Adam, A., and Moffatt, J. G. (1966). *J. Am. Chem. Soc.* **88**, 838.

Allende, J. E., and Bravo, M. (1966). *J. Biol. Chem.* **241**, 5813.

Anderson, D. T. (1967). *Australian J. Zool.* **15**, 47.

Barigozzi, C. (1957). *Annee Biol.* **33**, 241.

Bowen, S. T. (1962). *Biol. Bull.* **122**, 25.

Clegg, J. S. (1962). *Biol. Bull.* **123**, 295.

Clegg, J. S. (1964). *J. Exptl. Biol.* **41**, 879.

Clegg, J. S. (1965). *Comp. Biochem. Physiol.* **14**, 135.

Clegg, J. S. (1966). *Nature* **212**, 517.

Clegg, J. S. (1967). *Comp. Biochem. Physiol.* **20**, 801.

Clegg, J. S., Warner, A. H., and Finamore, F. J. (1967). *J. Biol. Chem.* **242**, 1938.

Cohn, W. E. (1955). *In* "The Nucleic Acids" (E. Chargaff and J. N. Davidson, eds.), Vol. I, p. 218. Academic Press, New York.

Dresden, M. H., and Hoagland, M. B. (1967). *J. Biol. Chem.* **242**, 1069.

Dutrieu, J. (1960). *Arch. Zool. Exptl. Gen.* **99**, 1.

Emerson, D. N. (1963). *Proc. S. Dakota Acad. Sci.* **42**, 131.

Fautrez-Firlefyn, N. (1951). *Arch. Biol. (Liege)* **42**, 391.

Finamore, F. J., and Warner, A. H. (1963). *J. Biol. Chem.* **238**, 344.

Goldschmidt, E. (1952). *J. Morphol.* **91**, 111.

Golub, A., and Clegg, J. S. (1968) *Develop Biol.* **17**, 644.

Green, J. (1965). *Biol. Rev.* **40**, 580.

Halvorson, H. O., Vary, J. C., and Steinberg, W. (1966). *Ann. Rev. Microbiol.* **20**, 169.

Iwasaki, T. (1964). *J. Radiation Res.* **5**, 91.

Kaulenas, M. S., and Fairbairn, D. (1966). *Develop. Biol.* **14**, 481.

Littlepage, J. L., and McGinley, M. N. (1965). *San Francisco Aquarium Soc. Spec. Publ. No. 1.*

Marcus, A, and Feeley, J (1964). *Proc. Natl. Acad. Sci. U.S.* **51**, 1075.

Marcus, A., and Feeley, J. (1965). *J. Biol. Chem.* **240**, 1675.

Monroy, A. (1965). *In* "The Biochemistry of Development" (R. Weber, ed.), Vol. I, pp. 73–139. Academic Press, New York.

Muramatsu, S. (1960). *Embryologia* **5**, 95.

Nakanishi, Y. H., Iwasaki, T., Okigaki, T., and Kato, H. (1962). *Annotationes Zool. Japan.* **35**, 223.

Nemer, M. and Infante, A. A. (1967). *In* "The Control of Nuclear Activity" (L. Goldstein, ed.), pp. 101–128. Prentice-Hall, Englewood Cliffs, New Jersey.

Oikawa, T. G., and Smith, M. (1966). *Biochemistry* **5**, 1517.

Provasoli, L., and Shiraishi, K. (1959). *Biol. Bull.* **117**, 345.

Spirin, A. S., (1966). *In* "Current Topics in Developmental Biology" (A. A. Moscona and A. Monroy, eds.), pp. 1–38. Academic Press, New York.

Tyner, E. P., Heidelberger, C., and Le Page, G. A. (1953). *Cancer Res.* **13**, 186.

Urbani, E. (1962). *Advan. Morphogenesis* **2**, 61.

Volkin, E., and Ruffilli, A. (1962). *Proc. Natl. Acad. Sci. U.S.* **48**, 2193.

Warner, A. H., and Finamore, F. J. (1965a). *Biochemistry* **4**, 1568.

Warner, A. H., and Finamore, F. J. (1965b). *Biochim. Biophys. Acta* **108**, 525.

Warner, A. H., and Finamore, F. J. (1967). *J. Biol. Chem.* **242**, 1933.

CHAPTER 13

The Program of Structural and Metabolic Changes following Fertilization of Sea Urchin Eggs

David Epel, Berton C. Pressman, Sigrid Elsaesser, and
Anthony M. Weaver ·

I. INTRODUCTION

The interactions between sperm and egg occurring upon fertilization result in new metabolic activities leading to cell division and differentiation. Studies on sea urchin eggs show that some of these new activities do not require synthesis of new enzymes since eggs fertilized and cultured in the presence of puromycin are not visibly affected until the time of the first cell division (Hultin, 1961; Wilt *et al.*, 1967; Black *et al.*, 1967). These new activities, moreover, do not result from provision of some new or limiting substance or enzyme from the sperm, since eggs can be artificially induced to develop by physical or chemical means (i.e., artificial parthenogenesis). It would appear, therefore, that the new postfertilization activities ensue from the activation of enzymes and metabolic pathways which are *already* present in the unfertilized egg.

Our analysis of this activation, to be described in this article, has been concerned with the description of the temporal sequence and mechanism of these metabolic changes. Our findings suggest that a programmed sequence of structural and metabolic events is initiated by fertilization, and that this program is remarkably similar in eggs of all species of sea urchins so far studied. In the following pages we shall review results from our own and other laboratories, and suggest some possible hypotheses to account for this remarkable program of structural and metabolic changes.

II. STRUCTURAL CHANGES

A. Cortical and Cytoplasmic Changes

The first observable changes following insemination are structural changes in the cortex involved in elevation of the fertilization membrane. This is a very complex process, involving the lysis of a peripheral ring of cortical granules, release of acidic sulfated mucopolysaccharides from these granules, and the subsequent elevation or unfolding of the fertilization membrane (reviewed by Runnström, 1966). Besides these surface changes, changes in "cytoplasmic granularity" also occur which can be seen *in vivo* in transparent eggs, or in histological sections of opaque eggs (Allen and Hagström, 1955).

1. *Kinetics of Structural Changes*

Since light-scattering can be a sensitive index of structural changes, the above structural changes might be monitored through measurement of the light-scattering of cell suspensions. Similarly, since the release of sulfated mucopolysaccharides from cortical granules results in increased acidity (Aketa, 1962; Ishihara, 1964), the time course of cortical granule lysis should be followable through simple measurement of the pH of the egg suspensions.

When such measurements are made, marked changes in both light-scattering and acidity are found within 1 minute after addition of sperm. Results of one such experiment are depicted in Fig. 1, and show that these two changes begin simultaneously at 45 seconds after sperm addition. As will be apparent later (Section VII), these two changes represent

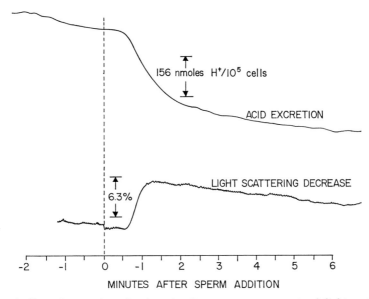

FIG. 1. Recorder tracing showing simultaneous measurement of light-scattering and pH changes following fertilization of eggs of *Strongylocentrotus purpuratus* at 17°C. Percentage light-scattering is a relative value, based on the total light-scattering of the egg suspension.

the earliest measurable response to fertilization, preceding changes in coenzyme content and respiratory rate by several seconds.

Although the pH change most probably results from the release of acidic mucopolysaccharides, the locus of the light-scattering change is not as easily defined. However, our experiments indicate that the light-scattering change does not result from the elevation of the fertilization membrane, since the identical scattering change occurs upon fertilization of eggs from which the membrane precursor has been removed by incubation in trypsin (D. Epel and B. C. Pressman, unpublished results).

Two other changes which could result in the observed alterations in light-scattering properties are the increases in "cytoplasmic granularity" and the lysis of the cortical granules. The relevance of the cytoplasmic changes is presently unclear, but two types of evidence suggest the cortical changes may be the major factor. First, the initiation of changes in light-scattering and pH are strikingly similar, suggesting that both result from the same event. If so, this event would be cortical granule lysis, since the pH change results from the release of acidic mucopolysaccharides from these granules. Secondly, a postfertilization light-scattering change can be seen in single cells under dark field microscope illumination (Runnström, 1928; Rothschild and Swann, 1949). This change is restricted to the cortex of the egg, suggesting that the light-scattering changes of cell suspensions may correspond to the cortical light-scattering changes of single cells.

2. Cortical Changes and Metabolic Activation

Cortical changes may be the first change of fertilization, but they do not appear to be prerequisites for the metabolic activation of the egg. Chase (1967; see also Whiteley and Chambers, 1961), in a series of elegant experiments, has shown that high hydrostatic pressure applied 20 seconds after sperm addition inhibits cortical granule breakdown, but does not affect sperm entry and subsequent development. These experiments, therefore, show that the cortical changes can be separated from those involved in activation of differentiation. They also suggest that the enzymes—or products of enzyme activity—which are involved in cortical granule breakdown do not subsequently activate or regulate other enzymes directly concerned with cell division and differentiation.

B. Subcellular Particles

Although the aforementioned cortical changes are not prerequisites of activation, structural changes in subcellular "particles" may be of paramount importance. The best evidence for such effects on metabolism comes from the work of Isono (Isono, 1963; Isono et al., 1963) who has studied the subcellular location of glucose-6-phosphate dehydrogenase. Before fertilization this enzyme is predominantly in a "pellet" fraction; after fertilization it is predominantly in the supernatant fraction. We have recently repeated and confirmed his findings, using eggs of Strongylocentrotus purpuratus.

Isono (1963) has found that the enzyme is released from the pellet fraction by either of its substrates, NADP or glucose 6-phosphate. Concentrations of both these compounds increase markedly after fertilization (see Sections IV,B, and V,B), suggesting that the translocation of this enzyme may result from an earlier increase in NADP or glucose 6-phosphate content.

III. PERMEABILITY CHANGES

A. Electrophysiological Evidence

Possibly related to the cortical and structural changes are marked increases in permeability and cation exchange. Thus, both exchange (Tyler and Monroy, 1959) and content (Hori, 1965) of sodium and potassium increase following fertilization. Temporally correlated with these changes in cations are changes in membrane potential (Tyler et al., 1956; Hiramoto, 1959), membrane resistance and capacitance (Hiramoto, 1959), and elevation of the fertilization membrane. Because ionic changes may be a critical factor in metabolic regulation (Kroeger and Lezzi, 1966), these changes may be of paramount importance in the metabolic activation of the egg.

B. Phosphate and Nucleoside Transport

Increases in rate of uptake of compounds directly utilized in macromolecular synthesis are also initiated by fertilization. Phosphate (Litchfield and Whiteley, 1959) and uridine (Piatigorsky and Whiteley, 1966) uptake, virtually undetectable in the unfertilized egg, begin 10–15 minutes after insemination. Transport of both of these compounds occurs against a concentration gradient, has some dependence on metabolic energy, and is apparently related to activation or synthesis of a surface-localized enzyme (Litchfield and Whiteley, 1959; Chambers and Whiteley, 1966; Whiteley and Chambers, 1966; Piatigorsky and Whiteley, 1966). The molecular mechanism of phosphate translocation is unclear (Griffiths and Whiteley, 1964), but uptake of uridine apparently involves a phosphorylation of this nucleoside (Piatigorsky and Whiteley, 1966).

C. Amino Acid Transport

Amino acid transport also increases after fertilization (Mitchison and Cummins, 1966; Gross and Fry, 1966), but unlike the transport of phosphate and nucleosides, an appreciable amount of uptake of amino acids also occurs in the *unfertilized* egg (Epel, 1967b, and unpublished data, 1967). This is indicated by the data in Table I, which shows the ratio between extracellular and intracellular leucine-^{14}C and valine-^{14}C after 5 minutes' incubation in the isotope. These data suggest "active" uptake since the exogenous amino acids are concentrated by the egg. The results also show that the transport system of the unfertilized egg, like that of the fertilized egg (Mitchison and Cummins, 1966), approaches saturation at micromolar levels. The mechanism of this uptake is unknown

TABLE I

UPTAKE OF ^{14}C-LABELED LEUCINE AND VALINE BY UNFERTILIZED EGGS[a]

Initial amino acid conc. $(10^{-6}\,M)$	Counts per minute (initial) in 0.014 ml seawater	Counts per minute (after 5 min) in 0.014 ml packed cells	Ratio intracellular/ extracellular	Uptake rate $(10^{-15}\,\text{mole}/ \text{min}/\text{cell})$
Leucine				
0.73	1,265	10,050	8.1	0.73
1.45	2,530	11,000	4.4	0.81
2.90	5,060	15,900	3.1	1.16
7.25	12,650	20,600	1.6	1.50
Valine				
0.53	816	4,500	5.5	0.30
1.06	1,632	5,900	3.6	0.43
2.12	3,264	9,000	2.8	0.66
5.25	8,160	10,230	1.3	0.72

[a] The packed cell volume (0.014 ml) is the volume of cells per milliliter of suspension used for these measurements. Uptake rate during the 5-minute period was linear, and at 5 minutes the extracellular isotope concentration was at least 88% of its initial level.

but does *not* involve any permanent chemical change in the amino acid moiety (Epel, 1967b).

D. KINETICS OF PERMEABILITY CHANGES

Since fertilization results in an increased rate of transport of such diverse compounds as sodium, phosphate, amino acids, and nucleosides, it is of interest to ascertain whether this permeability change occurs because of a generalized change in the cell membrane, or whether these permeability changes result from the individual activation of transporting enzymes specific for these compounds. No experiment specifically designed to answer this question has yet been done. The former hypothesis may have some merit, however, since increased rates of transport appear to begin between 10 and 20 minutes after insemination.

IV. CATION AND COENZYME CHANGES

A. CATION CONTENT

Changes in coenzyme and cation content are concomitants of fertilization and may be involved in controlling synthetic and enzymatic activity. For example, as the optimum potassium concentration for protein synthesis in sea urchin eggs is extremely high (Molinaro and Hultin, 1965), the postfertilization increase in K+ content (Hori, 1965) may be involved in regulating the rate of protein synthesis. An even more intriguing phenomenon is the release of bound Ca^{2+} after fertilization. Total Ca^{2+} does not change, but there occurs a redistribution between the free (dialyzable) and bound (nondialyzable) form of the cation (Mazia, 1937). This change could activate at least one enzyme, NAD kinase, as will be discussed later (Section VIII,C).

B. COENZYME CONVERSIONS

Krane and Crane (1960) observed that the NADPH content increased 6- to 10-fold after fertilization of sea urchin and pelycypod eggs. If unfertilized eggs were cultured in nicotinamide solutions, large amounts of NAD were formed, but there was no change in NADP or NADPH.

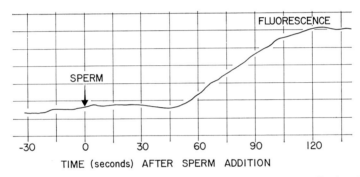

FIG. 2. Recorder tracing showing 366 mμ-induced fluorescence following fertilization of a suspension of eggs of *Strongylocentrotus purpuratus* at 17°C. Methods as described by Epel (1964a).

Following fertilization, however, this "extra NAD" was converted to NADP-NADPH (Krane and Crane, 1963). This strongly suggests that the mechanism of this change involves the activation, after fertilization, of the enzyme NAD kinase

$$NAD + ATP \xrightarrow[\text{cation}]{\text{divalent}} NADP + ADP \qquad (1)$$

Studies on the *in vivo* kinetics of these pyridine nucleotide changes indicate they are an early response to fertilization (Epel, 1964a). These changes can be followed by monitoring the 366 mμ-induced fluorescence of cell suspensions, since NADPH is fluorescent at this wavelength (Chance *et al.*, 1962). Such measurements, as depicted in Fig. 2, show that an increase in fluorescence begins at 45 seconds after insemination. This increase precedes the activation of respiration by 12–48 seconds, and coincides, or follows with a 3–5-second lag, the pH-light-scattering changes (Epel, 1964b, 1967a).

Chemical analyses of pyridine nucleotides at rapid intervals following insemination affirm the hypothesis of NAD kinase activation (Epel, 1964a). These studies show that the increase in product (NADP-NADPH) is stoichiometrically balanced by the decrease in reactants (NAD-NADH). This balance corresponds exactly to predictions derived from the proposed enzymatic reaction. Possible mechanisms for this remarkable activation will be discussed later (Section VIII,C).

Since NADPH is specific for biosynthetic reductions, the above alterations in pyridine nucleotides could profoundly modify metabolism. The redox potential ratio, NADPH/NADP, changes from unity in the unfertilized egg to 2–3 after fertilization (Epel, 1964a). As the direction and rate of NADPH-linked reductions is determined by this redox potential (Klingenberg and Bucher, 1960; Lowenstein, 1961), this change could also activate new synthetic pathways.

V. RESPIRATORY CHANGES

A. KINETICS

Historically, the postfertilization change first described was an increase in respiration rate. This 5-fold increase was first observed by Warburg in 1908 with the Winkler titration procedure and later corroborated with his newly developed "Warburg manometer" (Warburg, 1915). Subsequent manometric measurements by Laser and Rothschild (1939) indicated a transient postfertilization burst in respiration. This was later

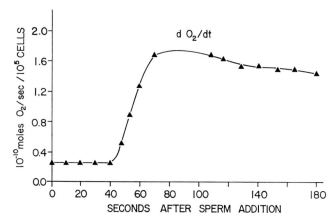

Fig. 3. Rate of respiration after fertilization of eggs of *Strongylocentrotus purpuratus* at 17°C. Respiration was determined polarographically with a Yellow Springs O_2 electrode, and rate was calculated by determination of slope.

affirmed by Ohnishi and Sugiyama (1963) and Epel (1964b), who measured rate directly with rapid-response O_2 electrodes. A typical determination, shown in Fig. 3, shows that the increase in rate begins at 60 seconds after insemination.

These respiratory changes always begin *after* the initiation of the structural and coenzyme changes. Significant lags between acid excretion

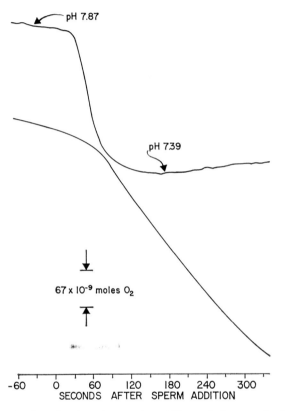

FIG. 4. Recorder tracing showing changes in acidity and decrease in O_2 tension following fertilization of eggs of *Lytechinus variegatus*. Temperature 29°C. Unpublished data of Epel and Iverson (see also Epel and Iverson, 1965).

(i.e., cortical granule lysis) and respiration have now been observed with eggs of five species of sea urchins (Ohnishi and Sugiyama, 1963; Epel, 1964b; Epel and Iverson, 1965; D. Epel, unpublished results). The lag is especially obvious in eggs of *Lytechinus variegatus*. As shown in Fig. 4, the pH changes begin at 18 seconds after insemination whereas respiratory activation is not apparent until 48 seconds later.

B. Mechanisms of Respiratory Activation

How the respiration rate is regulated in the unfertilized egg, and the mechanism of the respiratory activation after fertilization, is still far from clear. The most reasonable hypothesis, suggested by the control of mitochondrial respiration by ADP (Chance, 1959), is that the increased rate occurs in response to increased utilization of ATP and a

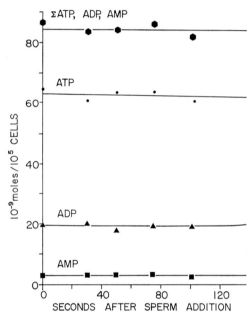

Fig. 5. Levels of adenine nucleotides following fertilization of eggs of *Strongylocentrotus purpuratus*. Samples of the egg suspension were added to 0.5 M perchloric acid at the indicated times, centrifuged, and the supernatants neutralized with K_2CO_3. The supernatants were then analyzed fluorimetrically by the procedure of Estabrook and Maitra (1962).

concomitant increase in ADP. This hypothesis can be tested by measurements of the ATP/ADP ratios during the periods of change in respiratory activity. Such measurements, shown in Fig. 5, indicate no changes in any of the adenine nucleotides during this period. This could mean that feedback control by ADP is not involved, or that local changes in ADP, undetectable by bulk analysis of adenine nucleotides, are the critical factor.

Alternate mechanisms have also been proposed. For example, respiratory substrate may be limiting in the unfertilized egg, and fertilization may activate pathways leading to breakdown of storage carbohydrate. Such a hypothesis has been proposed by Aketa *et al.* (1964), on the basis of their discovery that fertilization results in large increases in glucose 6-phosphate and other glycolytic intermediates (see also Epel and Iverson, 1965).

Activation of glycogen phosphorylase could account for the above increases. However, glucose 1-phosphate, the initial product of glycogen phosphorylase, is undetectable in fertilized eggs (Aketa *et al.*, 1964; D. Epel, unpublished results). Furthermore, there is no change in glycogen phosphorylase activity (D. Epel, unpublished results).

The aforementioned changes in the subcellular location of glucose-6-phosphate dehydrogenase (Section II,B) may also be critical in controlling respiratory rate. The binding of this enzyme to a particle could result in a block in carbohydrate flow through the pentose shunt (the primary pathway of glucose degradation in eggs). If so, translocation of this enzyme to the "soluble fraction" of the egg may result in large increases in pentose shunt metabolism and subsequent increases in respiratory rate (see also Isono, 1962; Isono and Ishida, 1962).

VI. CHANGES IN SYNTHESIS OF MACROMOLECULES

Many of the postfertilization changes, such as those in coenzyme and cation content, acquire additional interest because of their possible effects on control of rate and pattern of macromolecule synthesis. Some synthetic activities, such as DNA and RNA synthesis, are not even detectable until after fertilization. Other syntheses, such as protein and lipid synthesis, proceed in the unfertilized egg, but their rate increases after insemination.

A. PROTEIN SYNTHESIS

1. *Before Fertilization*

Rate of protein synthesis had previously been considered negligible in the unfertilized egg, and to be literally activated by fertilization (Nakano and Monroy, 1958; Hultin, 1961; Monroy and Vittorelli, 1962; Sofer *et al.*, 1966). Thus, comparison of incorporation rates before and after fertilization indicated relative differences as high as 300-fold. In view of the artifacts encountered in cell-free systems, it was perhaps not surprising that these differences were not reflected *in vitro*, where the incorporation rate of extracts from fertilized eggs was only 5-fold greater than from extracts of unfertilized eggs (Hultin, 1961; Hultin and Bergstrand, 1960; Nemer and Bard, 1963).

More recent studies, however, indicate that unfertilized eggs synthesize protein (Tyler *et al.*, 1966; Epel, 1967b; MacKintosh and Bell, 1967), and that the above discrepancies can be ascribed to differences in amino acid permeability between unfertilized and fertilized eggs (Epel, 1967b). Thus, if permeability changes are circumvented by preloading unfertilized eggs with ^{14}C-labeled amino acid and incorporation into hot-TCA-insoluble material determined before and after fertilization, the results shown in Fig. 6 are obtained. The data show that appreciable protein

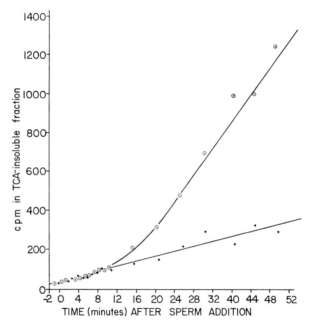

FIG. 6. Leucine-^{14}C incorporation in unfertilized (closed circles) and fertilized (open circles) eggs. Unfertilized eggs were preloaded with ^{14}C-labeled amino acid, the exogenous isotope was removed, and one portion of the eggs was fertilized. Incorporation into trichloroacetic acid-insoluble material was determined as described by Epel (1967b).

synthesis occurs in the unfertilized egg and that, in this particular experiment, the postfertilization increase is only 5-fold.

Several different lines of evidence indicate that the incorporation into TCA-insoluble material truly represents synthesis of protein. First, the TCA-insoluble material is released upon 16 hours' hydrolysis in 6 N HCl, and the ^{14}C-labeled amino acid can be recovered in an unmodified condition (Epel, 1967b). Second, greater than 90% of the TCA-insoluble material is rendered soluble in TCA upon incubation in pronase (D. Epel, unpublished results). Third, autoradiographic studies, done in collaboration with Dr. Meredith Gould, show that generalized cytoplasmic labeling occurs in all eggs of the population (Epel, 1967b). This latter result eliminates the possibility of bacterial contamination and also shows that the observed synthesis is a general property of all cells and is not restricted to a small proportion of immature oocytes or highly active eggs.

Other control experiments show that incorporation by unfertilized eggs is not affected by the experimental provision of a limiting amino acid. Thus, if cells are preloaded with small amounts of valine-^{14}C, washed

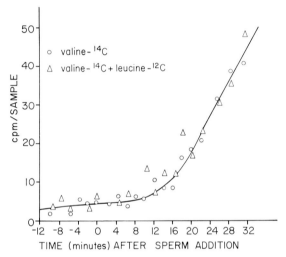

FIG. 7. Incorporation of limiting amounts of valine-^{14}C in the absence (circles) and presence (triangles) of large amounts of leucine-^{12}C Intracellular valine-^{14}C concentration was 1.3 nM/ml per milliliter of packed cells. Added intracellular leucine-^{12}C concentration was 114 nmole per milliliter of packed cells. Conditions as described by Epel (1967b).

free of this isotope, and then loaded with large amounts of leucine-^{12}C, the rate of valine-^{14}C incorporation is unaffected (Fig. 7).

2. After Fertilization

It is not yet known whether fertilization initiates synthesis of new types of protein, or whether there is, initially, an increased synthesis of the same proteins. Whichever the case, however, the insensitivity of early development to puromycin (Hultin, 1961; Wilt et al., 1967; Black et al., 1967) implies that the proteins made are nonessential until the time of the first mitosis.

The kinetics of protein synthesis after fertilization (Figs. 6 and 7) show that the rate does not increase immediately or within seconds after sperm addition. Rather, increased rate is a "late" response (compared to the events described earlier), which does not begin until 7–10 minutes after insemination. This lag provides strong evidence that activation of the egg occurs in a *sequential*, rather than a *simultaneous*, fashion.

B. NUCLEIC ACID SYNTHESIS

1. *DNA*

The initiation of new DNA and RNA synthesis represent other examples of "late" responses to fertilization. The elegant study of Hinegardner *et al.* (1964) has clearly shown that DNA synthesis does not begin until, or shortly after, pronuclear fusion (20–25 minutes in eggs of *S. purpuratus*). This S period is literally a frantic burst of activity, in which the DNA of 40 chromosomes is replicated in a period of 11 minutes.

2. *mRNA*

The timing of mRNA synthesis is more difficult to determine, largely because of the difficulty of detecting small amounts of synthesis in the face of extensive CCA turnover of tRNA and the possibility of bacterial contamination (Gross *et al.*, 1964; Glisin and Glisin, 1964). Recent studies of F. H. Wilt (personal communication), which circumvent the above difficulties, indicate that in eggs of *S. purpuratus* this synthesis does not begin until the fourth cell division.

C. LIPID SYNTHESIS

Rates of lipid synthesis, or at least rate of lipid interconversion, also increase after fertilization. There is a 5-fold increase in rate of acetate-^{14}C incorporation into the triglyceride-phospholipid fraction (Mohri, 1964). No data are presently available on the timing of this increased synthesis.

VII. THE PROGRAM OF FERTILIZATION CHANGES—1968

It is by now apparent that fertilization does not result in simultaneous activation or simultaneous increases in rate of enzymatic activity. Rather, there is a definite programmed sequence of temporally discrete events, which is very similar in eggs of different species of sea urchins. The present status of this sequence, as determined in eggs of *S. purpuratus* is shown in Fig. 8. The dotted lines indicate the estimated timing of changes whose exact temporal relationship to other changes is not yet known. For example, the exact time of initiation of phosphate transport is known, but experiments simultaneously comparing kinetics of

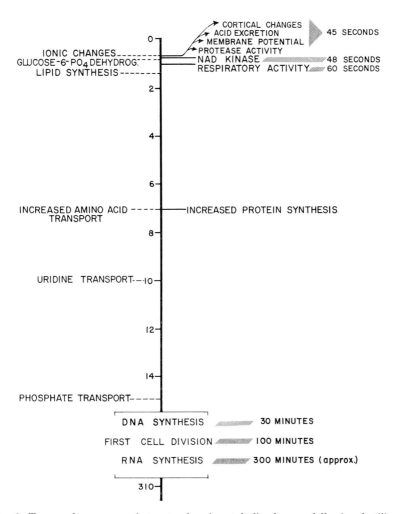

Fig. 8. Temporal sequence of structural and metabolic changes following fertilization of eggs of *Strongylocentrotus purpuratus* at 17°C. Established times are indicated by solid lines; estimated times, by dotted lines.

phosphate transport with other parameters (such as protein synthesis) have not been done.

VIII. INTERRELATIONSHIPS BETWEEN FERTILIZATION CHANGES

A. INFERENCES FROM KINETIC DATA

Although our present knowledge of the sequence permits a number of hypotheses to be made, it more importantly indicates which hypotheses no longer remain tenable. For example, the respiratory burst at 60 seconds after fertilization (Epel, 1964b) cannot result from increased energy utilization for peptide bond synthesis, since increased respiration begins 5 minutes before the onset of increased protein synthesis (Epel, 1967b). Similarly, NAD kinase activation precedes increased protein synthesis (Epel, 1964a; 1967b) indicating that new protein synthesis is not involved in this activation.

However, the temporal precedence of one reaction does not necessarily implicate it in regulating a later event. Thus, increased protein synthesis at 7 minutes has no relationship to initiation of DNA synthesis at 30 minutes, since DNA synthesis occurs in eggs fertilized and cultured in the presence of puromycin (Wilt et al., 1967; Black et al., 1967). Similarly, the cortical granule breakdown at 45 seconds is not required for metabolic activation since inhibition of granule lysis by high hydrostatic pressure has little effect on subsequent embryogenesis (Chase, 1967).

B. PROTEIN SYNTHESIS AND PROTEASES

Causal interrelationships surely do exist, however, and interaction between proteases and activation of protein synthesis may well become a classic example (Monroy et al., 1965; Mano, 1966). The proposed mechanism involves compartmentation changes via removal of a protein coat from mRNA complexes by the action of a protease which is activated at the time of membrane elevation. This mRNA complex may be synonymous with the heavy bodies described by Harris (1967; and this volume, Chapter 15). The removal of the protein of this complex allows interaction of all components of protein synthesis and is thus believed to be the controlling factor regulating protein synthesis after fertilization. This may indeed be the case, but the available data are not completely consistent with this time sequence since the proteases are apparently activated at the time of membrane elevation (Lundblad, 1954) and increased protein synthesis does not begin until 6–9 minutes later (Epel, 1967b).

C. NAD KINASE AND Ca^{2+}

Another possible interrelationship could involve the release of bound Ca^{2+} and the activation of NAD kinase. As noted, NAD kinase in vivo is inactive in the unfertilized egg and is activated after fertilization. The enzyme is equally active, however, in cell-free extracts prepared from fertilized or unfertilized eggs. Activation, then, occurs either upon cell disruption during homogenization or by provision of some substrate which is limiting in the unfertilized egg. However, both NAD and ATP are present in more than sufficient amount in the unfertilized egg. Mg^{2+}

or Ca^{2+} is required for activity (D. Epel, unpublished results) and since the free Ca^{2+} content increases after fertilization (see Section IV,A), this could be the factor controlling NAD kinase activity.

D. NAD KINASE AND GLUCOSE-6-PHOSPHATE DEHYDROGENASE

Another interrelationship may involve products of NAD kinase activity and glucose-6-phosphate dehydrogenase. As noted, the subcellular location of this enzyme changes, being found in the particle fraction in the unfertilized egg and in the soluble fraction in the fertilized egg (Isono *et al.*, 1963). *In vitro* experiments show that the substrates of

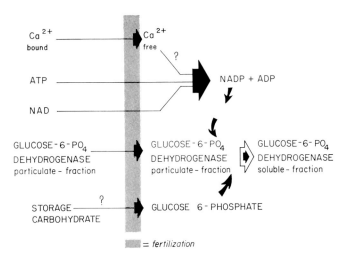

FIG. 9. Postulated reaction sequence resulting in activation of NAD kinase and translocation of glucose-6-phosphate dehydrogenase.

this enzyme, NADP and glucose 6-phosphate, can release the enzyme from the particle fraction (Isono, 1963). As concentrations of both of these substrates increase after fertilization (Epel, 1964a; Aketa *et al.*, 1964; Epel and Iverson, 1965), these could be the causal agents of the translocation. If this is the case, a chain-type reaction system is operative, as shown in Fig. 9, in which activation of one enzyme yields products causing changes in another enzyme.

IX. CONCLUDING REMARKS

The above overview of the fertilization reactions in the sea urchin egg points to a programmed response. As shown, various reactions occur sequentially rather than simultaneously and with a remarkably similar sequence in different species and genera. The major future problem will be to describe the mechanism of these reactions, and to ascertain if there are indeed causal relationships between them.

Although specific sets of metabolic responses to fertilization probably exist in eggs from other orders and phyla, it is not at all certain whether there is a set of universal responses. One would expect not, since fertilization can occur before meiosis (e.g., in pelecypod molluscs), during meiosis (as in some annelid eggs), or after meiosis is completed (as in sea urchin eggs). Supporting this concept of different types of metabolic responses are comparative data on respiratory changes after fertilization in the different phyla. These studies show that respiration can increase, remain unchanged, or even decrease (reviewed by Monroy, 1965). Unfortunately, little comparative data is available on the other metabolic responses, such as changes in ions, NADPH, or synthetic patterns. It is of interest in this connection, however, that NADPH has been found to increase in both sea urchin eggs (fertilized after meiosis) and in surf clam eggs (fertilized before meiosis) (Krane and Crane, 1960). Whether this is indicative of a general response to fertilization, or coincidental, will have to await further comparative investigations.

Programs involving transcription and translation of the genome have also been described for the cell cycle of *growing* cells (see e.g., Tauro *et al.* this volume, Chapter 5). Do these cell cycles also utilize programmed activation of preexisting enzymes? Such activation or control might be through allosteric mechanisms, through control of substrate levels, through changes in ions and hormones, or through compartmentational changes. Products of such controlled enzyme reactions may provide critical information to cytoplasm and nucleus for initiating new types of activity.

If such controls are found in the cell cycle, this need not imply that the cycle is independent of genetic control. Rather, it would suggest that some aspects of growth and division are dependent on earlier gene activity. Expression of these activities, as with the fertilization reactions in sea urchin eggs, would then be through the regulated activity of the preexisting enzyme molecule.

ACKNOWLEDGMENT

This work was supported by National Science Foundation Grant No. GB-4206.

REFERENCES

Aketa, K. (1962). *Embryologia* **7**, 223.
Aketa, K., Bianchetti, R., Marré, E., and Monroy, A. (1964). *Biochim. Biophys. Acta* **86**, 211.
Allen, R. D., and Hagström, B. (1955). *Exptl. Cell Res. Suppl.* **3**, 1.
Black, R. E., Baptist, E., and Piland, J. (1967). *Exptl. Cell Res.* **48**, 431.

Chambers, E. L., and Whiteley, A. H. (1966). *J. Cell. Physiol.* **68**, 289.
Chance, B. (1959). *Ciba Symp. Regulation Cell Metab.*, p. 91.
Chance, B., Cohen, P., Jobsis, F., and Schoener, B. (1962). *Science* **137**, 499.
Chase, D. (1967). Ph.D. thesis, University of Washington, Seattle, Washington.
Epel, D. (1964a). *Biochem. Biophys. Res. Commun.* **17**, 62.
Epel, D. (1964b). *Biochem. Biophys. Res. Commun.* **17**, 69.
Epel, D. (1967a). *In* "The Molecular Aspects of Biological Development" (R. A. Deering and M. Trask, eds.), pp. 17–34. NASA CR-673, Clearinghouse for Federal Scientific & Technical Information, Springfield, Virginia.
Epel, D. (1967b). *Proc. Natl. Acad. Sci. U.S.* **57**, 899.
Epel, D., and Iverson, R. M. (1965). *In* "Control of Energy Metabolism" (B. Chance, R. W. Estabrook, and J. R. Williamson, eds.), pp. 267–272. Academic Press, New York.
Estabrook, R. W., and Maitra, P. K. (1962). *Anal. Biochem.* **3**, 369.
Glisin, V. R., and Glisin, M. V. (1964). *Proc. Natl. Acad. Sci. U.S.* **52**, 1548.
Griffiths, W. M., and Whiteley, A. H. (1964). *Biol. Bull.* **126**, 69.
Gross, P. R., and Fry, B. J. (1966). *Science* **153**, 749.
Gross, P. R., Malkin, L. I., and Moyer, W. A. (1964). *Proc. Natl. Acad. Sci. U.S.* **51**, 407.
Harris, P. (1967). *Exptl. Cell Res.* **48**, 569.
Hinegardner, R. T., Rao, B., and Feldman, D. F. (1964). *Exptl. Cell Res.* **36**, 53.
Hiramoto, Y. (1959). *Exptl. Cell Res.* **16**, 421.
Hori, R. (1965). *Embryologia* **9**, 34.
Hultin, T. (1961). *Exptl. Cell Res.* **25**, 405.
Hultin, T., and Bergstrand, A., (1960). *Develop. Biol.* **2**, 61.
Ishihara, K. (1964). *Exptl. Cell Res.* **36**, 354.
Isono, N. (1962). *J. Fac. Sci. Univ. Tokyo Sect. IV*, **9**, 369.
Isono, N. (1963). *J. Fac. Sci. Univ. Tokyo Sect. IV*, **10**, 67.
Isono, N., and Ishida, J. (1962). *J. Fac. Sci. Univ. Tokyo Sect. IV*, **9**, 357.
Isono, N., Tsusaka, A., and Nakano, E. (1963). *J. Fac. Sci. Univ. Tokyo, Sect. IV*, **10**, 55.
Klingenberg, M., and Bucher, T. (1960). *Ann. Rev. Biochem.* **29**, 669.
Krane, S. M., and Crane, R. K. (1960). *Biochim. Biophys. Acta* **43**, 369.
Krane, S. M., and Crane, R. K. (1963). *In* "Control Mechanisms in Respiration and Fermentation" (B. Wright, ed.), p. 157. Ronald Press, New York.
Kroeger, H., and Lezzi, M. (1966). *Ann. Rev. Entomol.* **11**, 1.
Laser, H., and Rothschild, L. (1939). *Proc. Roy. Soc.* **B126**, 539.
Litchfield, J. B., and Whiteley, A. H. (1959). *Biol. Bull.* **117**, 133.
Lowenstein, J. M. (1961). *J. Theoret. Biol.* **1**, 98.
Lundblad, G. (1954). *Arkiv Kemi* **7**, 127.
MacKintosh, F. R., and Bell, E. (1967). *Biochem. Biophys. Res. Commun.* **27**, 425.
Mano, Y. (1966). *Biochem. Biophys. Res. Commun.* **25**, 216.
Mazia, D. (1937). *J. Cell. Comp. Physiol.* **10**, 291.
Mitchison, J. M., and Cummins, J. E. (1966). *J. Cell Sci.* **1**, 35.
Mohri, H. (1964). *Biol. Bull.* **126**, 440.
Molinaro, M., and Hultin, T. (1965). *Exptl. Cell Res.* **38**, 398.
Monroy, A. (1965). "Chemistry and Physiology of Fertilization." Holt, Rinehart, New York.
Monroy, A., and Vittorelli, M. (1962). *J. Cell. Comp. Physiol.* **60**, 285.

Monroy, A., Maggio, R., and Rinaldi, A. (1965). *Proc. Natl. Acad. Sci. U.S.* **54**, 107.

Nakano, E., and Monroy, A. (1958). *Exptl. Cell Res.* **14**, 236.

Nemer, M., and Bard, S. G. (1963). *Science* **140**, 664.

Ohnishi, T., and Sugiyama, M. (1963). *Embryologia* **8**, 79.

Piatigorsky, J., and Whiteley, A. H. (1966). *Biochim. Biophys. Acta* **108**, 404.

Rothschild, L., and Swann, M. M. (1949). *J. Exptl. Biol.* **26**, 164.

Runnström, J. (1928). *Protoplasma* **4**, 388.

Runnström, J. (1966). *Advan. Morphogenesis* **5**, 222.

Sofer, W. H., George, J. F., and Iverson, R. M. (1966). *Science* **153**, 1644.

Tyler, A., and Monroy, A. (1959). *J. Exptl. Zool.* **142**, 675.

Tyler, A., Monroy, A., Kao, C. Y., and Grundfest, H. (1956). *Biol. Bull.* **111**, 153.

Tyler, A., Piatigorsky, J., and Okazaki, H. (1966). *Biol. Bull.* **131**, 204.

Warburg, O. (1908). *Z. Physiol. Chem.* **57**, 1.

Warburg, O. (1915). *Arch. Ges. Physiol.* **160**, 324.

Whiteley, A. H., and Chambers, E. L. (1961). "Symposium on Germ Cells and Development," pp. 387–401. Inst. Intern. d'Embryologie and Fondazione A. Baselli, Pallanza.

Whiteley, A. H., and Chambers, E. L. (1966). *J. Cell. Physiol.* **68**, 309.

Wilt, F. H., Sakai, H., and Mazia, D. (1967). *J. Mol. Biol.* **27**, 1.

CHAPTER 14

Polysomes of Sea Urchins: Retention of Integrity

R. M. Iverson and Geraldine H. Cohen

I. INTRODUCTION

Much attention has focused on the biochemical aspects of sea urchin development. The organism is particularly suitable for the study of cell division (Mazia, 1961): upon fertilization of the eggs one obtains a large population of synchronously dividing cells. Our observations, however, have made us aware of a number of characteristics of the sea urchin as experimental material in the study of nucleic acid and protein synthesis during division and development, which we will discuss. As recent reviews (Gustafson, 1965; Løvtrup, 1966; Tyler and Tyler, 1966;

Monroy, 1967) on the biochemical studies of sea urchin development
have been published, we are presenting here primarily studies from this
laboratory.

Among the events that immediately follow fertilization (or artificial
activation) of the sea urchin egg is an increase in the rate of protein
synthesis (see reviews cited above). Insight into some of the mechanisms
for the control of protein synthesis during the ensuing cell divisions
and embryonic development may be gained through studies at the polysomal and translational levels. The validity of such an approach is
largely dictated by the degree to which certain obvious requirements
are successfully met: (1) the organisms should be cultured under growth
conditions conducive to normal synchronous division and development;
(2) the extraction procedure should preserve certain intrinsic structures
and functions which presumably relate as directly as possible to conditions inside the intact cell; and, finally, (3) the instrumentation and
analytical methods should be highly sensitive to the changes being
monitored.

II. ORGANISM AND CULTURE CONDITIONS

A. SELECTION OF OOCYTES

It is extremely important that the oocytes under study be uniformly
mature. We have found it necessary to examine the spawn from each
female to determine their experimental suitability. Immature oocytes
incorporate considerable amounts of labeled amino acids into hot trichloroacetic acid (TCA)-precipitable protein (Piatigorsky et al., 1967).
The presence of such cells among mature oocytes may thus lead to
incorrect conclusions in studies of events before or soon after fertilization.
Overripe (degenerating) eggs are discarded, particularly those from fe-
males collected at the end of their spawning season, as they are quite
fragile and easily cytolyzed. The oocytes of Lytechinus variegatus are
used within 2–3 hours from the time of shedding since they begin to
cytolyze and decline in fertilizability when held longer. Fertilizability
(greater than 95%) and synchrony of the cells at the first division
appear to be good indicators of the physiological health of the eggs.

B. GROWTH CONDITIONS

It has long been known that the respiration rate in the sea urchin
egg accelerates rapidly after fertilization. In other organisms, however,
there may be either no change, an increase, or a decrease in the respiration rate of the egg after fertilization (Monroy, 1967). This rapid metabolic enhancement in sea urchin zygotes may thus represent an event

secondary to the primary signal for the initiation and continuation of development. Our studies do indicate, however, that the incorporation of labeled amino acids into hot TCA-precipitable protein and the development of sea urchins zygotes (*L. variegatus*) is dependent upon the supply of oxygen. This is illustrated in Table I, which records results of studies on cultures of zygotes grown under three conditions of aeration: (1) not agitated at any time, (2) intermittently agitated, and (3) continuously agitated. The incorporation of a mixture of amino

TABLE I

EFFECTS OF VARIOUS CULTURE CONDITIONS ON THE INCORPORATION
OF LABELED AMINO ACIDS INTO PROTEIN[a]

Minutes after fertilization	Counts per minute per milligram protein		
	Not agitated	Agitated intermittently	Agitated continuously
10	7.3	45.0	46.6
20	16.7	91.6	112.8
25	10.6	164.2	184.5
27.5	23.4	217.4	50.2
30	13.4	92.2	65.2
35	11.8	87.4	98.2

[a] *Lytechinus variegatus* grown at 27°C. In the unagitated cells the streak stage is reached by 40% of the cells at 27.5 minutes, and by 45 minutes 30% of the cells commence the first division with abnormalities and cytolysis observable. In the intermittently agitated cells the streak stage is reached by 75% of the cells at 27.5 minutes and 40% have completed the first division by 45 minutes. In the continuously agitated cells 95% reach the streak stage by 27.5 minutes and 90% have completed the first division by 45 minutes. All cells were pulsed with 0.03 μc/ml of a mixture of L-amino acids-UL-[14]C (New England Nuclear Corp., No. 445) for 2.5 minutes at the times indicated. (Unfertilized cells incorporated 2.35 cpm per milligram of protein per 2.5-minute pulse.)

acids-UL-^{14}C into hot TCA-precipitable protein during a 2.5-minute pulse was followed as a function of time after fertilization. For maximal incorporation of amino acids into the hot TCA precipitate, a well-aerated sample or an intermittently (frequently) agitated sample surpasses the nonagitated culture. Under crowded growth conditions, such as in concentrated cultures not agitated frequently, highly aberrant sea urchin embryos arise at even the earliest cleavage stages (Sofer, 1966). With *L. variegatus* zygotes, however, agitation of the embryos in the first few minutes after insemination may partially rupture the newly forming fertilization membrane and so lead to anomalous forms of division and development.

Synchrony in the culture may easily be lost under the low aerobic (not agitated) conditions described above. In continuously aerated cultures, zygotes characteristically show a transient decrease in amino acid incorporation into protein during latter phases of the first synchronous division cycle (Sofer et al., 1966; Gross, 1967). But previous reports were conflicting (Gross and Fry, 1966; Timourian, 1966). This decline during division may not have been exhibited (Table I) if the aeration was inadequate, the cells asynchronous, or the overall rate of amino acid incorporation into protein very low.

Another possible source of conflicting results may be physiological differences between eggs obtained from sea urchins maintained for some time in the laboratory or by the commercial supplier, and eggs from recently collected animals (Harvey, 1956; Monroy and Maggio, 1966). Have changes occurred, for example, in the physiology of gametes obtained from sea urchins that have drawn on their gonads for nutrients during starvation (Giese, 1966a,b)?

To ensure adequate aeration, cultures of embryos are placed in brine shrimp hatcher bags No. 214 (Halvin Products Co., Brooklyn, New York) aerated by means of one or more fine air stones attached to an air pump. This system readily permits growth of the embryos in a reasonably synchronous manner to the prism stage at concentrations of about one volume of eggs to 50–100 volumes (2 liters total) of Millipore-filtered seawater. It is possible to use cultures of higher population densities for short-term developmental studies—radioactive tracer experiments, for example. Close attention must be paid, however, to the rate of aeration in order to prevent both the settling out of the embryos and excessive bubbling that may physically damage the eggs. Antibiotics (penicillin and streptomycin) added to the cultures inhibit bacterial growth during long-term development. The cultures may be kept at specified temperatures by suspending the bags in a constant-temperature water bath.

III. ANALYTICAL TOOLS

A. Continuous-Flow Optical Density Recording

One of the questions we have asked about sea urchin development is whether or not the unfertilized egg, which exhibits minimal incorporation of radioactive protein precursors into hot TCA precipitates, contains polysomes. For investigation of this problem we have used sucrose density gradient centrifugation of selected 15,000 g supernatants. It is critical to our evaluation that we know the consistency with which "blank" sucrose density gradients and flow artifacts (e.g., Schlieren patterns and absorbance by the sucrose) contribute to the overall absorption measurements. In this laboratory the gradients are analyzed with a Gilford multiple absorbance recorder (Model 2000) or with an ISCO density gradient fractionator (Model D). The gradients are scanned as quickly as possible upon completion of centrifugation. One obtains reasonably consistent tracings of blank sucrose gradients when monitored through the 1.0-cm (path length) flow cell on the ISCO unit (OD_{254}), as flow artifacts are virtually nonexistent in this system. A 1.0-cm flow cell

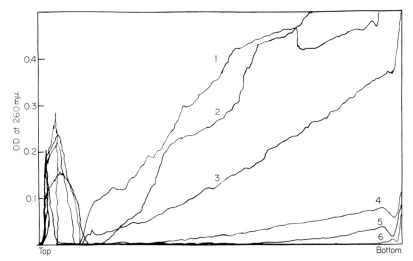

FIG. 1. Absorbance (OD_{260}) by uncentrifuged sucrose density gradients (14.6–50% sucrose, SW 25.1) layered with 0.2 ml of distilled water and scanned at a flow rate of 1.0 ml/min on a Gilford multiple sample absorbance recorder equipped with a Gilford 1.0-cm flow cell (tracings *1, 2, 3*) or with redesigned 1.0-cm flow cell (tracings *4, 5, 6*). The six gradients were made simultaneously.

distributed by Gilford is inadequate for our extended sucrose (15–50%) gradients (or for gradients made with a cushion of highly concentrated sucrose beneath the continuous gradient) when scanned from top to bottom at a reasonably fast flow rate (1.0 ml/min for SW 41 gradients and 1.0 or 1.9 ml/min for SW 25.1 gradients). Under these conditions, use of a flow cell supplied by Gilford produces blank sucrose gradient tracings such as those illustrated in Fig. 1, tracings *1, 2,* and *3*. A redesigned flow-cell spacer (1.0-cm path length) in which the inlet line has been placed at the bottom of the cell and the outlet line at the top, gives a maximum baseline shift of ±0.03 OD_{260} unit upon subtraction of the average optical density contribution of the blank sucrose density gradient (Fig. 1, tracings *4*, *5*, and *6*) from the experimental absorption profiles.

B. Sucrose Density Gradients

The sucrose gradients are made six at a time by a modification of a method recommended by Stafford and McCarty (1966) with a Technicon proportioning pump (Model 1). For 15–50% gradients a specified volume of buffered solution containing 50% sucrose (w/v) is placed in a mixing vessel with an outlet line that leads to six delivery lines. As sucrose is withdrawn from the mixing vessel (with continuous stirring by a magnetic bar) for delivery to the sucrose density gradient tubes, an equivalent volume of buffered medium lacking sucrose is introduced into the mixing vessel. The resulting gradients are slightly concave, which proves advantageous in that it is possible to resolve simultaneously both the ribosomal subunit particles and most of the heavier polysomes, depending upon the centrifugation conditions. Better resolution is obtained from freshly prepared and chilled SW 41 sucrose density gradients (15–50%) than from those stored in the refrigerator overnight.

IV. POLYSOME STUDIES

A. Choice of an Isolation Method

The choice of the initial isolation method is largely an arbitrary one. We have attempted to find an isolation method that (a) is rapid, (b) results in a high yield of ribosomal particles, (c) does not induce degradation of polysomes to smaller units, (d) yields reproducible results from one experiment to the next, (e) isolates polysomes that are functional under *in vitro* conditions, while (f) remaining susceptible to breakdown in the presence of low levels of added RNase, and finally, (g) does not itself cause aggregation or dissociation of ribosomal particles. Fulfilling criteria such as the above increases the likelihood that the analysis of polysomes during division and development directly relates to conditions within the intact cell.

B. Effects of Different Homogenization Media
on Isolation of Polysomes from Sea Urchins

The composition of the medium in which sea urchin embryos are homogenized has marked effects on the resulting polysomal sedimentation pattern. When the homogenization medium is extremely hypotonic with respect to the cells [0.01 M Tris-HCl, 0.01 M KCl, 0.0015 M MgCl$_2$, pH 7.8 at 25°C; designated RSB in Penman *et al.* (1963)], one consistently observes in sea urchin supernatants from late blastulae and gastrulae the predominance of the more slowly sedimenting polysomes (Cohen and Iverson, 1967a). The polysomes sedimenting in the gradient (for conditions see Fig. 2) from the ribosome (monomer) peak to two-thirds down the remainder of the tube are arbitrarily termed slow-sedimenting polysomes, while the polysomes in the lowest third are referred to as the fast-sedimenting polysomes. On the other hand, adjusting the ionic composition to near isotonicity (HERS buffer,* 0.01 M Tris-HCl, 0.43 M KCl, 0.018 M MgCl$_2$, pH 7.8 at 25°C) consistently results in the detection of a substantial increase, if not predominance, of faster-sedimenting polysomes from late blastula and gastrula embryos. HIS buffer* [0.01 M Tris-HCl, 0.24 M KCl, 0.01 M MgCl$_2$, pH 7.8 at 25°C (Stafford et al., 1964)], the tonicity of which is intermediate between those already cited, yields both types of profiles (Fig. 2C and 2E), as well as intermediate forms, but with no degree of predictability.

We have interpreted the isolation of predominantly slow-sedimenting polysomes from gastrulae under hypotonic buffer conditions as degradation of larger polysomes to smaller ones. This conclusion is supported by studies of optical density profiles and of radioactive amino acid incorporation into nascent proteins (Cohen and Iverson, 1967b). These show that when the isolations occur in hypotonic media (RSB or HIS buffer) and the slow-sedimenting polysomes predominate, radioactivity and OD$_{260}$ have shifted toward the slower-sedimenting polysomes and monosomes as compared to results from isolations in the near-isotonic HERS buffer. It is possible, but unlikely in view of the studies cited below, that the fast- and slow-sedimenting polysomes differ in their extractability under high- and low-ionic concentrations.

More evidence that polysomal breakdown has occurred in lower ionic strength media comes from the following experiments. The embryos are routinely washed in a hand centrifuge, aspirated, and 1–2 volumes of the desired buffer are added prior to homogenization (Cohen and Iverson, 1967a). Some degree of lysis is inevitable when the washing is carried out in hypotonic media. Unfertilized eggs are more likely to lyse than fertilized eggs, particularly when the RSB buffer is used. To eliminate the possibility of selective removal of the rapid-sedimenting polysomes during the wash procedure prior to homogenization, all the gastrula embryos were washed in HERS buffer. Half of these washed, packed embryos were homogenized in two volumes of HERS buffer (Fig. 2A) and the other half were homogenized in two volumes of a buffered medium that made the resulting ionic conditions comparable to homogenization in HIS buffer (Fig. 2E). The fast-sedi-

*Buffer HERS (higher-ionic strength) and buffer HIS (high-ionic strength) were referred to previously as buffer B and A, respectively (Cohen and Iverson, 1967a).

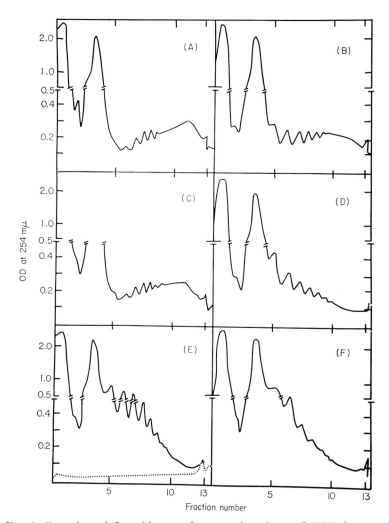

FIG. 2. Gastrulae of *Lytechinus variegatus* cultured at 26°–27°C for 12.5 hours were harvested and washed in HERS buffer (Cohen and Iverson, 1967a). Half of them were homogenized in two volumes of HERS buffer (A) and half in two volumes of a buffer that made conditions comparable to homogenization in HIS (E). In another series of similar experiments, the profile in (C) resulted when embryos were processed exactly as were those whose profile is shown in (E). A portion of the 15,000 *g* supernatants were also stored at refrigerator temperatures (3°–5°C) and then analyzed on sucrose density gradients, as follows: Supernatant shown in (A) after storage for 10 days resulted in (B); (C) stored 8 days resulted in (D); and (E) stored 10 days resulted in (F).

All operations from the time of washings were carried out at 0°–4°C. Four-tenths milliliter of the supernatants were layered onto 15–50% sucrose gradients in Beckman SW 41 tubes and centrifuged at 39,000 rpm for 80 minutes in a Spinco Model L–2 preparative centrifuge. Gradients were scanned in duplicate at full scales of menting polysomal units predominated in the former isolation (HERS buffer) while the slow-sedimenting polysomes predominated in the latter (Fig. 2E). Embryos were thus processed under identical conditions other than the ionic composition of the final homogenization medium, and exactly the same volume of the resulting 15,000 g supernatant was layered onto the gradients. The data (viz., Fig. 2A and 2E) clearly show that in isolations employing hypotonic buffers such as HIS buffer

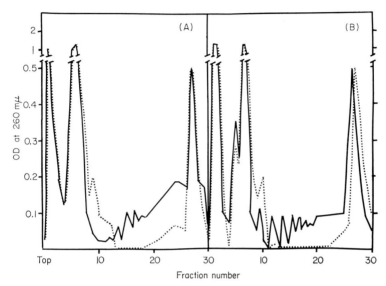

FIG. 3. Profiles (OD_{260}) of SW 41 sucrose density gradients scanned on Gilford unit equipped with redesigned 1.0-cm flow cell. Gradient and centrifugation conditions are described in Fig. 2. Supernatants (15,000 g) from embryos homogenized in HERS buffer were analyzed on day 0 (panel A) and after 10 days' storage (B). Dotted line represents profile (OD_{260}) of the same preparations, but treated for 30 minutes at 0°C and 5 μg of RNase per milliliter before layering onto the gradients.

there is a loss of material from the bottom regions of the gradient and an accumulation of material toward the top of the gradient. Homogenization in HIS buffer (Fig. 2C) sometimes results in sedimentation profiles indistinguishable from those obtained after homogenization in HERS (Fig. 2A). Upon storage in the refrigerator (3–5°C) for 8 days, however,

---

0.5 and 2.5 OD_{254} units with an ISCO Model D fractionator equipped with a 1.0-cm flow cell. The gradient in (C) was scanned only at a full scale of 0.5 OD_{254} unit. Dotted line represents "blank" absorbance from gradients layered with 0.4 ml of HERS buffer and analyzed after scanner unit was zeroed on distilled water.

most of the fast-sedimenting polysomes in the HIS supernatants break down, and smaller units accumulate toward the monosome peak (compare Fig. 2D to 2C and 2F to 2E). Interestingly, a comparable 15,000 g supernatant resulting from homogenization in HERS buffer shows only a small degree of breakdown after 8 or 10 days' storage in a refrigerator (compare Fig. 2B to 2A and Fig. 3B to 3A).

Interpretation of results from experiments in which the embryos are first washed in lower ionic strength media (RSB and HIS buffer), but not homogenized until after the ionic composition of the suspending media has been adjusted to that of the HERS buffer, is complicated by lysis of some of the cells when suspended in the low ionic strength media. In these cases, as well as in those where the 15,000 g supernatants from the gastrulae are homogenized in hypotonic RSB or HIS buffer and then adjusted to an ionic composition similar to HERS, the sedimentation profiles did not differ from those of embryos retained in the hypotonic media. We are now testing the possibility that exposure of the embryonic cells to hypotonic media brings about the release or decompartmentation of factor(s) (RNase?) leading to breakdown of the polysomes.

C. Preferred Method of Polysome Isolation

How well does the isolation method using HERS buffer conform to the criteria presented in Section IV,A? (1) The method is rapid. (2) The fact that 0.5% DOC (final conc.) added to the crude homogenates or the 15,000 g supernatants increases the yield of polysomes slightly or not at all over that obtained from HERS shows that recovery is high. (3) As discussed above, degradation of polysomes appears to be minimal. RNase inhibitors such as bentonite do not appear to be required when HERS is used, and the stability of the polysomes is not affected by as many as 40 passes of a fast-moving Teflon pestle in a Duall tube during homogenization. This stability is fortunate, for the embryonic cells retain their integrity so well in HERS buffer that many more homogenization strokes must be applied to rupture all of them. (4) The resulting sedimentation patterns are highly reproducible from one experiment to another. (5) The resulting 15,000 g supernatants, after the $MgCl_2$ concentration has been adjusted, are highly capable of incorporating amino acids into protein (Hudson, 1967). (6) The polysomes are sensitive to RNase (Fig. 3A and 3B) (Cohen and Iverson, 1967a).

D. Evidence for Polysomal Aggregates

It is unlikely that the ionic concentrations used in this study induce the aggregation of ribosomal particles not dependent on mRNA. We have frequently recovered the fast-sedimenting polysomes in the presence of high KCl (0.43 M) but low $MgCl_2$ (0.0015 M), but never with low concentrations of KCl (0.01 M) and increased $MgCl_2$ (0.01 M). The fast-sedimenting polysomes have been infrequently obtained (Cohen and Iverson, 1967b) in relatively low ionic strength buffers (0.01 M Tris-HCl, 0.05 M KCl, 0.01 M $MgCl_2$, pH 7.8). From preliminary studies it appears that the ribosomal subunits recovered in the buffers tested are quanti-

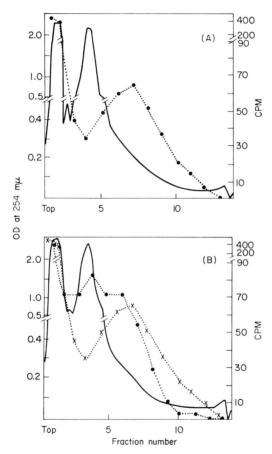

FIG. 4. Profiles (OD_{254}) of supernatants (15,000 g) from unfertilized eggs labeled *in vivo* for 30 minutes with L-phenylalanine-UL-^{14}C at 24°C and scanned with ISCO unit (A). Ribonuclease treatment (B) was with RNase at 5 μg/ml for 60 minues at 0°C. Solid lines represent OD_{254}; dashed lines, cpm. In (B), the cpm profile from (A) (x···x···x) has been included for comparison. Eggs were homogenized in HERS buffer and centrifuged for 70 minutes on 17–50% sucrose gradients. See legend of Fig. 2 for other details.

tatively comparable. Previously mentioned was the lack of an effect on the polysomal profile when the ionic concentration of 15,000 g supernatants obtained from embryos homogenized in RSB or HIS buffer was increased to that of the HERS. The polysomes prepared in HERS are sensitive to added RNase (Fig. 3A and 3B). The possibility of selective attachment of smaller polysomes to membranes or to DNA is eliminated in experiments with DOC or with DNase, for their addition produces

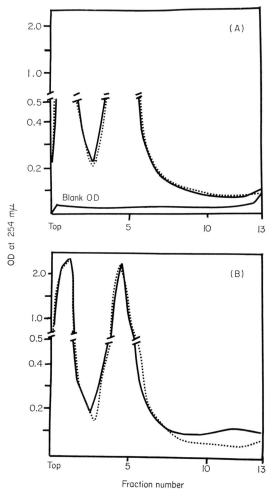

Fig. 5. Profiles (OD$_{254}$, ISCO) of untreated (———) and ribonuclease-treated (------) (5 μg/ml for 30 minutes at 0°C) 15,000 g supernatants from unfertilized eggs (A) and from the same eggs 15 minutes after fertilization (B). See legend of Fig. 2 for other details.

no observable changes in the sedimentation profiles (Cohen and Iverson, 1967a).

In spite of the abundance of single ribosomes (Fig. 4A and 4B), we have not detected discrete OD_{254} (OD_{260}) peaks sensitive to RNase in the "polysomal" region of sucrose gradients layered with supernatants from unfertilized eggs. We have, however, detected a shift of radioactivity from the lower region of the gradient toward the ribosome peak (fractions no. 3 and 4) of the gradient following RNase treatment of the 15,000 g supernatants from unfertilized eggs (cf. Fig. 4B to 4A). It should be noted that the time during which the unfertilized eggs were exposed to the isotope in this experiment is greater than we normally use for fertilized eggs or embryos. No immature oocytes, extraneous acellular material, or elevated fertilization membranes were observed among the eggs used. In contrast, UV-absorbing material removable by RNase is readily detected in the lower region of the gradient within 15 minutes after fertilization. This is illustrated in Fig. 5B where the 15-minute fertilized eggs had been initially withdrawn from the same pool of unfertilized eggs shown in Fig. 5A.

It should be noted that depending upon how heavily one chooses to load the gradient with 15,000 g supernatants from eggs, a small amount of ultraviolet-absorbing material, insensitive to RNase, sediments in the lower ("polysomal") regions of the gradient. The identity of this material is not yet known, but electron micrographs (Candelas, 1966; Stavy and Gross, 1967; Cayer and Iverson, 1967) show possible nonpolysomal material from sea urchin eggs sedimenting in the "polysomal region" of the sucrose density gradient.

V. CONCLUDING REMARKS

Particularly striking in this study is the effect of the homogenization medium on polysomal integrity. The ability to consistently obtain intact, presumably undegraded, polysomes appears to be dependent upon the experimental conditions, namely, the nearness to isotonicity of the homogenization medium. We do not know whether the increased potassium concentration functions solely as an osmotic factor, nor do we know whether the requirement for the monovalent species of ion is specific for potassium. A specific requirement for unusually high levels of potassium (0.25–0.40 M), apparently not just osmotic in nature, has been established for *in vitro* protein synthesis on sea urchin ribosomes-polysomes by Molinaro and Hultin (1965). This work on the *in vitro* synthesis of proteins, however, differs from ours in that the potassium was added to their low speed supernatants only *after* the initial homogenization took place in the absence of added potassium ions.

We know that after fertilization of sea urchin eggs a number of ionic changes occur, among which is an increase in potassium at 10 minutes after fertilization (Hori, 1965), an increase in the ratio of influx to efflux of potassium after insemination (Hori, 1965), and an increase in the fraction of potassium which is exchangeable following fertilization (Chambers and Chambers, 1949; Tyler and Monroy, 1959). It has been suggested that changes in the levels of exchangeable ions may affect the compartmentation of enzymes and other proteins within the cell (Monroy, 1967). In our case, increased potassium concentrations may in a number of possible ways prevent the action of RNase, and thereby preserve the polysomes *in vitro*.

Compared to the effects of increased potassium levels, the magnesium ion concentration appears to play a lesser role in the maintenance of polysomal integrity in cell-free systems. High magnesium concentrations, however, may contribute to polysomal integrity by inhibiting RNase activity or by impeding protein synthesis and hence preventing ribosomes from "running off" or being released from the mRNA.

We wish to emphasize that the true ionic composition of our 15,000 *g* supernatants from sea urchin embryos is undetermined. It is defined only in terms of the homogenization procedure employed. Dialysis of the supernatants against a specific buffer is thus a related but different line of study. The time required for dialysis necessarily entails "aging" of the preparation. It would be difficult also to compare polysomes isolated in homogenization media made isoosmotic primarily by the addition of sucrose to those isolated in media made near-isotonic by virtue of the ionic composition alone. The sucrose in the supernatants would have to be eliminated through dialysis (or considerable dilution) before layering the supernatants onto the gradient. Resuspension of high-speed ribosomal pellets into defined media is a questionable procedure since some of our findings (Cohen and Iverson, 1967b) indicate that one is likely to be studying "resuspendibility" of the pellets into the various media.

The HERS buffer has been successfully employed for the isolation of intact polysomes from three other species of sea urchins, *Arbacia punctulata*, *Trypneustes esculentes*, and *Echinometra lucunter*. In the case of *A. punctulata*, the large amount of pigment contained within the zygote did not noticeably interfere with the sedimentation analysis (Iverson, 1967).

ACKNOWLEDGMENT

The work presented in this paper has been supported by research and traineeship (G. H. C.) funds from the National Science Foundation, National Institutes of Health (Career Development Award GM 6809 to R.M.I.) and American Cancer Society. The assistance of Mr. E. E. Reeves and Mr. D. Young is gratefully acknowledged.

REFERENCES

Candelas, G. C. (1966). Ph.D. Thesis, University of Miami, Coral Gables, Florida.
Cayer, M. L., and Iverson, R. M. (1967). Unpublished observations.
Chambers, E. L., and Chambers, R. (1949). *Am. Naturalist* **83**, 269–284.
Cohen, G. H., and Iverson, R. M. (1967a). *Biochem. Biophys. Res. Commun.* **29**, 349–355.
Cohen, G. H., and Iverson, R. M. (1967b). Unpublished.
Giese, A. C. (1966a). *In* "Physiology of Echinodermata" (R. A. Boolootian, ed.), pp. 757–796. Wiley (Interscience), New York.
Giese, A. C. (1966b). *Physiol. Rev.* **46**, 244–298.
Gross, P. R. (1967). *In* "Current Topics in Developmental Biology" (A. A. Moscona and A. Monroy, eds.), Vol. II, pp. 1–46. Academic Press, New York.
Gross, P. R., and Fry, B. J. (1966). *Science* **153**, 749.
Gustafson, T. (1965). *In* "The Biochemistry of Animal Development" (R. Weber, ed.), Vol. I, pp. 140–202. Academic Press, New York.
Harvey, E. B. (1956). "The American Arbacia and Other Sea Urchins," pp. 52–55. Princeton Univ. Press, Princeton, New Jersey.
Hori, R. (1965). *Embryologia* **9**, 34–39.
Hudson, J. (1967). Personal communication.
Iverson, R. M. (1967). Unpublished observations.
Løvtrup, S. (1966). *Bull. Acad. Suisse Sci. Med.* **22**, 201–276.
Mazia, D. (1961). *In* "The Cell" (J. Brachet and A. E. Mirsky, eds.), Vol. III, pp. 80–412. Academic Press, New York.
Molinaro, M., and Hultin, T. (1965). *Exptl. Cell Res.* **38**, 398–411.
Monroy, A. (1967). *In* "Comprehensive Biochemistry" (M. Florkin and E. H. Stotz, eds.), Vol. 28, pp. 1–22. Elsevier, Amsterdam.
Monroy, A., and Maggio, R. (1966). *In* "Physiology of Echinodermata" (R. A. Boolootian, ed.), pp. 743–756. Wiley (Interscience), New York.
Penman, S., Scherrer, K., Becker, Y., and Darnell, J. (1963). *Proc. Natl. Acad. Sci. U.S.* **49**, 654.
Piatigorsky, J., Ozahi, H., and Tyler, A. (1967). *Develop. Biol.* **15**, 1.
Sofer, W. H. (1966). Personal communication.
Sofer, W. H., George, J. F., and Iverson, R. M. (1966). *Science* **153**, 1644.
Stafford, D. W., and McCarty, K. S. (1966). Personal communication.
Stafford, D. W., Sofer, W. H., and Iverson, R. M. (1964). *Proc. Natl. Acad. Sci. U.S.* **52**, 313.
Stavy, L., and Gross, P. R. (1967). *Proc. Natl. Acad. Sci. U.S.* **57**, 735.
Timourian, H. (1966). *Science* **154**, 1055.
Tyler, A., and Monroy, A. (1959). *J. Exptl. Zool.* **142**, 675–690.
Tyler, A., and Tyler, B. S. (1966). *In* "Physiology of Echinodermata" (R. A. Boolootian, ed.), pp. 683–742. Wiley (Interscience), New York.

CHAPTER 15

Relation of Fine Structure to Biochemical Changes in Developing Sea Urchin Eggs and Zygotes

Patricia Harris

I. INTRODUCTION

The synchronous early cleavage divisions of marine invertebrate eggs, especially those of echinoderms, provide excellent material for biochemical studies of cell division. Egg cells, however, are destined for more than a single division; they contain the information and much of the substance necessary for the development of an embryo, and it is not easy to distinguish those processes unique to mitosis from those related to later development. The first cleavage in the sea urchin egg, which has been overworked as a model for mitosis, is complicated by the fertilization process itself, which prior to any visible changes recognized as part of the division process, involves immediate cytoplasmic changes as well as the migration and fusion of the pronuclei.

At the chemical level there is an activation of metabolic processes, among these the stimulation of protein synthesis, and here we may ask whether this protein synthesis is related to the early cleavage divisions or whether it is destined for some later use in cell differentiation. There is evidence that preformed spindle protein is already present and in some excess in the unfertilized sea urchin egg (Went, 1959; Kane, 1965, 1967), yet there is also evidence that there is some early incorporation of amino acids in spindle proteins prior to first cleavage (Gross and Cousineau, 1963b; Stafford and Iverson, 1964; Mangan *et al.*, 1965). Hultin (1961a) and later Wilt *et al.* (1967), using puromycin, have demonstrated that without some protein synthesis, and specifically that which takes place between fertilization and pronuclear fusion, the cell cannot divide.

A morphologist's approach to the problems of cell division and cell differentiation is to ask what structural changes take place in the cell which can be related to concurrent biochemical changes. We have a voluminous literature from light microscopy to draw from, for indeed, it was the egg of the sea urchin, as well as that of *Ascaris megalocephala* which provided during the latter part of the 19th century the foundation for our knowledge of maturation, fertilization, and to a large degree, of mitosis. The classic compilation and review by Wilson (1925) of the work carried out during this "golden age" of light microscopy stands out as a monumental source book of observations and ideas that need reexamination with modern techniques now at our disposal. In the relatively few years that electron microscopy has been able to contribute significantly to our knowledge of cell ultrastructure, it has been most valuable primarily in those areas where structure could be related through biochemistry to function. Our increasing knowledge of the metabolic roles of various cell organelles allows us to interpret and to predict with some certainty from studies of general morphology the functions which a cell may be carrying out at any particular time, and these predictions, in turn, can be tested again at the biochemical level. In this way, a continuing dialogue between morphology and biochemistry has proved to be the most fruitful approach to our understanding of the living cell.

Thus, in regard to the question of early protein synthesis in the sea urchin embryo we are guided by our present knowledge of the roles of DNA, messenger RNA, and ribosomal RNA in this process, and we might naturally look for any suggestion of the movement of RNA from the nucleus to the cytoplasm. Early work by Afzelius (1957) on the basophilic structures of the sea urchin egg describes aggregates of granules identified as RNA, bounded by annulate lamellae, scattered in the cytoplasm, and also shows a close association of these structures with the nuclear membrane, as though formed at this site. Shortly afterward Merriam (1959) carried out a detailed study of the annulate lamellae in maturing sand dollar eggs and proposed that the "heavy bodies" of Afzelius, which are also present in this species, may contain the specificity necessary for synthesis of proteins in the cytoplasm. Subsequent biochemical studies concerning the site of the protein synthetic block seemed to preclude this conjecture, however, and the idea lay dormant for some time.

In recent years there has accumulated evidence that all the components of a protein-synthesizing system are present in the unfertilized mature sea urchin egg, yet these components remain for the most part inactive. Several recent reviews cover this literature in some detail (Grant, 1965; Monroy, 1965; Rünnstrom, 1966; Spirin, 1966). One of several hypotheses is that the messenger RNA is somehow "masked," presumably by protein, and that the "unmasking" is accomplished by the fertilization reaction (Monroy et al., 1965). Another explanation is that the ribosomes for some reason are unable to combine with messenger RNA and thus translate the maternal message (Candelas and Iverson, 1966; Stavy and Gross, 1967). A recent electron microscope study of sea urchin oogenesis has again brought up the question of the site of the stored messenger RNA, and the authors suggest that the annulate lamellae may play some role (Verhey and Moyer, 1967). In cell fractionation studies, the endogenous synthetic activity has invariably been found to reside in the heavy particulate fractions or "microsomes" (see reviews above). In the most recent work, Stavy and Gross (1967) have shown this fraction to contain ribosomal clusters, membranes, and various small particles, one of which they suspect may contain the messenger RNA. The biochemical data, however, do not at all preclude the possibility that mRNA is already bound to ribosomes, but in a defective way, perhaps spatially separated by protein, and for this reason the ribosomal and membrane constituents of the microsome fraction may have considerable significance.

From the morphologist's point of view, the cell structures that would be most likely to contain the messenger RNA are the heavy bodies, which contain RNA granules of ribosomal size. These granular bodies bounded by annulate membranes are apparently derived from the nucleus, and are unaccounted for in the biochemical preparations. Is it not possible that the microsome fraction may in fact contain fragmented heavy bodies? With this question in mind, and in the spirit of "picking up the dialogue" once more from the morphological point of view, the following studies were carried out to investigate in greater detail the nature of the heavy bodies and other nuclear membrane-associated structures. Some interpretations are offered, albeit farfetched, that might be helpful in the understanding of the current biochemical literature on this subject.

II. NUCLEAR-CYTOPLASMIC INTERACTION IN EARLY SEA URCHIN DEVELOPMENT

A. FORMATION AND BREAKDOWN OF "HEAVY BODIES"

Heavy bodies are structures characteristic of the mature unfertilized sea urchin egg, although occasionally they have been reported in immature oocytes. First described and named by Afzelius (1957) in unfertilized eggs of several species of European sea urchins, they appear as densely packed aggregates of ribosome-like granules bounded on several sides, but not completely enclosed by annulate lamellae. They are scattered widely throughout the cytoplasm and are often found attached to the nuclear membrane, suggesting at first the possibility that they are somehow extruded from the nucleus. By centrifuging whole cells, Afzelius found these rather angular structures situated in the most centrifugal part of the stratified cell; hence, the name "heavy bodies." Histochemical staining further indicated that these particles contained RNA. Heavy bodies have since been described in a number of other species of sea urchin (Merriam, 1959; Gross et al., 1960; Verhey and Moyer, 1967; Harris, 1967).

The urchins used in the studies described here were *Strongylocentrotus purpuratus* collected from the Oregon coast near Newport. Freshly shed eggs, obtained between January and March at the peak of the spawning season by injecting the urchins with 0.5 M KCl, showed great numbers of heavy bodies in the cytoplasm. The average number per egg is difficult to determine, for such a statistical study is a formidable one for electron microscopy. However, an estimate based on counts from median sections of several eggs showed the number to be roughly 1000–1500 per cell, about the same as that reported by Afzelius in *Echinus esculentus* and *Psammechinus miliaris*.

Many of these dense granular aggregates are found associated with the surface of the nuclear membrane, bounded on one side by the nuclear membrane and elsewhere by fragments of annulate lamellae similar in structure to the nuclear membrane (Fig. 1). The amount of annulate membrane associated with the heavy bodies varies from egg to egg, and occasionally granular aggregates may be seen without such membranes (Fig. 2).

The mode of formation of the heavy bodies still remains obscure. The granular minor nucleoli described by Afzelius in the germinal vesicles of the oocytes of several sea urchins, and by Merriam in the sand dollar oocyte nuclei, were noted to be very similar in appearance to the granular heavy bodies. Both authors have considered the possibility that this basophilic nucleolar material is somehow transferred from the nucleus to the cytoplasm. Afzelius shows aggregates located in outpocketings of the nuclear membrane or in the process of delamination from the germinal vesicle. Merriam, on the other hand, finds in *Dendraster* that these particles become associated with the annulate lamellae only after they have moved into the cytoplasm. It may be, since the heavy bodies are not usually seen free in the cytoplasm until after the maturation divisions (Afzelius, 1957; Merriam, 1959; Harris, 1967), that the minor nucleoli are released into the cytoplasm at the time of the germinal vesicle breakdown and acquire annulate membranes in the same manner as telophase chromosomes recruit nuclear membranes from cytoplasmic vesicles (Barer *et al.*, 1960; Harris, 1961).

This mode of formation, would not account for their occasional appearance before the maturation divisions, nor would it explain the continued association of granular aggregates with the outside of the nuclear membrane in the mature egg. It has been suggested that the association of the granular masses with annulate lamellae occurs in the cytoplasm as an accretion of granules between pairs of existing lamellae (Verhey and Moyer, 1967) or that the 150 Å particles of the heavy bodies are synthesized by the annulate membrane (Merriam, 1959) or are formed by delamination of membranes and nuclear material at the nuclear surface (Afzelius, 1957). Whatever the actual mode of formation may be, the timing of heavy body formation in *S. purpuratus* appears to extend from the time of the maturation divisions for an indefinite period thereafter, if their association with the nuclear membrane can be interpreted as a stage in their formation. This association of heavy bodies with the nuclear membrane seems to be unaffected by the fertilization process, for they are found attached to the nucleus in apparently undiminished numbers (five to ten at one time) up to the time of pronuclear fusion. After fusion of the pronuclei this association is no longer seen, but the heavy bodies remain widely scattered throughout the cytoplasm.

During the period between pronuclear fusion and the earliest manifestations of the first division, there is no apparent change in the number

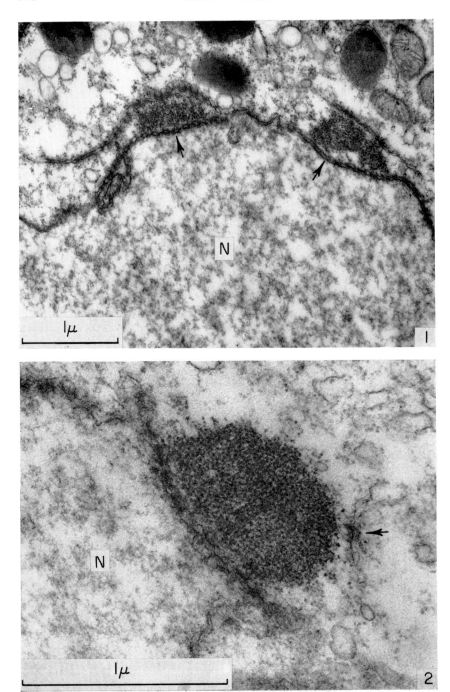

or distribution of the heavy bodies, although they appear to become less compact (Fig. 3). About 60 minutes after fertilization, a clear streak appears in the cytoplasm, extending on either side of the nucleus, and heavy bodies in this region are aligned parallel to it, but do not seem to play an active physical role in its formation.

As the asters begin to grow and the microtubular elements of the asters impinge on the nucleus, the nuclear membrane itself becomes quite irregular in profile and in many places the inner and outer membranes begin to separate to form large blebs. At the same time there is a corresponding change in appearance of the annulate lamellae of the heavy bodies. The paired membranes separate to form blebs or vesicles indistinguishable from the other vesicles and membranes present in the cytoplasm. Often these lamellae appear to break up into smaller fragments, the annuli become widely separated and may become associated with arrays of ribosomes (Fig. 4).

Heavy bodies present in the channels of the growing astral rays appear to contribute their annulate membranes to the massive membranous portion of the mitotic figure, while the granular aggregate breaks up and is indistinguishable from the great mass of ribosomes present in the cytoplasm. No heavy bodies were ever seen in the cytoplasm after the breakdown of the nuclear membrane in preparation for first cleavage, in this material about 70–75 minutes after fertilization.

B. INTRANUCLEAR CHANGES

The resting nucleus of the mature sea urchin egg has long been known for its homogeneous appearance in the light microscope and its apprently negative reaction to the Feulgen stain for DNA. In fact, this inability to demonstrate DNA with the Feulgen method in the egg nucleus led to the controversial proposal by Marshak and Marshak (1953) that indeed there was no DNA in the nucleus. Various investigators soon demonstrated that the nucleus did give a positive reaction, but that the stained chromatin was located around the inner surface of the nuclear membrane (Burgos, 1955; Brachet and Ficq, 1956; Agrell, 1959). Agrell's work especially demonstrated without doubt the location of condensed

FIG. 1. Heavy bodies (arrows) at the nuclear surface of an egg 15 minutes after fertilization. Note the extensive annulate membrane. ×26,000.

FIG. 2. A heavy body at the nuclear surface of an unfertilized egg. The amount of annulate membrane varies from cell to cell. In this case there is only an elongated vesicle with a single pore (arrow) in this section. Serial sections revealed only two other pores. ×55,000.

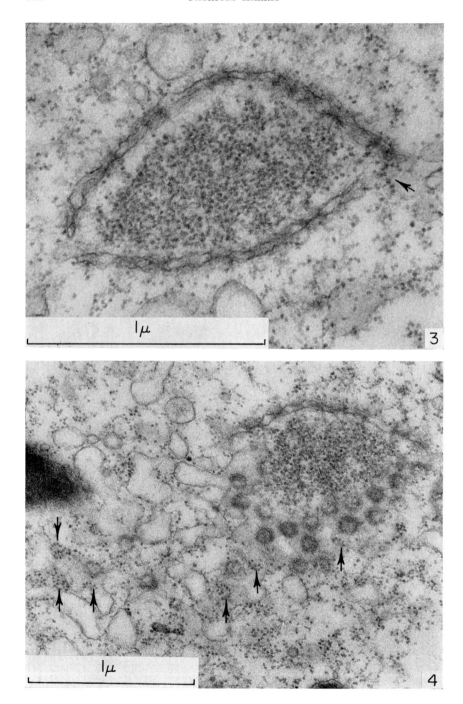

chromosomes on the pronuclear membrane, and followed their dissolution into threads before condensation into prophase chromosomes. More recent work, such as that of Simmel and Karnofsky (1961) and others, using tritium-labeled thymidine and autoradiography, has shown that the site of newly synthesized DNA is at the nuclear membrane.

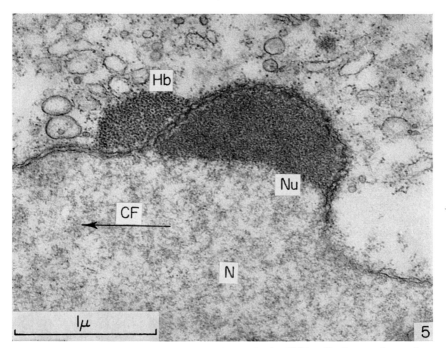

Fig. 5. Nucleolus-like body (*Nu*) and heavy body (*Hb*) adhering to the nuclear membrane of a centrifuged unfertilized egg. Neither structure was displaced by the centrifugation. Note the difference in granule size. Arrow indicates the direction of centrifugal force (*CF*) *N*, nucleus. ×38,000.

Nucleolus-like bodies associated with the nuclear membrane have been demonstrated in the pronuclei of various metazoans (Austin, 1965), and

---

Fig. 3. A heavy body in the cytoplasm of an egg 61 minutes after fertilization. The aggregate is less compact than those found earlier at the nuclear surface. Note the open ended nature of the enclosure, with granules apparently free to migrate through the opening (arrow). ×65,000.

Fig. 4. A heavy body just prior to nuclear membrane breakdown at 74 minutes after fertilization. The annulate membranes also begin to break down at this time, and arrays of ribosomes are often seen at the sites of the annuli (arrows). ×45,000.

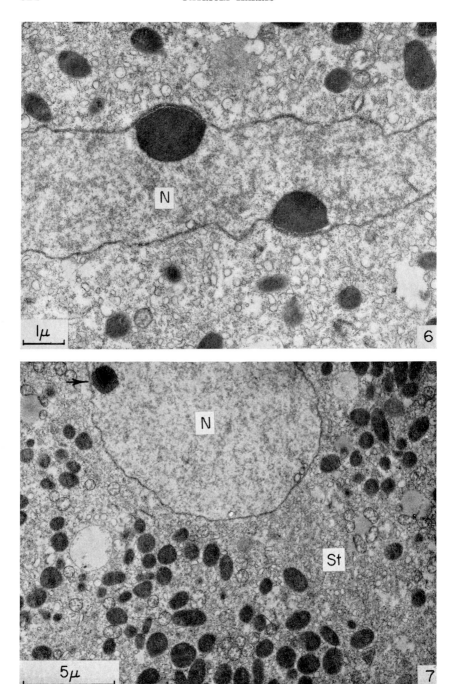

their developmental history has been described in some detail in the rat by Austin (1951) using phase optics on living material, and by Szollosi (1965) using electron microscopy. Mercer and Wolpert (1962) noted the presence of a small nucleolus in the unfertilized egg of *Psammechinus miliaris*, but did not comment further.

Freshly shed mature eggs of *Strongylocentrotus purpuratus* showed one or two fairly large dense aggregates of finely granular material similar to that shown in *Psammechinus* and as many as ten smaller ones adhering to the inner surface of the nuclear membrane. They are very similar in appearance to the secondary nucleoli of mouse pronuclei described by Szollosi, but unlike those in the mouse, these appear to be Feulgen positive and are probably akin to what Wilson (1925) referred to as "chromatin nucleoli." It is not certain whether these condensations represent entire chromosomes or only parts of them. Agrell's (1959) preparations show condensed regions in each of the chromosomes at the nuclear surface. These might serve as attachment points for the chromosomes at the nuclear surface.

The nucleus of the unfertilized egg is often very irregular in shape and the nucleolus-like bodies occupy smoothly rounded cuplike protrusions, making very close contact with the nuclear membrane. The difference in granule size between these structures and the heavy bodies adhering to the outer surface of the nuclear membranes is quite apparent when the two lie in close proximity (Fig. 5). Very often fragments of annulate lamellae or nuclear membrane within the nucleus may be associated with the nucleolar bodies.

The closeness of the association of these nucleolus-like structures with the nuclear membrane diminishes during the course of their existence. In the unfertilized and newly fertilized eggs they are extremely flattened and present the greatest possible area in contact with the nuclear surface. They persist through the period of pronuclear fusion, though at this time they are somewhat more rounded and have less contact surface with the membrane (Fig. 6). By streak stage they are essentially spherical and though they occupy a position close to the nuclear membrane, they appear to have no very intimate contact with it (Fig. 7). What happens to these structures when the nuclear membrane breaks down, whether they contribute in some way to chromosome condensation or whether they are released intact into the cytoplasm, is not known. They were

FIG. 6. Fusion nucleus of a zygote 32 minutes after fertilization. Note the two dense nucleolus-like structures. ×11,000.

FIG. 7. Nucleus of streak stage zygote with rounded nucleolus near the nuclear membrane. The clear streak (*St*) extends from the nucleus (*N*) to lower right. Dense particles in the cytoplasm are yolk. ×5000.

not seen associated with any mitotic figures, nor were similar bodies present in the newly re-formed daughter nuclei after first division.

III. CHARACTERIZATION OF NUCLEUS-RELATED STRUCTURES

A. CENTRIFUGATION OF WHOLE CELLS

The classic work of Harvey (reviewed 1956) on centrifugation of whole sea urchin eggs, the subsequent stratification of cell components and the separation of an egg into light and heavy, nucleate and anucleate fragments, has been an invaluable technique in studies of sea urchin development. The first study of such a stratified egg at the electron microscope level was the work of Afzelius (1957) in which he described the heavy bodies as sedimenting at the centrifugal pole of the eggs of several species of urchins, and showed the vital staining of this layer in a light micrograph of a stratified *Echinus esculentus* egg. A later study of Gross *et al.* (1960) on centrifuged eggs of *Arbacia punctulata* showed these so-called heavy bodies to occur in the clear upper zone along with numerous annulate lamellae, and not at the heavy pole. More recently Geuskens (1965) investigated the fine structure of light and heavy fragments of *Arbacia lixula* eggs produced by centrifugation. This author made no mention of heavy bodies either in the light or heavy halves, but does describe parallel lamellae covered with ribosomal granules in the clear zone of the light fragment. The results on *Strongylocentrotus purpuratus* are in close accord with those obtained by Gross *et al.* (1960) with *Arbacia punctulata*. The heavy bodies are found in the heavy end of the clear zone and are accompanied by large masses of lamellar systems, many with annuli. In eggs broken into light and heavy fragments, the heavy body layer was divided between the two halves.

Eggs of *S. purpuratus* were centrifuged according to the protocol outlined by Harvey (1956). Some modifications were necessary, since higher centrifuge speeds than those adequate for *Arbacia* were needed to stratify and to separate the eggs into heavy and light fragments. As shown by Harvey in *Arbacia*, lower centrifuge speeds at longer times left more of the heavier particles in the light fragment. Higher speeds broke the eggs more rapidly and provided a cleaner separation, with all the yolk contained in the heavy fragment. This was also true with *S. purpuratus*. Figures 8 and 9 show portions of the heavy body layer of the light and heavy fragments obtained by centrifuging the cells at 11,000 g for 25 minutes. Although fragments obtained at higher speeds were not examined with the electron microscope, one might guess that with proper manipulation of centrifuge speeds and times one could place the heavy body layer either in the light fragment, the heavy fragment, or as in this case distributed between the two. A knowledge of the location of these particles obviously would be of some importance if this method were used to test their developmental role.

The stratification of the organelles in *S. purpuratus* is somewhat different from that found in *Arbacia*. Immediately below the oil cap is a wide layer of large vesicles, which does not exist in *Arbacia,* and there is no layering of pigment at the heavy pole. The large vesicles may contain pigment. Their general distribution throughout the cytoplasm of the uncentrifuged unfertilized egg and their migration to the cortex after fertilization (to be reported elsewhere) is identical to the behavior of the pigment vesicles in *Arbacia*. As found by Geuskens (1965) and Gross *et al.* (1960), the mitochondria do not layer well, but are found interspersed with yolk, membranes, and ribosomes in the heavy part of the egg, and may be found in the heavy part of the clear zone. In an egg stratified by centrifuging at 11,000 *g* for 25 minutes, the layers from top to bottom consist of (1) lipid with associated mitochondria; (2) large vesicle layer (pigment?); (3) upper clear zone with nucleus at top, very small vesicles, ribosomes, and Golgi elements; (4) lower clear zone with ribosomes, small vesicles, masses of very long stacks of lamellae, some with annuli, heavy bodies with associated membranes, some scattered yolk and mitochondria; (5) yolk, mitochondria, membranes, and ribosomes.

The nucleus of the mature egg, which is small and without any well-defined nucleolus, has not received much attention, while the oocyte germinal vesicle with its large nucleolus has been the object of a number of centrifugation studies. In centrifuged oocytes the nucleolus is always found at the bottom of the germinal vesicle and in fact has been found to descend slowly by gravity alone (discussed by Harvey, 1956). The nucleolus-like bodies of the mature egg, however, remain tightly attached to the nuclear membrane, as do any heavy bodies that may be adhering to the outer surface. Figure 5 shows a nucleus of an egg centrifuged at 11,000 *g* for 25 minutes. The arrow indicates the direction of centrifugal force. This close association of the heavy body and the intranuclear structure is rarely seen, but is pictured here to show more clearly the difference in granule size of the two structures. Usually they adhere at quite separate points and there is no visual evidence, at least, of any large-scale transfer of material from one to the other.

B. CYTOCHEMICAL STUDIES

Afzelius (1957), in his study of basophilic structures of the sea urchin egg, reported that histochemical staining of the heavy bodies for light

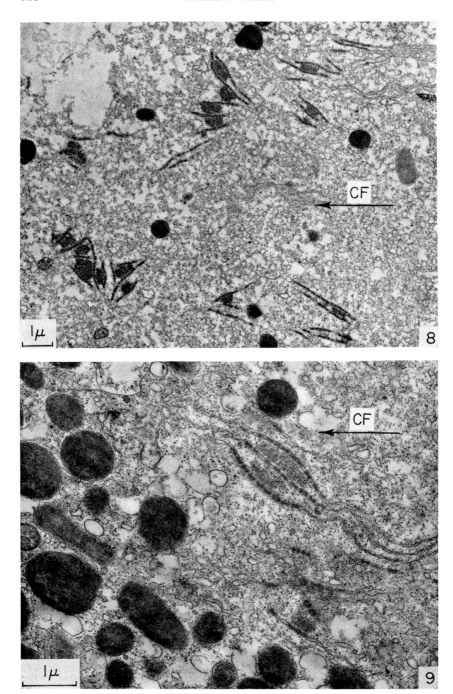

microscopy indicated that they contained RNA. Heavy bodies, whose size range is about 0.5–3 μ, are rather small entities to stain individually with an absolute degree of certainty. The mass staining of the heavy body layer of a centrifuged egg is more convincing, but in the light of the possible developmental importance of these bodies it would be well at this time to verify these findings. To date this has not been done. Histochemistry at the electron microscope level could be done without too much difficulty. Ideally, an analysis should be made on an isolated purified preparation, but this of course presents many problems. Until further verification we can only say that these structures are probably RNA.

In the present study, unfertilized eggs were fixed in 10% neutral formalin or in Carnoy's fixative, embedded in paraffin, and sectioned at 5 μ. They were then stained for DNA with Feulgen technique after acid hydrolysis, or with Azure B for nucleic acids in the manner used by Szollosi (1965). Nuclear membrane-associated material stained in both cases, but the small size of the structures in question and the lack of resolution due to section thickness did not warrant further enzymatic extraction studies on paraffin-embedded material. The close association of two different structures with the nuclear membrane and yolk granules of approximately the same size in the surrounding cytoplasm further complicated the interpretation.

Although cytochemical studies with the electron microscope were not deliberately undertaken, the fortuitous extraction of condensed chromosomes by neutral osmium fixatives containing sodium chloride or seawater provided some clue to the composition of the membrane-associated structures. The effect of neutral osmium in seawater on condensed chromosomes in sea urchin eggs was first noted by Gross et al. (1958) and later by Harris (1962). Figure 10 shows the effect of a pH 7.4 osmium fixative containing seawater on prophase chromosomes artificially condensed by mercaptoethanol. Figure 11 shows the effect of the same fixative on the nucleolus-like structure. The heavy bodies, on the other hand, appear unaffected by this fixation.

These results show that there is a chemical as well as a structural difference between the heavy bodies and nucleoli, and that the heavy bodies are not formed by simple nucleolar extrusion.

FIG. 8. Heavy body layer of the light half of an unfertilized egg fragmented by centrifugation. Arrow indicates direction of centrifugal force (CF). ×8000.

FIG. 9. Heavy body layer of the heavy half of an egg from the same batch as Fig. 8. Heavy bodies lie just above the yolk layer. (CF) Centrifugal force. ×15,000.

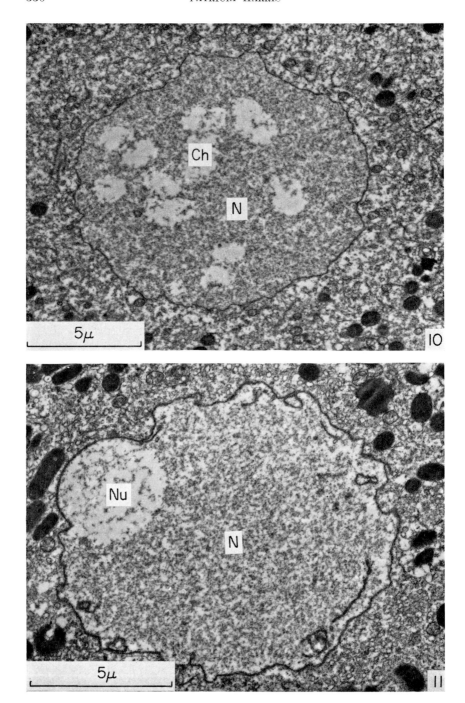

IV. TIMING OF DEVELOPMENTAL STAGES AND BIOCHEMICAL SAMPLING

One serious impediment to the study of sea urchin development is the difficulty of comparing work from different laboratories. The great number of investigations on sea urchin development have been carried out on a wide number of species and at varying temperatures with the same species. In some cases the mature "unfertilized egg" is not distinguished from the oocyte, and the "fertilized egg" can mean anything from 10 seconds after fertilization to the completion of first cleavage or even later embryonic stages.

The normal cleavage time of different species of urchins varies from about 50 minutes in the case of *Arbacia* to 2 hours in *S. purpuratus,* and each species in turn will show a difference in division time when the temperature is varied within its range of tolerance. The effect of temperature on cell division in the sea urchin has been studied in some detail (Tyler, 1936; Agrell, 1958). In a study of the mitotic dynamics of fish eggs, Detlaff (1963) has stressed the importance, when studying the duration of stages of mitosis, of expressing this duration in terms of the ratio of a particular stage to the total length of the division cycle. It is only in this way that a meaningful comparison can be made between species with different division times, or between experiments with the same species at different temperatures. This holds true not only for the actual division stages, but also for other events that occur between fertilization and first division, i.e., pronuclear fusion, streak stage, nuclear membrane breakdown, as well as stages in differentiation of the cortical layers. When a sample is taken 10 minutes after fertilization, one must know whether these eggs will divide in 50 minutes or 150 minutes. In other words, were these samples of eggs taken before or after pronuclear fusion, before or after the onset of prophase or nuclear membrane breakdown, before or after DNA synthesis.

Strongylocentrotus purpuratus, the common intertidal urchin of the North American Pacific Coast spawns during the winter months, beginning in December and continuing into March. Eggs often can be obtained later in the spring, but their ability to be fertilized and to develop into normal larvae may be somewhat impaired. Cleavage time can vary

Fig. 10. Extraction effect of neutral osmium tetroxide in seawater on chromosomes (*Ch*) in a sea urchin blastomere. The cell was blocked in prophase with mercaptoethanol to condense the chromosomes. *N*, nucleus ×6300.

Fig. 11. Extraction of nucleolus-like body (*Nu*) with the same fixative as that used in Fig. 10. Female pronucleus (*N*) 12 minutes after fertilization. ×7800.

from one batch of eggs to another even under identical conditions of temperature and aeration. Variation within batches, resulting in extreme asynchrony even in the first cleavage division, is most frequently found very early or very late in the spawning season. Fine structure studies of such batches of eggs (P. Harris, unpublished) show that in a great many of these eggs only a portion of the cortical granules have moved to the surface of the egg.

The wide variation in temperature used in experiments with the same species also calls for a word of caution. The annual mean water temperature recorded at several shore stations along the Oregon coast is between 11° and 12.5°C, with a range of 5 to 7 degrees, and a low of about 9°C in January during the spawning season (J. Gonor, unpublished). Corresponding seasonal temperatures are several degrees higher in California, where most of the studies of gametogenesis and early development of this urchin have been carried out. Although there is no evidence of geographic races with different temperature tolerances within this species, it would seem wise nevertheless, to use experimental temperatures well within the normal environmental temperature range of the animal where collected. While it may be expedient to use somewhat higher temperatures in the laboratory for convenience to shorten the division time or to bring experimental conditions closer to those used in some other laboratory, these temperatures should allow the normal development of the larva at least to the pluteus stage. A ciliated blastula may form in spite of abuses, but preparations for later differentiation may be seriously impaired. It would be decidedly ill-advised, for example, in a study comparing northern species with tropical ones, to keep them all at the same temperature and to expect the larvae to develop equally well.

With the many species of urchin and the wide range of temperatures used, there should be some indication in published reports of the length of the division cycle or the length of time from fertilization to first cleavage. The timing of specific events during this period should also be known; for example, when does pronuclear fusion occur, or streak stage, beginning condensation of the chromosomes, breakdown of the nuclear membrane, and so on. With further investigations of fine structure it may be that many other such events will become known on a much finer scale. The fact that we cannot see them with the light microscope, or cannot centrifuge them, does not mean they are unimportant.

V. DISCUSSION: A MORPHOLOGIST'S POINT OF VIEW

As noted earlier, a voluminous literature on fertilization and activation of early synthetic processes in egg cells has without much question established the fact that the egg already possesses the components necessary for these activities and only awaits some trigger to set the developmental process in motion. This implies that during oogenesis information from the nucleus is made potentially available to the cytoplasm in a form unusable until fertilization or artificial activation. Morphological studies at the fine structural level give convincing evidence of a transfer of RNA containing material from the nucleus to the cytoplasm, though the mode of this transfer may vary drastically from one organism to another and may occur at slightly different times. There may be a direct extrusion of material through the nuclear membrane of the oocyte, as in the case of *Thyone* (Kessel and Beams, 1963), there may be a delamination of heavy bodies from the nuclear surface (Afzelius, 1957), or the extrusion of RNA containing "tertiary nucleoli" by budding of the nuclear membrane as in the pronuclei of the mouse (Szollosi, 1965). It has also been suggested that such transfer may occur in conjunction with the formation of annulate lamellae at the nuclear surface, with RNA becoming bound in some way to the annuli (Merriam, 1959; Verhey and Moyer, 1967). There is also the obvious time of nuclear-cytoplasmic exchange during the maturation divisions when the germinal vesicle breaks down and releases its contents.

The primary question, however, is how the components of the protein synthesizing system are kept inactive, for obviously the answer to this question may give some important clues to the regulation of protein synthesis in later development. Several theories have been proposed on the basis of current biochemical data, but none of these is completely satisfactory. It has been suggested that the ribosomes of the unfertilized egg are somehow defective, or that the necessary messenger RNA is synthesized immediately after fertilization, or that the messenger RNA is already present, but somehow masked and unable to combine with the ribosomes, or that it is already combined with the ribosomes and somehow held inactive. It is obvious that not all of these proposals can be correct, so let us add to them still another based on the present morphological studies with the electron microscope.

It may well be that in the echinoids we have a very fortunate situation in that the information from the nucleus exists in the cytoplasm in morphologically identifiable packages, the heavy bodies. These structures, apparently derived from the nucleus and present in the cytoplasm of mature unfertilized sea urchin eggs have a structure best described as an aggregate of dense particles of ribosomal size incompletely bounded by annulate membranes similar to the nuclear envelope. Histochemical studies (Afzelius, 1957) indicate that these structures contain RNA, though admittedly the nature of this RNA requires further elucidation. Let us suppose, for the sake of argument, that there are present in unfertilized sea urchin eggs two morphologically distinct populations of ribosomes: free ribosomes without messenger RNA and thus unable to engage in protein synthesis, and ribosome-mRNA complexes in the form of large polyribosomal aggregates bounded by annulate membranes and somehow rendered inactive—the "heavy bodies." Let us see how this concept fits the biochemical data.

A. Evidence for the Presence of Two Morphologically Distinct Populations of Ribosomes

The presence of two populations of ribosomes was first noted by Hultin (1961b) in homogenates of unfertilized *Paracentrotus* eggs: free ribosomes and those bound in what he called "endoplasmic particles." These particles, when isolated from fertilized eggs, were shown to be synthetically active even after removal of the associated membranes. In more recent work Hultin (1964) attempted to rule out the possibility that these membrane-associated RNA aggregates were the site of the messenger RNA by demonstrating that those RNA aggregates isolated from unfertilized eggs, when released from their membranes by deoxycholate, do not incorporate labeled amino acids. He assumed, however, that the membranes do the masking. Evidence from work of Monroy et al. (1965) is that the masking is due to a protein that is removable with trypsin.

There is without doubt a progressive binding of ribosomes to membranes, a process most striking at the time of nuclear membrane breakdown prior to first division. At this time the heavy bodies are also beginning to disintegrate and clusters or strings of ribosomes can be seen at the sites of the annuli (Fig. 4). The total population of membrane-associated ribosomes may thus contain the ribosomal aggregates of the heavy bodies (if these are ribosomes) as well as the newer association of polyribosomes with membranes, and the ratio of the two may vary from the former immediately after fertilization to the latter at first division. It would be difficult to determine simply in terms of sedimentation rate which of these forms is the active component in early protein synthesis.

Stafford et al. (1964) noted an association of ribosomes with membranes and that 30 minutes after fertilization of *Lytechinus* eggs about 10% of the active polyribosomes are bound to membranes. These authors do not say what stage of development 30 minutes represents in this species. Presumably the fertilized eggs were sampled before the onset of the actual division process, and one might guess that heavy bodies were still present in this case. The association of active ribosomes with membranes should not necessarily be regarded as equivalent to the later association of ribosomes with the forming endoplasmic reticulum in the

blastula, since Hultin (1961b) has shown that membranes are not required for protein synthesis.

B. Unfertilized Eggs

Several experimental conditions have been imposed on unfertilized
eggs to test their ribosomal capacity for protein synthesis. In each case
the presence of two differently reacting ribosomal populations could explain the results.

1. Untreated Eggs

The unfertilized egg itself shows no protein synthesis (Gross, 1964;
Gross and Cousineau, 1964; Monroy, 1960; Nakano and Monroy, 1958),
although a low level synthesis has been reported in homogenates of unfertilized eggs (Hultin, 1961b; Hultin and Bergstrand, 1960; Tyler, 1963;
Wilt and Hultin, 1962). Whether this latter synthesis is real or whether
it is an artifact due to the disruption of structural compartments within
the cell is difficult to assess, but it does not appear to be necessary
for development of the blastula. Thus in the untreated, unfertilized egg
both ribosome populations remain inactive; the polyribosomes are
masked, and the free ribosomes are without mRNA.

2. Exogenous Templates

Exogenous templates such as polyuridylic acid stimulate amino acid
incorporation in nonfractionated homogenates (Wilt and Hultin, 1962)
as well as in ribosome preparations from unfertilized eggs (Brachet *et
al.*, 1963a; Monroy *et al.*, 1965; Nemer, 1962; Nemer and Bard, 1963;
Tyler, 1963; Wilt and Hultin, 1962). In this case the polyribosomes
remain masked and inactive, while the free ribosomes are activated by
exogenous templates.

3. Trypsin Treatment

Trypsin treatment of ribosomes from unfertilized eggs stimulates their
ability to incorporate amino acids into proteins in the absence of exogenous templates (Monroy *et al.*, 1965). Here the polyribosomes are released and active, while the free ribosomes remain without mRNA and
are inactive.

4. Exogenous Templates with Trypsin Treatment

Polyuridylic acid greatly increases synthetic activity of trypsin-
treated ribosomes of unfertilized eggs (Monroy *et al.*, 1965). Again the
polyribosomes are released and active and in addition the free ribosomes
are activated by poly U. It should be noted that the combined activity

is much greater than that induced by either of the treatments alone. The activity induced by poly U also increases over that in the non-trypsin-treated preparations. A possible explanation suggested by the authors is that poly U may displace natural messenger because of its greater affinity for binding with the ribosomes.

C. Fertilized Eggs

1. Untreated Eggs

Soon after fertilization ribosomal aggregates can be seen in the cytoplasm and there is a rise in protein synthesis (Monroy and Tyler, 1963; Stafford et al., 1964). It has also been demonstrated that there is an early synthesis of mRNA after fertilization and that this messenger soon becomes attached to preexisting free ribosomes (Wilt, 1963, 1964; Spirin and Nemer, 1965; Infante and Nemer, 1967). While the appearance of the polyribosomes could be interpreted to mean that newly synthesized mRNA combines with free ribosomes to form functional units, there is a wealth of information to show that this is not the case.

2. Enucleation

Nonnucleate fragments of unfertilized eggs obtained by centrifugation can be fertilized or activated and will begin to synthesize protein and will cleave (Brachet et al., 1963b; Denny, 1963; Denny and Tyler 1964; Harvey, 1936; Rinaldi, 1967). Since no nucleus is required for synthesis, the messenger RNA must already be present in the system. It might be recalled here that the heavy bodies are distributed between the light and heavy fragments and thus could be the site of the messenger.

3. Treatment with Actinomycin D

In the presence of actinomycin D, protein synthesis and cleavage continue into the blastula stage (Brachet et al., 1963b; Gross and Cousineau, 1963a, 1964; Gross et al., 1964). Actinomycin D, which blocks protein synthesis at the level of DNA-dependent RNA synthesis, acts essentially in the same way as enucleation. As in the nonnucleate merogones, this system is also complete and need not rely on synthesized mRNA from the nucleus.

4. Exogenous Templates

There has been reported a progressive decrease in ability of exogenous templates to stimulate protein synthesis after fertilization (Nemer, 1962), although this observation has been challenged by later work

(Tyler, 1963). In part, the discrepancy may lie in the method of breaking up the cells. Homogenates show a decrease in stimulation of the blastula compared to the unfertilized egg, while in sonicated preparations there is no significant difference. In fact, Tyler observes that in sonicates the endogenous activity can be reduced considerably, while the ability to respond to poly U remains as high as in the homogenate, or even higher. These results would suggest a preparation artifact whereby free ribosomes may be released from disrupted polyribosomes.

With the increasing evidence of early mRNA synthesis and binding to ribosomes in an inactive form, a decrease in the ability of embryo preparations to respond to poly U may simply reflect a decreasing number of uncommitted ribosomes. There is ample evidence that a cell becomes committed to the synthesis of a specific protein some time before the synthesis actually takes place. In the synthesis of embryonic hemoglobin, for example, treatment with actinomycin D is effective only if applied in earlier stages of development, not immediately prior to the appearance of hemoglobin (Wilt, 1965). Likewise, in developing sea urchin embryos, errors in the message due to transcription from faulty DNA as early as the eight cell stage become manifest only later at the time of pigment formation in the gastrula (Gontcharoff and Mazia, 1967). Since no new ribosomes are formed this early, a removal of free ribosomes from an uncommitted to a committed, but synthetically inactive, state could account for a decreased response to such exogenous templates as poly U.

5. Origin of Polyribosomal Aggregates

There are several structures in the fertilized egg which contain messenger RNA, but only one of these, the heavy polyribosomes containing preexisting or "maternal" mRNA accounts for the bulk of the protein synthesis (Spirin and Nemer, 1965). What is the origin of these heavy polyribosomes: Are they assembled after fertilization by the binding of preformed messenger to existing ribosomes, or are they released already formed from larger aggregates? Hultin (1964) presents evidence that the former is the case. Treatment of fertilized eggs with chloramphenicol, which inhibits the binding of mRNA to ribosomes, produces a decrease in the rate of amino acid incorporation. The conclusion drawn from this result was that the appearance of active polyribosomes in the fertilized egg was not due to the breakdown of larger aggregates, but rather to the net formation of such aggregates under the influence of mRNA.

The morphological evidence presented in this paper shows that aggregates which could be mRNA ribosomal complexes continue to be formed at the surface of the nuclear membrane for some time after fertilization. Also, the dense chromosomal material so closely applied to the nuclear membrane at this time may be instrumental in the movement of already formed mRNA within the nucleus to the cytoplasm, where it is bound to ribosomes. This activity could very well be inhibited by chloramphenicol, producing the results shown by Hultin. At the same time, already existing aggregates in the cytoplasm may be releasing active polyribosomes under the influence of the proteolytic activity which accompanies the fertilization reaction, and are unaffected by chloramphenicol.

D. SOME UNANSWERED QUESTIONS

The relationship of the hypothetical ribosomal populations based on the preceding evidence can be summarized as follows: (1) The unfertilized egg contains free ribosomes as well as intact heavy bodies that presumably contain a ribosomal mRNA complex. (2) Soon after fertilization the heavy bodies release active polyribosomes, but the latter remain associated with membranes. (3) After first cleavage heavy bodies no longer exist, but are dispersed as active polyribosomes. Free ribosomes still exist, but some have now become "committed" for future synthesis.

This is a morphologist's oversimplified version of a much more complicated situation, yet it does raise some questions that should be answered. One might ask where the heavy bodies are in the biochemical preparations. Are they broken up by the homogenization and "seen" in the centrifuge studies as heavy polyribosomes or "microsomes"? Are they discarded with the cellular "debris" in the initial centrifugation to get rid of yolk and other particles? Since they stratify at the interface of yolk and clear zone in whole centrifuged unfertilized eggs, are some perhaps retained with the ribosome-containing supernatant and the remainder discarded? If the heavy bodies do not contain mRNA already bound to ribosomes, what do they contain? Are the membranes that are associated with active polyribosomes after fertilization the remnants of annulate lamellae of the heavy bodies, or is this a new association with smooth-surfaced vesicles in the cytoplasm? Some of these are questions that could be answered if complementary biochemical and morphological studies were carried out, yet up to this time there has been no attempt to section the centrifuge pellets for examination with the electron microscope, and the few negative stained preparations that have been made are exceedingly difficult to interpret.

VI. CONCLUSION

The main purpose in presenting the foregoing arguments based on morphological studies was to offer alternate explanations to those based on physical and chemical methods alone. They were not meant to solve all the problems of protein synthesis in early development, nor at the same time were they completely an exercise of the imagination. The ultracentrifuge and the electron microscope are two very powerful tools for the study of the cell and its metabolic processes, but neither one is sufficient alone. In broader terms, the dialogue between morphology and biochemistry is important and necessary for our understanding of the living cell, whether we are studying mitosis or syntheses necessary for cellular differentiation in embryonic development. It is hoped the present study will contribute something useful to this dialogue.

Acknowledgments

This work was supported by a grant from the USPHS National Institutes of Health (GM-12963). The author also wishes to acknowledge the use of laboratory facilities of the Oregon State University Marine Science Center, Newport, Oregon.

References

Afzelius, B. A. (1957). *Z. Zellforsch. Mikroskop. Anat.* **45**, 660.

Agrell, I. (1958). *Arkiv. Zool.* **11**, 383.

Agrell, I. (1959). *Arkiv. Zool.* **12**, 95.

Austin, C. R. (1951). *J. Roy. Microscop. Soc.* **71**, 295.

Austin, C. R. (1965). "Fertilization." Prentice-Hall, Englewood Cliffs, New Jersey.

Barer, R., Joseph, S., and Meek, G. A. (1960). *Proc. Roy. Soc. (London)* **B152**, 353.

Brachet, J., and Ficq, A. (1956). *Arch. Biol. (Liége)* **67**, 431.

Brachet, J., Decroly, M., Ficq, A., and Quertier, J. (1963a). *Biochim. Biophys. Acta* **72**, 660.

Brachet, J., Ficq. A., and Tencer, R. (1963b). *Exptl. Cell Res.* **32**, 168.

Burgos, M. H. (1955). *Exptl. Cell Res.* **9**, 360.

Candelas, G. C., and Iverson, R. (1966). *Biochem. Biophys. Res. Commun.* **24**, 867.

Denny, P. C. (1936). *Am. Zoologist* **3**, 505.

Denny, P. C., and Tyler, A. (1964). *Biochem. Biophys. Res. Commun.* **14**, 245.

Detlaff, T. A. (1963). *Exptl. Cell Res.* **29**, 490.

Geuskens, M. (1965). *Exptl. Cell Res.* **39**, 413.

Gonor, J. Unpublished. Calculated from data collected by the Oregon State University Department of Oceanography, Corvallis, Oregon.

Gontcharoff, M., and Mazia, D. (1967). *Exptl. Cell Res.* **46**, 315.

Grant, P. (1965). *In* "The Biochemistry of Animal Development" (R. Weber, ed.), Vol. I, pp. 483–593. Academic Press, New York.

Gross, P. R. (1964). *J. Exptl. Zool.* **157**, 21.

Gross, P. R., and Cousineau, G. H. (1963a). *Biochem. Biophys. Res. Commun.* **4**, 321.

Gross, P. R., and Cousineau, G. H. (1963b). *J. Cell Biol.* **19**, 260.

Gross, P. R., and Cousineau, G. H. (1964). *Exptl. Cell Res.* **33**, 368.

Gross, P. R., Philpott, D. E., and Nass, S. (1958). *J. Ultrastruct. Res.* **2**, 55.

Gross, P. R., Philpott, D. E., and Nass, S. (1960). *J. Biophys. Biochem. Cytol.* **7**, 135.

Gross, P. R., Malkin, L. I., and Moyer, W. A. (1964). *Proc. Natl. Acad. Sci. U.S.* **51**, 407.

Harris, P. (1961). *J. Biophys. Biochem. Cytol.* **11**, 419.
Harris, P. (1962). *J. Cell Biol.* **14**, 475.
Harris, P. (1967). *Exptl. Cell Res.* **48**, 569.
Harvey, E. B. (1936). *Biol. Bull.* **71**, 101.
Harvey, E. B. (1956). "The American Arbacia and Other Sea Urchins." Princeton Univ. Press, Princeton, New Jersey.
Hultin, T. (1961a). *Experientia* **17**, 410.
Hultin, T. (1961b). *Exptl. Cell Res.* **25**, 405.
Hultin, T. (1964). *Develop. Biol.* **10**, 305.
Hultin, T., and Bergstrand, A. (1960). *Develop. Biol.* **2**, 61.
Infante, A., and Nemer, M. (1967). *Proc. Natl. Acad. Sci. U.S.* **58**, 681.
Kane, R. E. (1965). *Biol. Bull.* **129**, 396.
Kane, R. E. (1967). *J. Cell Biol.* **32**, 243.
Kessel, R. B., and Beams, H. W. (1963). *Exptl. Cell Res.* **32**, 612.
Mangan, J., Miki-Noumura, T., and Gross, P. R. (1965). *Science* **147**, 1575.
Marshak, A., and Marshak, C. (1953). *Exptl. Cell Res.* **5**, 288.
Mercer, E. H., and Wolpert, L. (1962). *Exptl. Cell Res.* **27**, 1.
Merriam, R. W. (1959). *J. Biophys. Biochem. Cytol.* **5**, 117.
Monroy, A. (1960). *Experientia* **16**, 114.
Monroy, A. (1965). "Chemistry and Physiology of Fertilization." Holt, New York.
Monroy, A., and Tyler, A. (1963). *Arch. Biochem. Biophys.* **103**, 431.
Monroy, A., Maggio, R., and Rinaldi, A. M. (1965). *Proc. Natl. Acad. Sci. U.S.* **54**, 107.
Nakano, E., and Monroy, A. (1958). *Exptl. Cell Res.* **14**, 236.
Nemer, M. (1962). *Biochem. Biophys. Res. Commun.* **8**, 511.
Nemer, M., and Bard, S. G. (1963). *Science* **140**, 664.
Rinaldi, R. A. (1967). *J. Cell Biol.* **35**, 113A.
Rünnstrom, J. (1966). *Advan. Morphogenesis* **5**, 221–325.
Simmel, E. A., and Karnofsky, D. A. (1961). *J. Biophys. Biochem. Cytol.* **10**, 59.
Spirin, A. S. (1966). *In* "Current Topics in Developmental Biology" (A. Monroy and A. A. Moscona, eds.), Vol. 1, pp. 1–38. Academic Press, New York.
Spirin, A. S., and Nemer, M. (1965). *Science* **150**, 214.
Stafford, D. W., and Iverson, R. M. (1964). *Science* **143**, 580.
Stafford, D. W., Sofer, W. H., and Iverson, R. M. (1964). *Proc. Natl. Acad. Sci. U.S.* **52**, 313.
Stavy, L., and Gross, P. R. (1967). *Proc. Natl. Acad. Sci. U.S.* **57**, 735.
Szollosi, D. (1965). *J. Cell Biol.* **25**, 545.
Tyler, A. (1936). *Biol. Bull.* **71**, 59.
Tyler, A. (1963). *Am. Zoologist* **3**, 109.
Verhey, C. A., and Moyer, F. H. (1967). *J. Exptl. Zool.* **164**, 195.
Went, H. A. (1959). *J. Biophys. Biochem. Cytol.* **6**, 447.
Wilson, E. B. (1925). "The Cell in Development and Heredity," 3rd ed. Macmillan, New York.
Wilt, F. H. (1963). *Biochem. Biophys. Res. Commun.* **11**, 447.
Wilt, F. H. (1964). *Develop. Biol.* **9**, 299.
Wilt, F. H. (1965). *J. Mol. Biol.* **12**, 331.
Wilt, F. H., and Hultin, T. (1962). *Biochem. Biophys. Res. Commun.* **9**, 313.
Wilt, F. H., Sakai, H., and Mazia, D. (1967). *J. Mol. Biol.* **27**, 1.

CHAPTER 16

Essential Biosynthetic Activity in Synchronized Mammalian Cells*

D. F. Petersen, R. A. Tobey, and E. C. Anderson

I. INTRODUCTION

The biochemical processes occurring within cells can be conveniently divided into two classes: "cyclic," that is, those that appear only at certain definite times in each life cycle (such as the replication of DNA during the S period); and "continual," those that are more or less persistent (such as ATP production or total protein synthesis). The former are presumably under close genetic control, are sequentially ordered, and succinctly define the biochemical age of the cell. The latter comprise the housekeeping and support operations and provide the basic structures, precursors, and energy sources needed for viability. Any regulatory scheme describing the mechanisms of cell growth and division must account for the periodicity of the former processes and define their inter-

* This work was performed under the auspices of the U.S. Atomic Energy Commission.

relationships with the latter. While the continual reactions may be largely regulated by classical mechanisms through the laws of chemical kinetics and equilibria, evidence is now accumulating which indicates that temporal control of the cyclic processes is effected through sequences of transcriptions and translations (see reviews by Halvorson et al., 1966; Prescott, 1968) which provide the necessary pathways from gene to functional polypeptide.

Our studies have included mathematical analysis of the relation between these two classes of events as it is manifest in the growth and division of cultures (Anderson et al., 1968; Bell, 1968), as well as the mapping of cyclic events for use as markers to determine the biochemical age of the cell in its life cycle (Tobey et al., 1966a,b). This chapter is concerned only with some of the latter results. Data will be presented that are relevant to the question of transcription and translation as control mechanisms, and several markers will be described that are points of termination of biosynthetic activities essential for cell division.

II. LIFE-CYCLE ANALYSIS

The process of locating the times of occurrence of cyclic events can be termed "life-cycle analysis," which has as its aim the mapping of the times of synthesis of specific macromolecules (DNA, RNA's, proteins). The several experimental techniques which have been used in such studies are based on the fact that, if an event occurs at some specific point in the life cycle, the consequences of this event will appear in mitotic cells after a time delay equal to the time required for the affected cells to reach mitosis. This delay will, therefore, indicate the age in the life cycle at which the event occurred. The event can be the introduction of a radioactive label or the imposition of a block by an inhibitor of a specific reaction. The test culture can be random or synchronized, the latter offering better temporal resolution but sometimes at the risk of biochemical imbalance of the test culture (Anderson et al., 1967a). The use of mitosis as the fiducial mark is dictated primarily by the experimental accessibility of this unique event, on the basis of either the specific morphology of mitotic cells or the increase of cell number at division.

The first such analysis was accomplished by Howard and Pelc (1953) using ^{32}P-labeled Vicia faba. They identified a period which they called S, during which the DNA of the cell was replicated. The adjacent periods, separating S from mitosis M, were labeled G_1 and G_2—"gaps" in which unknown processes were occurring (Fig. 1). In studies of the life cycle, the comparatively long lifetime and precise mitotic reproductive mechanism of mammalian cells are distinct advantages, and the first work with such cells was done by Lajtha *et al.* (1954). A very important advance was the introduction by Puck and Steffan (1963) of the use of the Colcemid block to accumulate mitotic figures with a resultant great increase in precision and ease of scoring. In our use of the method, we have found total cell number to be a useful alternative to mitotic fraction. The general equations developed by Puck and Steffan (1963; see also Trucco, 1965) for the accumulation of mitosis in a Colcemid

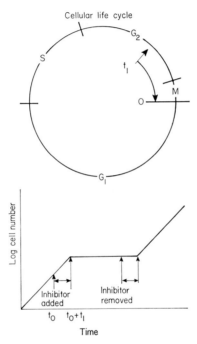

FIG. 1. Cellular life cycle showing the cell growth curve following addition and subsequent removal of an inhibitor capable of preventing progress in the life cycle reversibly.

block are directly applicable to the increase in cell number if the total cell concentration is substituted for $(1 + N_M)$, where N_M is the mitotic fraction.

The method of analysis is indicated schematically in Fig. 1 as applied to exponentially growing (i.e., nonsynchronized) cultures. The logarithm of the total cell number is a linear function of time prior to inhibition. If an inhibitor added at time T_0 halts progress of cells in the life cycle at a point T_1 (measured *backward* along the life cycle from division) while those *beyond* T_1 continue to divide, then the log cell number would continue to increase linearly from T_0 to $T = T_0 + T_1$ and thereafter remain constant. If the inhibitor failed to stop traverse but changed the rate, the point of inhibition would still be discernible as an intersection, but the slope of the second component would be greater than zero. If progress is halted at more than one point in the life cycle, only the terminal point (that is, the one nearest the impending mitosis) will be defined. In the event of reversible inhibition, removal of the inhibitor would result in resumption of growth with typical changes in slope, depending upon the nature of inhibition. Figure 1 depicts the situation upon release from an inhibitor causing a continuum of block points beyond the final measured point. All cells halt traverse, remain in register, and resume division at a rate identical to that observed before addition of the inhibitor. However, if the block point proved unique, a steeper slope indicating progress and accumulation of some cells at the point of blockade during inhibition would be expected, as has been demonstrated in the case of DNA synthesis inhibitors (Klenow, 1959; Xeros, 1962; Mueller *et al.*, 1962; Bootsma *et al.*, 1964; Petersen and Anderson, 1964; Puck, 1964). In this general case, exponential cultures could be used, but the number of cells dividing per unit time can be greatly increased by employing a synchronized culture, thus providing sharper delineation of the point of action. A more detailed account of these aspects of the life-cycle analysis by enumeration of the cell concentration may be found in an earlier discussion (Tobey *et al.*, 1966a).

III. CONDITIONS AFFECTING QUALITY OF CELL SYNCHRONY

In order to determine optimum experimental conditions for production of populations with maximum quality of synchrony, we have investigated the effects of prolonged thymidine exposure on cell survival. The rates of cell division and cell yield were determined for a series of cultures of Chinese hamster ovary (CHO) cells (Tjio and Puck, 1958) maintained in 10 mM thymidine for periods ranging from 3 to 33 hours, then resuspended in normal growth medium.

The expected pattern of division of thymidine-synchronized cells and means of estimating the fraction of cells in the S and synchronous division waves have been described in detail (Tobey *et al.*, 1967a). Briefly, when randomly growing cultures of CHO cells are treated with 10 mM thymidine for sufficient periods of time, cells in the S phase of the life cycle halt progress toward division, while cells in all other phases accumulate at the G_1/S boundary. Upon resuspension of cells in normal growth medium, a biphasic division curve results composed of cells trapped in the S phase during blockade which exhibit a growth rate approximately equivalent to that of the original random population, and cells collected at the G_1/S boundary during blockade, which divide at a significantly faster rate. Although dispersion (Anderson and Petersen, 1964; Engelberg, 1964) tends to blur the boundary between the S and synchronous portions of the division wave, straight-line approximations to the data do not introduce significant errors in determining the point of intersection, and estimation of the number of cells in each segment is possible.

The doubling times of the synchronous segments were calculated from the slopes for all cultures held from 3 to 33 hours in 10 mM thymidine prior to release and are shown in Fig. 2A. The rate of cell division was greatest in cultures held in thymidine for periods of 12–24 hours.

The fraction of the population in S and in the synchronized segment was calculated for the cultures held in thymidine for varying periods of time before release and are shown in Fig. 2B.* The cell yield was reduced in cultures held in thymidine for periods in excess of 18 hours.

We have shown elsewhere (Tobey et al., 1967a) that the proper duration of the thymidine blockade for optimum synchrony induction in cultures of CHO cells is independent of generation time but is, instead, approximately equal to the sum of the $G_2 + M + G_1$ phases in a rapidly growing culture. From that observation and the results presented in Fig. 2, we suggest that the block period be kept as short as possible (about 12 hours for our CHO culture), consistent with complete collection of the population at the G_1/S boundary and a minimum loss in cell viability.

A comparison of cell count and mitotic index data is shown in Fig. 3 for a double-thymidine block experiment. The common logarithm of total cell concentration (open points) is plotted against time in hours from the beginning of the experiment, and the magnitude of statistical

---

* The maximum increase in cell number (in the combined S and synchronous division waves) for the cultures in Fig. 2B was 74% rather than the 100% expected if all cells divided. Similar reduced yields have been observed in different lines where cells were synchronized by temperature shock (Newton and Wildy, 1959; Mayhew, 1966) or by amethopterin (Mueller et al., 1962), indicating that the reduced yield is not peculiar to the thymidine synchronization technique. A second thymidine block does not further reduce the dividing fraction, and thymidine-synchronized CHO cultures and nonblocked control cultures yield similar results for the trypan blue permeability test (unpublished observations). The reduced number of dividing cells is, therefore, not attributable to a selective killing of part of the population by thymidine. Experiments in progress suggest that the reduced fraction may be due, in part, to delayed separation of daughter cells (Anderson et al., 1967b); a nonseparated pair of daughter cells is enumerated as a single event in an electronic particle counter.

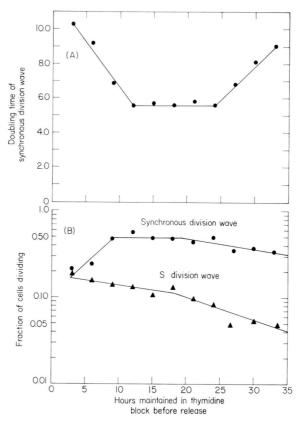

FIG. 2. The effects of duration of thymidine blockade on the subsequent behavior of partially synchronized populations of CHO cells. (A) The slopes of synchronous populations are plotted against time in blockade. (B) The fraction of cells dividing in the S and in the synchronous division waves is plotted as a function of block period for cultures held in 10 mM thymidine for different periods before release into fresh medium. The S division wave is composed of cells which were stopped in the S phase of the life cycle during thymidine block, while the synchronous division curve is composed of cells accumulated at the G_1/S boundary during blockade.

error is indicated (1 $\sigma = 0.6\%$). The mitotic index is directly proportional to the rate of growth and can be integrated numerically to give comparable data (solid points in Fig. 3). The arbitrary constant of integration is related to the duration of the scored figure as recognizably mitotic; it was chosen to normalize the slope of the curve after release from the second thymidine blockade and corresponds to 0.5 hour. Figure 3 clearly demonstrates that agreement between mitotic index and total cell concentration determinations is good, the principal discrepancy occurring at about 42 hours. At this time the counting data indicate no division, while the mitotic index data imply continued growth. A few percent of unusually persistent mitotic figures could account for the disagreement which, in any case, is not large.

An important factor in improving the precision of our measurements has been the design and construction of cell-counting equipment which

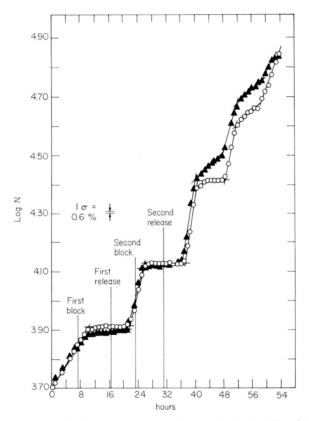

FIG. 3. Comparison of cell counts (open circles) and mitotic index determinations (solid triangles) as methods for evaluation of thymidine-synchronized cell populations. The mitotic index data have been integrated numerically to permit direct comparison using as an integration constant 0.5 hour, the approximate duration of a microscopically recognizable mitotic figure.

approaches the maximum attainable counting precision (Anderson *et al.*, 1966). The minimum error obtainable is set by counting statistics and has a value (1 σ) equal to the square root of the total counts recorded. Thus, for the total cell concentration, a coefficient of variation of 1% can be obtained from 10^4 counts and 0.3% from 10^5 counts. For the calculation of division rates, however, an order of magnitude is lost in the process of differentiating the primary data (i.e., in subtraction of two large numbers, the total counts, to get the comparatively small net increase). The longer the time interval over which the increase is taken, the higher the precision in rate but the poorer the time resolution. If the total counts are of the order of 10^4 and the time interval is

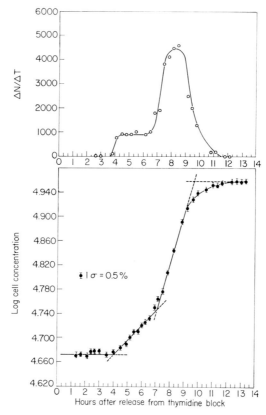

FIG. 4. Quality of data obtained from automated cell counting equipment capable of frequent sampling over prolonged periods. The sample is a Chinese hamster cell culture grown in suspension and counted at 20-minute intervals after release from a 9-hour blockade in 10 mM thymidine. The top curve is a differential plot of primary data obtained (bottom curve) from the automated counter.

0.14 T_G (T_G = generation time), the counting statistical precision in the determination of rate of increase would be 10%. That an automated sampling and counting system closely approximates these minimum values is demonstrated in Fig. 4. Here the data show $\sigma = 0.5\%$ for the results of a typical single-thymidine block experiment. In another test, samples were measured every 4 minutes over a 4-hour period of exponential growth of a random culture, 25,000 to 30,000 counts being taken per sample. The data were averaged in groups of four, and division rates were calculated for the resulting 16-minute intervals $(0.025\ T_G)$. The average of the rates observed in the five determinations was 0.062 with a standard deviation of ± 0.005 (σ_m) per hour calculated from the variance of the individual determinations about the mean.

IV. TIMING OF BIOSYNTHETIC EVENTS

Once the technical requirements for high precision population monitoring were attained, the problem of establishing additional markers could be pursued meaningfully. The basic idea was simply that progress

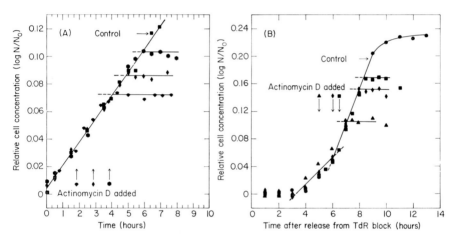

Fig. 5. Terminal point of action of actinomycin D in random and synchronized cultures of Chinese hamster cells. (A) Three random cultures were treated at varying intervals with 2 μg of actinomycin D per milliliter. (B) Three thymidine-synchronized cultures were treated just prior to and during the parasynchronous burst of division.

implied macromolecular synthesis and that times existed in the cellular life cycle when essential macromolecular synthesis was complete. Cells in this category should be insensitive to effects of various inhibitors. To test the hypothesis, several inhibitors with known specific effects upon macromolecular synthesis were studied under conditions which employed the minimum inhibitor concentrations shown maximally effective in arresting biosynthesis. These experiments resulted in data of the kind shown in Figs. 5 and 6, where 2 μg of actinomycin D or cycloheximide were added per milliliter. Cells poisoned with actinomycin (Fig. 5) continued to divide for 1.9 hours after addition of the drug and then stopped abruptly. We interpret this finding as indicating that cells in the last 1.9 hours of the life cycle were insensitive to actinomycin, while cells at earlier stages in the life cycle were unable to complete processes essential for division. The lag between addition of the inhibitor and

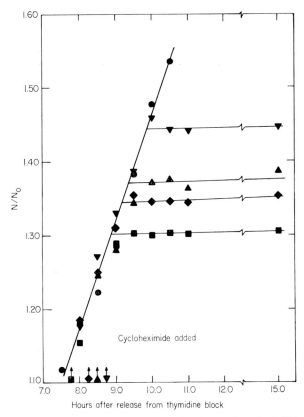

Fig. 6. Terminal point of action of cycloheximide in synchronized Chinese hamster cells. Cycloheximide (2 μg/ml) was added at varying times during the parasynchronous burst of division.

cessation of growth is then an accurate measure of the terminal point of action of actinomycin. Cells poisoned by cycloheximide (Fig. 6) continued to divide for 1 hour, indicating that cells in the last hour of their life cycle (late G_2 and the mitotic period) no longer require the biosynthetic machinery affected by cycloheximide and divide at the normal rate (no change in slope) despite its presence. Similar data obtained from experiments with a variety of inhibitors* are summarized in Table I. In each case, a discrete portion of the terminal region of the life cycle is insensitive to inhibition, while all cells prior to the block point fail to proceed.

From the data in Table I, it is evident that actinomycin and the other inhibitors differ by approximately 1 hour in their respective terminal blocking points. In view of the known primary effects of actinomycin and puromycin on the synthesis of RNA and protein synthesis, respectively, we proposed two additional markers (Tobey *et al.*, 1966a): "end of essential RNA synthesis" and "end of essential protein synthesis" on the original map of Howard and Pelc (1953).

It was possible to subdivide the temporal marker describing the end

---

* The time between puromycin addition and subsequent cessation of division measured in 30 experiments with CHO cells was 0.70 ± 0.15 hour, an increase of 6 minutes over our earlier estimate of this temporal marker (Tobey *et al.*, 1966a). Results from more than 50 such experiments with another inhibitor of protein synthesis (i.e., cycloheximide) have yielded a value of 0.98 ± 0.08 hour. The explanation for the differential time of action of the two inhibitors may lie in their method of disruption of protein synthesis.

Incomplete polypetides terminating in a puromycin moiety are released from polyribosome complexes in puromycin-treated cells (Allen and Zamecnik, 1962). High concentrations of puromycin bring about an accelerated rate of polyribosome breakdown (Wettstein *et al.*, 1964; Trakatellis *et al.*, 1965; Colombo *et al.*, 1965). Puromycin treatment should, therefore, lead to an immediate cessation of synthesis of *functional* proteins. [In our experiments with CHO cells, from 42 minutes onward after addition of 50–100 μg/ml of puromycin, there is no further increase in cell number over many hours, suggesting that functional protein synthesis is indeed completely inhibited.]

In contrast, cycloheximide presumably causes a reduction in the rate of translation of mRNA information into protein (Wettstein *et al.*, 1964) by preventing the attachment of activated amino acids onto the developing polypeptide. This decrease in rate of readout in cells treated with cycloheximide appears to be concentration dependent, but it is difficult to inhibit readout completely even with extremely high concentrations of the drug (Wettstein *et al.*, 1964). If, therefore, functional protein synthesis continues (albeit at a reduced rate) in CHO cells treated with cycloheximide in our studies, cells with nearly complete complements of essential division protein could finish synthesis and divide in the presence of the drug, resulting in a larger number of cells dividing in medium containing cycloheximide than in puromycin-containing medium. Since division is our fiducial mark and the time of action is calculated as the time *back* in the life cycle preceding division at which the block became effective, if more cells divide in the presence of cycloheximide than in puromycin, the time of action for cycloheximide would appear *earlier* than for puromycin. Attempts to detect a slight decrease in the slope of the growth curve in cycloheximide-treated cultures after the first 0.7 hour in the drug (i.e., assuming a slowdown of the rate of readout and rate of division for cells 0.7 hour or more before division) have been inconclusive because of the inherent scatter of the cell count data.

of essential protein synthesis further. In following the mitotic index of cells poisoned with puromycin, it was noticed that the mitotic index did not drop to zero as we expected from a block point in G_2. Rather, a small fraction of the mitotic cells persisted in metaphase for periods up to 12 hours, indicating that a sequence of biosynthetic events affecting more than one protein might be involved. Based upon the number of cells trapped in metaphase, it was calculated that a second protein permitting cells to *enter* mitosis is synthesized approximately 8 minutes prior to synthesis of a final protein which permits cells to *leave* mitosis and divide successfully. It should be noted that both protein synthesis

TABLE I

TIME OF ACTION OF VARIOUS INHIBITORS OF CELL DIVISION

Inhibitor	Concentration (μg/ml)	Apparent delay (hours)[a]
Actinomycin D	2	1.87 ± 0.08
Puromycin	50	0.70 ± 0.15
Cycloheximide	2	0.98 ± 0.08

[a] Average $\pm$ standard deviation of the mean. The minor differences between the values presented here and those published previously are due largely to accumulation of more data. The differences in delay produced by the protein synthesis inhibitors are discussed in footnote (see p. 351).

markers appear to be located earlier in the life cycle than the beginning of prophase.

From data obtained with inhibitors of known specific action, we have established a sequence of biochemical events that transpire prior to division. Neither the order nor timing of these events is altered in cultures of Chinese hamster cells with grossly different generation times, in which the variation in generation time is solely dependent upon variations in the length of G_1.

V. STABILITY OF mRNA ASSOCIATED WITH CELL DIVISION

The cells located between the RNA and protein synthesis markers constitute a unique population which has completed synthesis of all species of mRNA required for division but not the associated protein synthesis. By simultaneously treating these cells with inhibitors of both RNA and protein synthesis for varying periods of time, then releasing them into medium in which protein but not RNA synthesis can resume, the stability of mRNA species coding for functional division proteins can be determined by measuring the fraction of cells able to complete division. The simultaneous addition of actinomycin and cycloheximide to a culture produces three classes of cells: (1) cells closer to division than 1 hour which will divide in the presence of inhibitors and disappear rapidly into G_1; (2) cells trapped between 1 and 1.9 hours prior to division (i.e., trapped between the actinomycin and cycloheximide markers); and (3) cells farther from division than 1.9 hours with incomplete complements of both RNA and protein. In view of the irreversibility of actinomycin at the concentrations employed, these latter cells will never be able to divide. If the cycloheximide block is removed, only those cells of class (2) which possess functional mRNA species capable of coding for division protein will be able to divide, since actinomycin continues to prevent synthesis of new mRNA species.

Random cultures could be used for such an experiment, but by utilizing synchronized cells the number trapped between the actinomycin and cycloheximide markers can be increased and the precision of the experiment appreciably improved. Accordingly, cells were synchronized with 10 mM thymidine, and cycloheximide and actinomycin (2 μg/ml of each) were added simultaneously at the beginning of the synchronous division wave. After varying periods in blockade, aliquots were resuspended in normal medium, and the fraction of cells which subsequently divided (relative to the number of cells contained in the same segment of the life cycle of a control culture) was then calculated. In Fig. 7 (open points), the fraction of trapped cells that divided is presented as a function of time in both cycloheximide and actinomycin prior to resuspension in normal medium. The fraction of dividing cells dropped to 50% in 2.9 hours.

A variation of the above experiment was performed to investigate possible synergistic effects of the two drugs. Synchronized cells were blocked with only cycloheximide for varying periods of time, then released into medium containing actinomycin. The number of dividing cells was again expressed as the fraction of cells trapped between the actinomycin and cycloheximide markers (Fig. 7, solid points). The effect is indistinguishable from that observed in the preceding experiment in which both drugs were present during the block periods. These results eliminate the possibility of untoward side effects from actinomycin but, far more important, demonstrate that *no RNA coding for division protein was synthesized in the presence of cycloheximide alone.* Such synthesis would have allowed cells earlier in the life cycle than the actinomycin marker to cross that marker, resulting in an increase in the number of dividing cells. The dividing fraction would then have exceeded 100%. Thus, by inhibiting protein synthesis, functional mRNA synthesis is

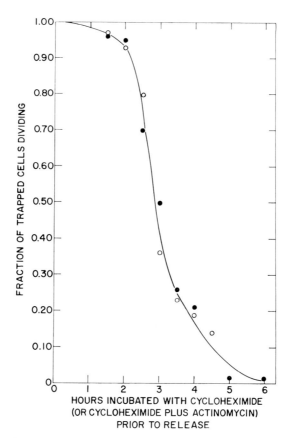

FIG. 7. Fraction of expected number of cells dividing after different periods of time in cycloheximide blockade. The open points represent cultures incubated in cycloheximide and actinomycin (final concentration of both drugs 2 μg/ml) and then released into drug-free medium. The solid points represent cultures incubated in cycloheximide only prior to release into medium containing actinomycin.

concomitantly shut off, under conditions in which bulk RNA synthesis continues at a nearly normal rate (Tobey *et al.*, 1966b; Enger *et al.*, 1968).

VI. SUITABILITY OF SYNCHRONIZED CELLS FOR MARKER CHARACTERIZATION

While it has been possible to provide new markers to characterize the G_2 phase of the life cycle, G_1 remains essentially featureless. Part of the difficulty in establishing markers in G_1 is the poor quality of synchrony across G_1 afforded by synchronization techniques that act by reversibly inhibiting DNA synthesis, due to synchrony decay arising from variations in the rate of traverse around the life cycle of individual cells in the population. It is necessary, therefore, to obtain a technique capable of producing synchronized populations close to G_1.

The variation of Terasima and Tolmach's method (1963), suggested by Robbins and Marcus (1964) for selective removal of mitotic cells from monolayer cultures, proved to be most valuable because selection is close to G_1 and the degree of synchrony obtained by our modification (Tobey *et al.*, 1967b) is exquisitely sharp. We obtain mitotic populations by shaking glass-grown cultures mechanically at specific intervals (10 minutes between successive treatments for CHO cells, 15-minute intervals for HeLa cells). By the time the monolayers have been agitated 7 times, 90% or more of the detached cells are mitotic, and mitotic populations may be collected from subsequent treatments over the ensuing 8 hours. Because we depend upon a new crop of cells to enter mitosis between successive treatments, we are selecting for biochemically competent cells (i.e., cells continuing traverse of the life cycle during the course of the experiment). Detached cells may be allowed to complete mitosis without interruption by placing them in a spinner flask immediately after collection. None of these cells has been forced to delay traverse of the life cycle, in contrast to other methods of synchronization employing chemical agents. Thus, criticisms (which may or may not be justified) relating to possible biochemical changes resulting from use of chemical agents to induce synchrony are avoided.

When material (medium plus detached cells) from the seventh harvest was placed in a spinner flask immediately after collection, the fraction of mitotic cells in the population (open points in Fig. 8) dropped from 0.95 to less than 0.05 in 19 minutes, approximately 12 times the rate of division observed in thymidine-synchronized cells which had undergone appreciable synchrony decay by the time they reached division. With cells prepared by this technique we have obtained information on the postdivision period for use in our growth analysis program (Bell and Anderson, 1967; Anderson and Petersen, 1967), as well as a better understanding of the process of division as it relates to particle counting methodology (Anderson *et al.*, 1967b). We anticipate that this technique will be extremely useful in the search for G_1 markers.

Large quantities of synchronized cells may be prepared for biochemical studies by chilling material from a series of successive harvests. The entire collection will divide synchronously after resuspension in warm medium (Fig. 8, solid points). Although the cells collected by this method are prevented from traversing the life cycle during cold storage, it is unlikely that the cells are accumulating macromolecules at this time

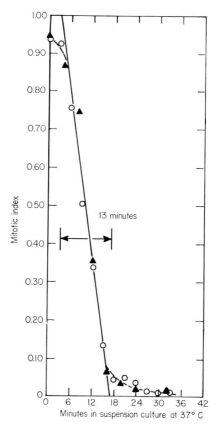

Fɪɢ. 8. Comparison of the rate of completion of mitosis by a culture placed in a spinner flask immediately following collection by the mechanical selection technique (open points) and another culture containing cells obtained from a series of harvests over a 4-hour period, chilled immediately after collection, and resuspended in warm medium in a suspension culture at $T = 0$ (solid points).

and thereby entering an abnormal biochemical state, since the biosynthetic activity of the mammalian cell is minimal during mitosis (Prescott and Bender, 1962).

VII. DISCUSSION

We have provided evidence for three temporal markers of biological significance which subdivide the G_2 phase of the life cycle and which are related to impending mitosis and division. Biochemically, these markers represent the end of essential RNA synthesis* and end of essen-

* It should be emphasized that the temporal marker for end of essential RNA

tial protein synthesis, the latter further divided into completion of synthesis of a protein which allows the cell to *enter* mitosis and approximately 8 minutes later a time at which a protein is synthesized which allows the cell to *complete* mitosis. The protein synthesis markers are located late in the life cycle and are all clearly in G_2 rather than in mitosis. If the final RNA species is a messenger coding for the last essential species of protein, then the interval between the actinomycin and puromycin markers could represent the time from completion of transcription to appearance of functional protein.

Half of the cells which possessed a complete set of RNA species but incomplete protein species are unable to divide if held in cycloheximide for 2.9 hours, then released into medium containing actinomycin (Fig. 7). This result could be explained by the following model. After completion of synthesis of RNA required for division, cistron(s) containing information for essential division protein may become inaccessible for further transcription. Upon reversal of prolonged cycloheximide blockade, the cells trapped between the actinomycin and cycloheximide markers would be unable to synthesize the proteins required for division, due to loss of mRNA, and would further be unable to retranscribe the last species of essential mRNA, due to inacessibility of a specifically related segment of the genome. Thus, as the cycloheximide block is maintained for progressively longer periods, progressively larger numbers of cells would be incapacitated by loss of division-related mRNA and would be unable to divide.

In view of the gross reduction in rate of protein synthesis and low polyribosome content of mitotic cells (Prescott and Bender, 1962; Johnson and Holland, 1965), it is possible that in the normal cell the mRNA species under observation might be broken down at the start of mitosis (i.e., the lifetime for division-associated mRNA might be less than the 2.9-hour value reported here). In any event, the messenger(s) coding for final division protein(s) is functionally stable only during the portion of the life cycle in which it is needed.

Finally, our observation that inhibition of synthesis of essential division protein by cycloheximide may be accompanied by an inhibition of an associated mRNA species may imply that, for a cell to proceed around the life cycle, RNA and protein synthesis must continue concomitantly. Results such as these are expected from the models proposed

---

synthesis refers only to the time prior to division at which the last species of RNA essential for division is synthesized. RNA synthesis can indeed continue throughout interphase, and one class of RNA which is synthesized in late G_2 is ribosomal RNA (Enger *et al.*, 1968).

by Stent (1966) and by Cline and Bock (1966) for control of genetic transcription at the translational level. With the available method for evaluating differential sensitivity, more work in the mitotic region of the life cycle will be initiated to define more clearly a regulatory role of translation.

VIII. SUMMARY AND CONCLUSIONS

There is little doubt that the ability of a cultured mammalian cell to traverse its life cycle is dependent upon a sequence of transcriptions and translations. The degree to which these events are coupled has been investigated with the result that cells were found to be unable to traverse their life cycle when protein synthesis is inhibited. This conclusion, based on data which demonstrate that cells cannot cross a transcription marker when translation is inhibited, is consistent with proposed models for control at the translation rather than at the transcription level. Since the same markers occur in a number of stable cell lines, we view the sequence as a fundamental property and consider that traverse of the life cycle (the reproductive phases S, G_2, and M in particular) is rigidly impressed in a tightly coupled transcription-translation sequence with continuous translation operating as a major control feature.

REFERENCES

Allen, D., and Zamecnik, P. (1962). *Biochim. Biophys. Acta* **55**, 865–874.
Anderson, E. C., and Petersen, D. F. (1964). *Exptl. Cell Res.* **36**, 423–426.
Anderson, E. C., and Petersen, D. F. (1967). *Biophys. J.* **7**, 353–364.
Anderson, E. C., Carlson, D. L., Glascock, R. B., Larkins, J. H., Perrings, J. D., and Walters, R. A. (1966). *Los Alamos Sci. Laboratory Rept. LA-3610-MS,* pp. 44–49.
Anderson, E. C., Petersen, D. F., and Tobey, R. A. (1967a). *Nature* **215**, 1083–1084.
Anderson, E. C., Petersen, D. F., and Tobey, R. A. (1967b). *Biophys. J.* **7**, 975–977.
Anderson, E. C., Bell, G. I., Petersen, D. F., and Tobey, R. A. (1968). *Biophys. J.,* in press. (Abstr.)
Bell , G. I. (1968). *Biophys. J.* **8**, 431–444.,
Bell, G. I., and Anderson, E. C. (1967). *Biophys. J.* **7**, 329–351.
Bootsma, D., Budke, L., and Vos, O. (1964). *Exptl. Cell Res.* **33**, 301–309.
Cline, A. L., and Bock, R. M. (1966). *Cold Spring Harbor Symp. Quant. Biol.* **31**, 321–333.
Colombo, B., Felicetti, L., and Baglioni, C. (1965). *Biochem. Biophys. Res. Commun.* **18**, 389–395.
Engelberg, J. (1964). *Exptl. Cell Res.* **36**, 647–662.
Enger, M. D., Tobey, R. A., and Saponara, A. G. (1968). *J. Cell Biol.* **36**, 583–593.
Halvorson, H. O., Bock, R. M., Tauro, P., Epstein, R., and LaBerge, M. (1966). *In* "Cell Synchrony—Studies in Biosynthetic Regulation" (I. Cameron and G. Padilla, eds.), pp. 102–116. Academic Press, New York.
Howard, A., and Pelc, S. R. (1953). *Heredity Suppl* **6**, 261–273.
Johnson, T. C., and Holland, J. J. (1965). *J. Cell Biol.* **27**, 565–574.

Klenow, H. (1959). *Biochim. Biophys. Acta* **35**, 412–421.

Lajtha, L. G., Oliver, R., and Ellis, F. (1954). *Brit. J. Cancer* **8**, 367–379.

Mayhew, E. (1966). *J. Gen. Physiol.* **49**, 717–725.

Mueller, G. C., Kajiwara, K., Stubblefield, E., and Reuckert, R. (1962). *Cancer Res.* **22**, 1084–1099.

Newton, A. A., and Wildy, P. (1959). *Exptl. Cell Res.* **16**, 624–635.

Petersen, D. F., and Anderson, E. C. (1964). *Nature* **203**, 642–643.

Prescott, D. M. (1968). *Cancer Res.* **28**, 1815.

Prescott, D. M., and Bender, M. A. (1962). *Exptl. Cell Res.* **26**, 260–268.

Puck, T. T. (1963). *Cold Spring Harbor Symp. Quant. Biol.* **29**, 167–176.

Puck, T. T., and Steffan, J. (1963). *Biophys. J.* **3**, 379–397.

Robbins, E., and Marcus, P. I. (1964). *Science* **144**, 1152–1153.

Stent, G. (1966). *Proc. Roy. Soc. London* **B164**, 181–197.

Terasima, T.. and Tolmach, L. J. (1963). *Exptl. Cell Res.* **30**, 344–362.

Tjio, J. H., and Puck, T. T. (1958). *J. Exptl. Med.* **108**, 259–268.

Tobey, R. A., Petersen, D. F., Anderson, E. C., and Puck, T. T. (1966a). *Biophys. J.* **6**, 567–581.

Tobey, R. A., Anderson, E. C., and Petersen, D. F. (1966b). *Proc. Natl. Acad. Sci. U.S.* **56**, 1520–1527.

Tobey, R. A., Anderson, E. C., and Petersen, D. F. (1967a). *J. Cell Biol.* **35**, 53–59.

Tobey, R. A., Anderson, E. C., and Petersen, D. F. (1967b). *J. Cell. Physiol.* **70**, 63–68.

Trakatellis, A., Heinle, E., Montjar, M., Axelrod, A., and Jensen, W. (1965). *Arch. Biochem. Biophys.* **112**, 89–97.

Trucco, E. (1965). *Biophys. J.* **5**, 743–753.

Wettstein, F., Noll, H., and Penman, S. (1964). *Biochim. Biophys. Acta* **87**, 525–528.

Xeros, N. (1962). *Nature* **194**, 682–683.

CHAPTER 17

Markers in the Cell Cycle

J. M. Mitchison

I. CYCLE HISTORY

Any reader of this book will appreciate the present flourishing state of work on the cell cycle, especially since the use of synchronous cultures is only one of several methods for studying the cycle. Over the last fifteen years the number of published papers has risen along a roughly exponential curve with a doubling time of four years (Fig. 1A). But this boom will continue only if we set ourselves questions which are intellectually profitable and technically answerable. There is a moral here which comes from a review on the Golgi apparatus by Hibbard (1945). The number of papers rose even faster to a peak in 1929, but it was then followed by an equally rapid decline (Fig. 1B).

It is interesting to look back at the early 1950's when today's river was only a trickle. This was when the first cell cultures were being synchronized (*Tetrahymena* by Zeuthen and Scherbaum, 1954; *Chlorella* by Tamiya *et al.*, 1953), when the first autoradiographs were being made (Howard and Pelc, 1951), and when work was starting on Feulgen and ultraviolet measurements of DNA during the cycle (Swift, 1950; Walker and Yates, 1952). Hughes published a book in 1952, "The Mitotic Cycle," which set out very clearly the views of many cell biologists at the time. The title itself is indicative, for the timing of the mitotic events was well known and the whole process of mitosis and cleavage was by far the most dramatic part of the cell cycle—as indeed it still is at the level of visible structure. But the rest of the cycle was treated as "interphase" and though half the book was concerned with the interphase cell, the treatment was almost entirely a static one. The cell was regarded as the same throughout interphase, and there was practically no discussion of the time patterns of growth. It was still possible to view interphase as a period of continuous steady growth where all the chemical components of the cell increased in equal proportions. The resulting lack of differentiation at the chemical level would then parallel the lack at the morphological level.

Since those early days, work on the cycle has spread out in two broad streams. In the one which is well represented in this book, the main experimental tool is the synchronous culture. James (1966) has produced a most useful table of the techniques for synchronization in which there are eight general methods for "induced synchrony" (where all the cells of a culture are induced into synchrony) and four for "selection synchrony" (where cells at a particular stage of the cycle are separated off from the rest and grown as a synchronous culture). All the early synchronous cultures were produced by induction and the first reactions were of surprise and fascination at the phenomenon itself. This led on to the question of how the cultures could be used; from this two lines developed. In one, the main emphasis was on the mechanism of synchronization and on the stimulus to division. In the other, the emphasis was on the events of growth and synthesis in the cycle between divisions. These two approaches can be called "division-oriented" and "cycle-oriented" to use the nomenclature of James (1966). The latter approach has often been criticized on the grounds that the cycle after induction is unnatural, abnormal, or distorted. "Unnatural" is an unsuitable word to use both because anything that a cell can do is natural and also because in some cases the conditions that induce synchrony may resemble conditions in nature, for instance, the 12 hours cycle of light and dark with photosynthetic microorganisms (Edmunds, 1965). But it is possible to call a cycle abnormal provided we set the standard of normality, a point rightly emphasized by Zeuthen (1964). The best standard of normality is probably the cycle of the average single cell growing exponentially in constant conditions. It is an ideal rather than a practical standard because of the difficulty of measuring any of its parameters except with single-cell methods and the impossibility of getting completely constant conditions of the medium in any batch culture. Even so, it does have value because it can be used to make a rough scale of the degrees of abnormality produced by the different synchrony techniques. The low end of the scale would include the selection methods and some of the induction methods that depend on starvation and re-growth (e.g., Williamson and Scopes, 1960). The high end of the scale would include the heat shock induction of *Tetrahymena* (Zeuthen, 1964). Having made this scale, we should not then convert it into a scale of cellular or experimental virtue. The features about the *Tetrahymena* system that made the cycles abnormal are also those that make it very valuable. It provides a clear demonstration that division can be disso-ciated from growth, and it is in this system that there has been most progress in discovering the mechanism of synchronous induction.

The second broad stream has been concerned with measurements on single cells from asynchronous growing cultures. The basic tool here

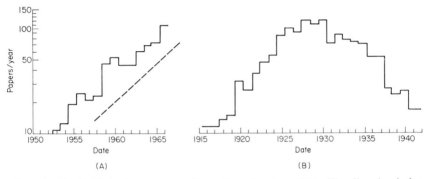

FIG. 1. (A) Published papers on the cell cycle since 1951. The line in dashes represents a doubling time of 4 years. These papers, which total 612, come from my card index. This index is not comprehensive, but it is a representative sample made by someone interested in the cycle for the last 15 years. (B) Published papers on the Golgi apparatus between 1914 and 1942. Data from Hibbard (1945) replotted on a logarithmic scale.

has been the microscope either in its normal form for measuring volume and scoring autoradiographs, or adapted for measuring dry mass by inter-ferometry and nucleic acid content by microspectrophotometry. Far the most important technique has been labeling DNA with a pulse of tritiated thymidine and then determining the timing of DNA synthesis by following in autoradiographs the appearance of labeled mitoses. Na-ture has been good to us because the success of this method depends on a number of lucky circumstances. Thymidine not only enters cells easily, but it is also a specific label for a single type of macromolecule even though it is not on the normal pathway for DNA synthesis. In most cases the thymidine pool is small so a short pulse can be given. Finally, DNA synthesis is a discontinuous event in the cycle of many cells. If things had not been this way many scores of papers would not

have been written. As it is, thymidine labeling has been the most reveal-
ing single method in all the work on the cell cycle. Whether it will remain
so is another matter. The single-cell methods have the collective disadvantage that only a limited number of cell properties can be measured.
By contrast, the powerful separation and assay techniques of modern
bulk biochemistry can be used with synchronous cultures.

It is obviously impossible in this short article to go in any detail
into the results that have come from the last fifteen years of work.
But we can look in a broad and general way at some of the patterns
of growth and synthesis that have emerged. In most cell cycles the
overall properties such as volume and dry mass increase steadily. The
rate of increase may double slowly through the cycle, giving an approximate exponential curve or it may double abruptly at some point (e.g.,
division) and then remain constant until the same point in the next
cycle. There are two striking exceptions to this general pattern which
should be remembered. One is a pattern of growth where the rate of
increase starts at a high value at the beginning of the cycle and then
falls until it is at or near zero at the end of the cycle (giving a growth
curve which is convex to the time abscissa). This has been shown for
reduced weight (equivalent to dry mass), volume, and total protein in
Amoeba in the pioneer paper by Prescott (1955) and also for total
dry mass in *Streptococcus* by Mitchison (1961). The other interesting
pattern is the change in the total acid-soluble pool of *Schizosaccharo-
myces* (Mitchison and Cummins, 1964). This is a case where the relative
size of one of the bulk components of a cell fluctuates during the cycle
by a factor of two. Turning to total protein and RNA, the two main
chemical components (apart from water) in animal cells and many
microorganisms, the patterns are very similar to those of dry mass and
volume except at the time of mitosis. Here there is good evidence of
a sharp drop in RNA synthesis and in many cases a reduction of protein
synthesis in cells that have a true mitotic division. But with DNA the
position is totally different. Its synthesis is discontinuous in all cells
except bacteria [and even here there is evidence for a sharp rate change
(Clark and Maaløe, 1967) and of discontinuous synthesis in the two
genomes of slow growing cells, (Lark, 1966)]. In many cells, the synthesis (S) period is about half to two-thirds of the way through the cycle
with a gap (G_1) before it and a gap (G_2) after. There is, however,
a select and interesting group of cells with no G_1, a short S and then
a long G_2 occupying most of the cycle (references in Cummins, this
volume).

This periodic synthesis of DNA, which has been known since the
early days of cell cycle work, is a matter of great importance. Until

recently, it was the only clear piece of evidence against the idea of continuous growth during interphase. But even more important was its practical use in providing two markers (the beginning and end of the S period) to chart the progress of a cell through the cycle. Other events in the cycle, for instance radiation resistance, could be positioned in relation to the G_1-S-G_2-M sequence. This sequence has become of such importance in higher cells that it has sometimes been identified as the cell cycle itself, forgetting the other things that happen.

II. CYCLE MARKERS

The DNA markers serve as a convenient starting point for exploring the general principles of cell cycle markers. We should be clear initially as to what they are. A reasonable definition is that they are discrete events which take place at a particular point in the normal cell cycle. The events can be of any measurable or detectable kind—chemical,

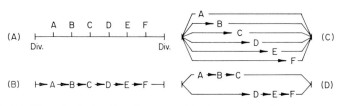

FIG. 2. (A) Hypothetical cell cycle map with six markers, A to F. These markers as (B) a fixed sequence, (C) a variable sequence, (D) a partially fixed sequence.

structural, or physiological. The standard of normality will in most cases follow the definition given above: the cycle of the average single cell growing exponentially in constant conditions. Exactly what these conditions are will vary, but they will usually be taken as those that produce the fastest growth in a particular medium.

Let us suppose that six markers have been found spread evenly through a cell cycle. They are shown in the cell cycle map in Fig. 2A by the letters A to F, with F as the last marker before division. This map can be used as it stands for a number of purposes. If a new marker is discovered, it can be placed accurately in the cycle by its relation to the existing ones. If a synchronous culture is to be compared with this normal cycle, the degree of abnormality can be judged by the extent to which this sequence has been displaced or distorted. One of the questions with mammalian tissues such as liver is whether the majority of the cells are very slowly moving through G_1 or whether, as seems more likely, they are stopped at some point early in the cycle. The point could be either early in G_1 or in a pre-G_1 stage that has been called G_0. This question could be answered with a cell cycle map like Fig. 2A providing the markers could be identified. If the cells were slowly moving through G_1, there should be a fairly even distribution of cells which had not reached A, of cells which had passed A, and of cells which had passed B. If the cells had stopped between A and B, then all of them should have passed A and none should have reached B. If the cells had stopped in a G_0, then none of them should have reached A. Another similar question about tissues which could be answered in the same way is whether the "G_2 cells" of Gelfant (1962) are stationary in G_2 or are moving through it very slowly. For these questions to be answered in practice, more markers would be needed at either end of the cycle than are shown in Fig. 2A.

The natural history of detecting the markers is a vital initial step, but we shall get more out of the marker maps if we are prepared to experiment with the cycle. In particular, we should be able to determine some of the causal connections between the markers. Two extreme cases of causality are shown in Fig. 2 in diagrammatic form. In Fig. 2B there is a direct causal connection between each marker and the one that follows it. As a result it is a "causal fixed sequence" in which the spacing between the markers can be altered, but their order is fixed and the omission of any one of them will stop further progress through the cycle. Another way of producing a fixed sequence is to have the markers without a direct causal chain but depending instead on signals from a single internal timing system, as with the hour bells on a church clock. This can be called a "noncausal fixed sequence" and differs from the causal one in allowing a marker to be omitted without stopping the cycle. The order, however, cannot be altered. The other extreme is a "variable sequence" (Fig. 2C). Here each of the markers is the product of an independent chain of events within the cell, and there is no direct causal connection or single timing mechanism between them. In the normal cycle they will occur in a given order, but if the cycle is changed or distorted not only their spacing but also their order can be varied. All of them have to occur at some point in the cycle but it may not matter to the cell when this point is. The church clock analogy can be extended to this variable sequence by considering six separate clocks each with bells. Normally they will strike in a particular sequence, but if some of them start to run fast or slow the sequence will change. Between these two extremes there are many compromise situations which can be called "partially fixed sequences." One of them is shown in Fig. 2D with two sets each with three markers. The order between the sets is variable, but the order within the sets is fixed. As well as the order

ABCDEF it would be possible to have *DEFABC* or *ADBECF*, but not *FEDCBA*. This is not of course the only possible terminology. Instead of "fixed" and "variable," one could use "series" and "parallel." I prefer, however, the former terms mainly because a "parallel sequence" is an awkward concept.

How could we apply these concepts to real cell cycles? There are two obvious points to start with: that almost all sequences will be partially fixed; and that no distinction can be made between the types of sequence unless the cell cycle is altered either deliberately or accidentally by changing the growth conditions. Although most of the work lies in the future, certain things can be said now. The two markers of the beginning and end of the S period must be a fixed sequence since it is logically impossible that they could be reversed. The main morphological stages of mitosis (prophase, metaphase, etc.) are also a fixed sequence, probably causal, whereas the two markers of mitosis and division are a noncausal fixed sequence. Mitosis can happen without division (e.g., syncytia) and division without mitosis (activated enucleate sea urchin eggs), but in no circumstances does division precede mitosis. On the other hand, the division marker is variable compared to the doubling point marker—the point in the cycle where the size of the cell is double what it was at the beginning of the cycle. There are many cases, especially in batch culture, where cells become larger or smaller in successive cycles.

The cycle can be altered in a variety of ways. One is to increase the time of the cycle by growing at lower temperatures or in a different medium. If this only altered the time scale but did not affect the spacing or order of the markers, it would not be very informative. But there is some encouraging evidence that at least the spacing of one marker can be changed. Varying the length of the cycle alters the position of the S period in mammalian cells (e.g., Sisken and Kinosita, 1961), the position of the rate change point of DNA synthesis in bacteria (Clark and Maaløe, 1967; Helmstetter, 1967), and the pattern of dry mass increase in fission yeast (Mitchison *et al.*, 1963). Another interesting way of increasing the length of the cycle would be to make use of the division delay produced by radiation.

Specific inhibitors are potentially useful tools for distorting the cycle. Fluorodeoxyuridine inhibits DNA synthesis in *Bacillus subtilis*, yet a marker event, the synthesis of ornithine transcarbamylase, occurs after the inhibition (Masters and Donachie, 1966). Another DNA inhibitor, mitomycin C, allows protein and RNA synthesis to continue in yeast for some time after DNA synthesis has been stopped (Williamson and

Scopes, 1962). With these and other inhibitors, it should be possible to separate those markers which depend on the completion of a synthetic event from those which do not.

One of the most interesting situations to look for markers is the heat shock method of synchronizing *Tetrahymena* which was mentioned earlier. Let us suppose there were twelve markers during the normal cycle. If four of these markers were absent during the heat treatment (where there is no division) and only reappeared after the end of the heat treatment and before the synchronous burst of division, then we would have a case for regarding these markers as a fixed sequence culminating in division. If this same sequence appeared in other synchronous systems, the case would be stronger still. One or more of these markers could of course be the "division proteins" suggested by Williams and Zeuthen (1966). This kind of analysis seems a profitable way at the moment of tackling the old, important, and still unsolved problem of the trigger to division. It also shows a way of getting round, at any rate partially, one of the logical difficulties about marker analysis. The difficulty is this. It is easy to show that a sequence is variable if the order of the markers is changed in an altered cycle: but it is hard to prove the existence of a small fixed sequence in the presence of a large variable one until many alterations have been tried without affecting the order (or unless there is an obvious causal connection between the markers). On the other hand, if a "division sequence" is absent from growing nondividing cells and present in nongrowing dividing cells and if the same thing occurs with one or two other systems, e.g., repetitive synchronization of *Tetrahymena* (Padilla *et al.*, 1966), most people would accept this as a strong indication of a fixed sequence.

Another promising situation for marker analysis is at the end of the cycle in cells such as *Amoeba* and *Streptococcus* which have convex growth curves, as was mentioned earlier. Division is preceded by a period in which there is little or no growth. It is therefore possible that a division sequence might be found fairly easily in this period since it would not be encumbered with growth markers. The same freedom from growth also occurs in early embryos.

Until recently, marker analysis would have seemed utopian because of the very few markers that were known. But the situation has changed so radically in the last four years that it has become a practical possibility in some synchronous systems and could easily be made so in others. There has been the least progress in structural markers. Apart from the well-known changes at mitosis, no other ones have been found in most cells (e.g., Robbins and Scharff, 1966). There is an exception in *Tetrahymena* where a new oral apparatus develops before division, and

Williams and Zeuthen (1966) have suggested that the oral fibers may be an important division protein.

It is quite a different story with chemical or synthetic markers, where the progress has been rapid. DNA, RNA, and protein have become so deeply embedded in cell biology that they are sometimes felt to form a trio of equal members. Of course they are not. Most of the DNA has only one function and is composed of a relatively small number of very large molecules. There are at least three kinds of RNA of different size and function, and probably another three. One of these kinds of RNA (the messenger fraction) should be subdivisible into as many fractions as there are cellular proteins, though this is as yet technically impossible. The proteins of a cell number many hundreds, perhaps thousands, and differ greatly in size and function. We might therefore have guessed that the place to look for discontinuous synthesis like that of DNA was among individual proteins and not in the total cell protein; and so it has turned out to be. Enzyme assays on synchronous cultures of a number of microorganisms (bacteria, budding yeast, fission yeast, and *Chlorella*) have shown that enzyme synthesis rarely follows the typical smooth exponential increase of total protein but instead falls into one of three main patterns:

1. *Step enzymes.* These show periodic synthesis at a characteristic point in the cycle, and the position of this point varies among enzymes. The pattern is just like DNA with a G_1, S, and G_2 (Masters and Pardee, 1965; Tauro and Halvorson, 1966; Bostock *et al.*, 1966).

2. *Rate doubling enzymes.* These show continuous synthesis at a constant rate until a characteristic point in the cycle. At this point the rate doubles and then stays constant until the same point in the next cycle (Donachie, 1965; Kuempel *et al.*, 1965).

3. *Peak enzymes.* These show a peak in activity at a characteristic point in the cycle (Sylvén *et al.*, 1959; Shen and Schmidt, 1966; Johnson and Schmidt, 1966; Eckstein *et al.*, 1967; see Schmidt, this volume, Chapter 8).

These are three patterns shown in normal undisturbed cycles. But an additional way of using enzyme assays is to remove aliquots at intervals through the cycle and induce or derepress an inducible enzyme. In most of the systems on which this has been tried there are sharp changes in the rate of production of the enzyme from aliquots at different stages of the cycle (Masters and Pardee, 1965; Knutsen, 1965; Kuempel *et al.*, 1965; Donachie and Masters, 1966; Donachie and Masters, this volume, Chapter 3). There is no need to consider the explanations of these patterns or the relation between enzyme activity and enzyme quantity because the main point to be emphasized here is that all the patterns provide excellent cycle markers. The assays are sensitive and fairly simple, and it is certainly time they were applied more widely to higher cells (e.g., Littlefield, 1966). Another possibility is measurement of the total quantity of a specific protein. This is technically more difficult but does get over the difficulties of relating enzyme activity to enzyme quantity and can also be used on nonenzyme proteins. Robbins and Borun (1967) have recently shown that two groups of histones are synthesized at the same time as DNA in HeLa cells.

RNA is less promising as a source of markers. In some cases RNA synthesis is approximately exponential, except for the fall at mitosis (Scharff and Robbins, 1965; Kimball and Perdue, 1962; Mitchison and Lark, 1962; Hermann and Schmidt, 1965; Cummings, 1965; Tauro *et al.*, this volume, Chapter 5). In other cases there are rate changes which could be used as markers (Zetterberg and Killander, 1965; Stubblefield and Draven, 1966; Mittermayer *et al.*, 1965; Vincent, personal communication). The position seems a little obscure at the moment though it might clarify if the RNA components were separated. It is interesting that there appears to be a change in the base composition of the newly transcribed RNA during the cycle of *Physarum* (Cummins, this volume, Chapter 7).

The third kind of marker is a "physiological" one. This marks the stage of the cycle at which there is a sharp change in the response of the cell to some agent. For instance, a common marker of this sort is the point in the cycle before which an inhibitor stops the subsequent division and after which it does not. This is often called a "critical point" or a "transition point." These markers have been found in many cell systems: e.g., *Tetrahymena* (Zeuthen, 1964, Table VII; Mazia and Zeuthen, 1966; Stone and Prescott, 1964; Lazarus *et al.*, 1964): *Paramecium* (Rasmussen, 1967), *Physarum* (Mittermayer *et al.*, 1965; Cummins, this volume), sea urchin eggs (Swann, 1953), and mammalian cells (Tobey *et al.*, 1966; Petersen *et al.*, this volume, Chapter 16). The agents in these cases were inhibitors or changes in media and temperature. There is another agent, radiation, which also gives physiological markers defined by the resistance of cells to division delay, chromosome damage, or death (reviewed for mammalian cells by Sinclair, 1968; also Swann, 1962). The only general difficulty with the physiological markers is that they tend to be concentrated in the later parts of the cycle. This is not surprising since most of them are with agents that affect division. If there was more work done with agents that affect DNA synthesis (as in Stone and Prescott, 1964), it might be possible to fill in the empty G_1 region at the beginning of the cycle.

The method that I have advocated here of searching for markers and then distorting the cycle in order to find their causal relations, is not of course the only method of attack. Another is to concentrate on one particular event, such as the initiation of DNA synthesis (Lark, 1966), and follow it in depth. But I believe this is a promising moment to start marker analysis because of the recent emergence of many new markers particularly in the enzyme field. These markers are certain to be of use in the future, and even now they demonstrate one point very clearly—that we must completely abandon the old idea of interphase being a period of steady growth with no change in the relative proportions of the molecules of the cell. Instead we must picture the cell cycle as a complex sequence of changes with the chemical composition of the cell varying from moment to moment. This microcosm of growth coupled with differentiation at the molecular level is not only much more complex than it appeared to be fifteen years ago, but also that much more fascinating.

REFERENCES

Bostock, C. J., Donachie, W. D., Masters, M., and Mitchison, J. M. (1966). *Nature* **210**, 808.
Clark, D. J., and Maaløe, O. (1967). *J. Mol. Biol.* **23**, 99.
Cummings, D. J. (1965). *Biochim. Biophys. Acta* **85**, 341.
Donachie, W. D. (1965). *Nature* **205**, 1084.
Donachie, W. D., and Masters, M. (1966). *Genet. Res. (Cambridge)* **8**, 119.
Eckstein, H., Paduch, V., and Hilz, H. (1967). *European J. Biochem.* **3**, 224.
Edmunds, L. N. (1965). *J. Cell. Comp. Physiol.* **66**, 159.
Gelfant, S. (1962). *Exptl. Cell Res.* **26**, 395.
Helmstetter, C. E. (1967). *J. Mol. Biol.* **24**, 417.
Hermann, E. C., and Schmidt, R. R. (1965). *Biochim. Biophys. Acta* **95**, 63.
Hibbard, H. (1945). *Quart. Rev. Biol.* **20**, 1.
Howard, A., and Pelc, S. R. (1951). *Exptl. Cell Res.* **2**, 178.
Hughes, A. F. (1952). "The Mitotic Cycle." Butterworths, London and Washington, D.C.
James, T. W. (1966). *In* "Cell Synchrony–Studies in Biosynthetic Regulation" (I. L. Cameron and G. M. Padilla, eds.), pp. 1–13. Academic Press, New York.
Johnson, R. A., and Schmidt, R. R. (1966). *Biochim. Biophys. Acta* **129**, 140.
Kimball, R. F., and Perdue, S. W. (1962). *Exptl. Cell Res.* **27**, 405.
Knutsen, G. (1965). *Biochim. Biophys. Acta* **103**, 495.
Kuempel, P. L., Masters, M., and Pardee, A. B. (1965). *Biochem. Biophys. Res. Commun.* **18**, 858.
Lark, K. G. (1966). *In* "Cell Synchrony–Studies in Biosynthetic Regulation" (I. L. Cameron and G. M. Padilla, eds.), pp. 54–80. Academic Press, New York.
Lazarus, L. H., Levy, M. R., and Scherbaum, O. H. (1964). *Exptl. Cell Res.* **36**, 672.
Littlefield, J. W. (1966). *Biochim. Biophys. Acta.* **114**, 398.
Masters, M., and Donachie, W. D. (1966). *Nature* **209**, 476.
Masters, M., and Purdee, A. B. (1965). *Proc. Natl. Acad. Sci. U.S.* **54**, 64.
Mazia, D., and Zeuthen, E. (1966). *Compt. Rend. Trav. Lab. Carlsberg* **35**, 341.
Mitchison, J. M. (1961). *Exptl. Cell Res.* **22**, 208.

Mitchison, J. M., and Cummins, J. E. (1964). *Exptl. Cell Res.* **35,** 394.

Mitchison, J. M. and Lark, K. G. (1962). *Exptl. Cell Res.* **28,** 452.

Mitchison, J. M., Kinghorn, M. L., and Hawkins, C. (1963). *Exptl. Cell Res.* **30,** 521.

Mittermayer, C., Braun, R., and Rusch, H. P. (1964). *Biochim. Biophys. Acta* **91,** 399.

Mittermayer, C., Braun, R., and Rusch, H. P. (1965). *Exptl. Cell Res.* **38,** 33.

Padilla, G. M., Cameron, I. L., and Elrod, L. H. (1966). *In* "Cell Synchrony–Studies in Biosynthetic Regulation" (I. L. Cameron and G. M. Padilla, eds.), pp. 269–288. Academic Press, New York.

Prescott, D. M. (1955). *Exptl. Cell Res.* **9,** 328.

Rasmussen, L. (1967). *Exptl. Cell Res.* **45,** 301.

Robbins, E., and Borun, T. W. (1967). *Proc. Natl. Acad. Sci. U.S.* **57,** 409.

Robbins, E., and Scharff, M. (1966). *In* "Cell Synchrony–Studies in Biosynthetic Regulation" (I. L. Cameron and G. M. Padilla, eds.), pp. 353–374. Academic Press, New York.

Scharff, M., and Robbins, E. (1965). *Nature* **208,** 464.

Shen, S. R-C., and Schmidt, R. R. (1966). *Arch. Biochem. Biophys.* **115,** 13.

Sinclair, W. K. (1968). *Radiation Res.* **33,** 620.

Sisken, J. E., and Kinosita, R. (1961). *J. Biophys. Biochem. Cytol.* **9,** 509.

Stone, G. E., and Prescott, D. M. (1964). *J. Cell Biol.* **21,** 275.

Stubblefield, E., and Deaven, L. (1966). *J. Cell Biol.* **31,** 114A.

Swann, M. M. (1953). *Quart. J. Microscop. Sci.* **94,** 369.

Swann, M. M. (1962). *Nature* **193,** 1222.

Swift, H. H. (1950). *Physiol. Zool.* **23,** 169.

Sylvén, B., Tobias, C. A.. Malmgren, M., Ottoson, R., and Thorell, B. (1959). *Exptl. Cell Res.* **16,** 75.

Tamiya, H., Iwamura, T., Shibata, K., Hase, E., and Nihei, T. (1953). *Biochim. Biophys. Acta* **12,** 23.

Tauro, P., and Halvorson, H. O. (1966). *J. Bacteriol.* **92,** 652.

Tobey, R. A., Anderson, E. C., and Petersen, D. F. (1966). *Proc. Natl. Acad. Sci. U.S.* **56,** 1120.

Walker, P. M. B., and Yates, H. B. (1952). *Proc. Roy. Soc.* **B140,** 274.

Williams, N. E., and Zeuthen, E. (1966). *Compt. Rend. Trav. Lab. Carlsberg* **35,** 101.

Williamson, D. H., and Scopes, A. W. (1960). *Exptl. Cell Res.* **20,** 338.

Williamson, D. H., and Scopes, A. W. (1962). *Proc. 22nd Intern. Congr. Physiol. Leiden, 1962,* p. 759.

Zetterberg, A., and Killander, D. (1965). *Exptl. Cell Res.* **39,** 22.

Zeuthen, E. (1964). *In* "Synchrony in Cell Division and Growth" (T. Zeuthen, ed.) pp. 99–158. Wiley (Interscience), New York.

Zeuthen, E., and Scherbaum, O. H. (1954). *In* "Recent Developments in Cell Physiology" (J. A. Kitching, ed.), pp. 141–156. Butterworths, London and Washington, D.C.

Author Index

Numbers in italics refer to the pages on which the complete references are listed.

A

Abbo, F. E., 38, 53, 56, *74*
Abe, M., 17, 28, *33*, 45, 47, *74*
Adam, A., 256, *278*
Adler, H. I., 73, *74*
Afzelius, B. A., 317, 318, 319, 326, 327, 333, *339*
Agrell, I., 321, 325, 331, *339*
Aketa, K., 281, 289, 295, *296*
Alderton, G., 86, *98*
Allen, D., 351, *358*
Allen, M. B., 197, *201*
Allen, R. D., 229, 237, *246*, 280, *296*
Allende, J. E., 276, *278*
Anders, M., 150, *157*
Anderson, D. T., 252, *278*
Anderson, E., 154, *158*
Anderson, E. C., 342, 344, 345, 347, 351, 354, 355, *358*, *359*, 370, *372*
Anderson, W. A., 228, 230, 231, 235, *246*
André, J., 231, 233, *246*
Arnand, M., 82, *97*
Aronson, A. I., 82, *95*, *96*
Aschoff, J., 137, *138*
Attardi, G., 111, *117*
Aubert, J. P., 82, 83, 84, *95*, *99*
Austin, C. R., 323, 325, *339*
Axelrod, A., 351, *359*

B

Bach, J. A., 83, *95*
Baechtel, F. S., 164, *176*
Baglioni, C., 351, *358*
Baillie, A., 83, *95*
Baker, A. L., 160, 162, *176*, 181, 182, *201*
Balassa, G., 78, 79, 82, 83, 84, 89, *95*, *99*

Baldwin, H. H., 142, *157*
Baptist, E., 280, 291, 294, *296*
Bard, S., 289, *298*
Bard, S. G., 335, *340*
Barer, R., 319, *339*
Barigozzi, C., 250, *278*
Barlow, J. J., 105, 109, *117*
Barner, H. D., 28, *33*, 70, *74*, 85, 88, *95*, *96*
Barnicot, N. A., 228, 231, 233, *246*
Batty, I., 83, *100*
Baumann-Grace, J. B., 83, *100*
Bautz, E., 155, *157*
Beach, D., 245, *248*
Beams, H. W., 333, *340*
Beck, J. V., 82, *97*
Becker, J., 153, *158*, 304, *313*
Beckwith, J., 51, *76*
Beers, R. J., 82, *98*
Behnke, O., 227, 230, 231, 233, 234, 245, *247*
Bell, E., 290, *297*
Bell, G. I., 342, 355, *358*
Bender, M. A., 356, 357, *359*
Bendigkeit, H. E., 22, *34*
Bendix, S., 197, *201*
Bennett, E. L., 194, 200, *202*
Bennett, L. L., 145, *157*
Bensch, K. G., 236, *247*
Berg, C. M., 17, *33*, 45, 47, *74*
Berger, L. R., 204, *224*
Bergére, J. L., 84, *95*
Bergmann, L., 184, *201*
Bergstrand, A., 289, *297*, 335, *340*
Berliner, M. D., 121, 137, *138*
Bernlohr, R. W., 82, 84, *95*, *96*, *98*
Berrah, G., 94, *96*
Beskid, G., 84, 88, *98*
Bianchetti, R., 289, 295, *296*

373

Hartman, P. E., 3, *13*
Harvey, E. B., 302, *313*, 326, 327, 336, *340*
Hase, E., 198, 199, *201, 202,* 361, *372*
Haselbrunner, H., 112, *118*
Hashimoto, T., 79, 87, 88, *96, 97*
Hauschild, A. H. W., 200, *201*
Hawkins, C., 367, *372*
Hawley, E. S., 136, *138*
Hawthorne, D. C., 101, 103, *118*
Haxo, F. T., 192, *201*
Hay, J., 152, *158*
Hayes, W., 20, *34*
Heidelberger, C., 262, *278*
Heinle, E., 351, *359*
Helmstetter, C., 45, 66, 67, *74, 75*
Helmstetter, C. E., 16, 17, 18, 19, 21, 22, 28, 33, *33, 34,* 39, 45, 68, *74,* 367, *371*
Hempstead, P., 73, *74*
Henrici, A., 86, *97*
Herman, E. C., 370, *371*
Herman, L., 230, 235, *247*
Hermier, J., 84, *95*
Hermolin, J., 209, 210, 212, *224*
Herrmann, E. C., 167, *176*
Hershey, A., 86, *97*
Hewitt, R., 28, 30, 31, *33, 34*
Hibbard, H., 361, 363, *371*
Higa, A., 56, 61, *75,* 90, *97*
Hilz, H., 369, *371*
Hinegardner, R., 143, *157*
Hinegardner, R. T., 292, *297*
Hiramoto, Y., 204, *224,* 237, 242, *248,* 283, *297*
Hirose, S., 56, 65, 66, *75*
Hirota, Y., 73, *74*
Hirshfield, H. I., 242, *247*
Hoagland, M. B., 272, *278*
Hobbs, D. G., 70, 71, *74*
Hodson, P. H., 82, *97*
Hoffman, E. J., 16, 18, 30, 31, *34,* 45, *75*
Hogness, D. S., 65, *75*
Holland, J. J., 357, *358*
Holmes, P. K., 89, 91, *97*
Holz, G. G., 241, 245, *247, 248*
Hori, R., 283, 284, *297,* 312, *313*
Horiuchi, T., 39, 42, 43, 45, 48, *75*
Horowitz, M. H., 2, 3, *13*

Howard, A., 22, *34,* 342, 351, *358,* 361, *371*
Howitt, C. J., 86, *97*
Hsu, T. C., 236, *247*
Huang, P. C., 111, *117*
Hudson, J., 308, *313*
Hughes, A. F., 3, *371*
Hultin, T., 280, 284, 289, 291, *297,* 311, *313,* 316, 334, 335, 337, *340*
Hunter, J. R., 83, *98*
Huntington, E., 86, *97*
Hurwitz, J., 150, *157*
Hutchison, D. J., 31, *35*

I

Igarashi, R. T., 82, 84, *96*
Ikeda, M., 217, 220, 221, *224*
Imanaka, H., 88, *97*
Imperato, S., 83, *96*
Infante, A., 336, *340*
Infante, A. A., 276, *278*
Inoué, S., 236, *247*
Ionesco, H., 78, 79, 82, 83, 84, 92, *95, 99*
Ishida, J., 289, *297*
Ishida, M. R., 166, *177*
Ishihara, K., 281, *297*
Isono, N., 282, 289, 295, *297*
Iverson, R., 317, *339*
Iverson, R. M., 287, 289, 295, *297, 298,* 302, 305, 306, 308, 309, 310, 311, 312, *313,* 316, 334, 336, *340*
Iwamura, T., 193, 200, *201,* 361, *372*
Iwasaki, T., 250, 252, 254, 260, 278, *278*

J

Jacob, F., 2, 6, *13,* 18, 19, *34, 35,* 49, 73, *74,* 79, 94, *97,* 171, *176*
Jacobs, P., 148, *157*
Jahn, T. L., 229, *247*
James, T. W., 235, *247,* 362, *371*
Jayaraman, J., 112, *118*
Jenkins, R. A., 243, *247*
Jensen, W., 351, *359*
Jobsis, F., 285, *297*
Johnson, R. A., 18, *34,* 164, 165, *176,* 369, *371*
Johnson, T. C., 357, *358*

Montjar, M., 351, *359*
Moore, C., 112, *117*
Morgan, R., 143, *157*
Morimura, Y., 187, *201, 202*
Morrison, J., 152, *158*
Mortimer, R. K., 101, 103, *118*
Moustacchi, E., 115, *118*
Moyer, W. A., 292, 297, 317, 318, 319, 333, 336, *339, 340*
Mueller, G., 144, 148, 154, *157*
Mueller, G. C., 344, 345, *359*
Mulders, P. F. M., 166, *177,* 200, *202*
Murakami, T. H., 204, *224*
Muramatsu, S., 251, *278*
Murrell, W. G., 77, 78, 79, 83, 85, 86, 92, *96, 98, 100*
Myers, J., 160, *177,* 193, *201*

N

Nagata, T., 16, *35,* 171, *176*
Nags, E. H., 90, *96*
Nakada, D., 20, 28, *35,* 67, *74*
Nakai, T., 73, *75*
Nakanishi, Y. H., 250, 251, 252, 254, 260, *278*
Nakano, E., 282, 289, 295, *297, 298,* 335, *340*
Nakata, D., 82, *98*
Namboodiri, A. N., 137, *138*
Nandi, U. S., 109, 116, *118*
Nason, A., 127, *138*
Nass, S., 318, 326, 327, 329, *339*
Nelson, C. D., 200, *201*
Nelson, D. L., 77, 79, *97*
Nemer, M., 276, *278,* 289, *298,* 335, 336, 337, *340*
Neurath, P. W., 121, 137, *138*
Newman, A., 17, 28, *35*
Newton, A. A., 345, *359*
Nihei, T., 361, *372*
Nilsson, J. R., 229, 237, *247*
Nishi, A., 39, 42, 43, 45, 48, 56, 65, 66, *74, 75*
Noll, H., 351, *359*
Norris, J. R., 83, *95, 98*
Novelli, G. D., 82, *96*
Novick, A., 9, *13*
Nygaard, O., 142, 148, *157, 158*

O

Ohnishi, T., 287, *298*
Ohye, D. F., 78, 83, 92, *98, 100*
Oikawa, T. G., 256, *278*
Oishi, M., 90, *98,* 166, *177*
Okazaki, H., 290, *298*
Okigaki, T., 250, 251, 252, 254, 260, *278*
Oliver, R., 343, *359*
O'Sullivan, A., 19, 22, *35,* 90, *100,* 166, *177*
Otsuka, H., 199, *201*
Ottoson, R., 101, *118,* 163, *177,* 369, *372*
Owen, R. D., 122, *138*
Owens, R. G., 128, *138*
Ozahi, H., 300, *313*

P

Padilla, G. M., 86, 88, 89, *96, 98,* 121, *138,* 209, 212, 214, *224,* 237, *247,* 268, *372*
Paduch, V., 369, *371*
Palay, S. L., 227, 230, *247, 248*
Pardee, A., 86, *98,* 369, *371*
Pardee, A. B., 18, 33, *34, 35,* 38, 39, 40, 41, 42, 45, 51, 53, 56, 57, 58, 60, 63, 64, 65, 67, *74, 75, 76,* 163, 166, 171, *176*
Parsons, J. A., 115, *118*
Peabody, R. A., 204, *224*
Pease, D. C., 228, 231, 233, *247*
Pecora, P., 242, *247*
Pelc, S. R., 22, *34,* 342, 351, *358,* 361, *371*
Penman, S., 109, *118,* 304, *313,* 351, *359*
Pepper, R. E., 82, *98*
Perdue, S. W., 370, *371*
Perrings, J. D., 347, *358*
Perry, J. J., 82, 83, 87, *96, 98*
Petersen, D. F., 342, 344, 345, 351, 354, 355, *358, 359,* 370, *372*
Peterson, D., 154, *158*
Phillips, D. M., 233, *248*
Philpott, D. E., 318, 326, 327, 329, *339*
Piatigorsky, J., 283, 290, *298,* 300, *313*
Pickett-Heaps, J. D., 230, 235, 236, *248*
Pickett, J. M., 200, *201*
Pierucci, O., 18, 28, *34*

382

Schmidt, R. R., 18, *34*, 160, 161, 162, 163, 164, 165, 166, 167, 168, 170, 175, *176, 177*, 181, 182, 183, *201*, 369, 370, *371, 372*
Schmitt, F. O., 227, 230, *248*
Schneider, W. C., 127, *138*
Schoener, B., 285, *297*
Schoser, G., 180, 181, 191, 201, *202*
Scopes, A. W., 106, 107, *118*, 363, 368, *372*
Scott, D. B. M., 88, *99*
Scott, N. S., 169, 174, *177*
Sebald, M., 82, *99*
Senger, H., 180, 181, 182, 183, 184, 186, 187, 190, 191, 192, 194, 195, 196, 197, 198, 200, 201, *201, 202*
Setlow, R. B., 28, *35*
Shah, V. C., 174, *177*
Shannon, C. E., 11, *13*
Shelanski, M. L., 233, 234, 236, 245, *248*
Shen, S. R. C., 164, 165, *177*, 369, *372*
Shepherd, C. J., 127, *138*
Shibata, K., 361, *372*
Shihira, I., 180, *202*
Shiraishi, K., 254, *278*
Siegel, A., 111, *118*
Sikyta, B., 63, *76*
Silberman, L., 204, *225*
Silver, S., 94, *99*
Simmel, E. A., 323, *340*
Simmons, P. J., 83, *99*
Simpson, R. E., 221, *224*
Sinclair, W. K., 370, *372*
Sisken, J. E., 367, *372*
Slautterback, D. B., 227, 230, 231, *248*
Sleigh, M. A., 228, *248*
Slepecky, R. A., 78, 79, 80, 87, 88, 89, 91, 92, 94, 95, *96, 97, 98, 99, 100*
Slezak, J., 63, *76*
Slonimski, P. O., 115, *118*
Smillie, R. M., 169, 174, *177*
Smith, D., 109, 115, *118*
Smith, D. W., 19, *35*
Smith, E. L., 1, *13*
Smith, M., 256, *278*
Smithers, D., 145, *157*
Snell, N. S., 86, *98*
Soeder, C. J., 180, 186, 187, *202*
Sofer, W. H., 289, *298*, 301, 302, 305, 313, 334, 336, *340*

Sokawa, Y., 198, *202*
Solari, A. J., 228, 231, 233, *247*
Sorokin, C., 160, *177*, 187, *202*
Sorokin, S., 228, *248*
Sotelo, C., 230, *248*
Sotelo, J. R., 228, *248*
Spencer, D., 174, *177*
Spencer, H. T., 161, *177*
Spiegelman, S., 109, 111, *117, 118*
Spirin, A. S., 276, *278*, 317, 336, 337, *340*
Spizizen, J., 79, 82, 84, *99*
Spotts, C. R., 82, *99*
Spudich, J. A., 77, 79, 82, 87, *97, 99*
Srb, A. M., 122, *138*
Srinivasan, V. R., 82, 83, 84, 88, *97, 99*
Stadler, D. R., 137, *138*
Stafford, D. W., 304, 305, *313*, 316, 334, 336, *340*
Stahl, F. W., 16, *35*, 45, *75*
Stange, L., 194, 200, *202*
Stanier, R. Y., 84, *99*
Starka, J., 92, *99*
Stavy, L., 311, *313*, 317, *340*
Steffan, J., 343, *359*
Steinberg, W., 56, 59, 61, *75, 76*, 77, 78, 90, 91, *97, 99*, 276, *278*
Stent, G., 358, *359*
Stern, H. J., 65, *76*
Stewart, B. J., 83, *99*
Stine, G. J., 121, 123, 126, 127, 128, 133, 135, 136, *138*
Stone, G. E., 370, *372*
Stonehill, E. H., 31, *35*
Strange, R. E., 83, *98, 99*
Strotmann, H., 186, *202*
Stubblefield, E., 148, *157*, 228, 235, 236, *247, 248, 344*, 345, *359*, 370, *372*
Stull, H. B., 79, *100*
Subak-Sharpe, H., 152, *158*
Sueoka, N., 17, 19, 22, *35*, 39, *76*, 86, 90, 94, *98, 99, 100*, 166, 171, 174, *176, 177*
Sugiyama, M., 287, *298*
Suit, J. C., 31, *35*
Sussman, A. S., 89, *99*, 137, *138*
Sussman, M., 65, *76*
Swann, M. M., 282, *298*, 370, *372*
Swenson, P. A., 28, *35*
Swift, H., 228, *248*, 361, *372*
Sylvén, B., 101, *118*, 163, *177*, 369, *372*

Subject Index

A

Acetate
  conidia formation and, 123
  incorporation, fertilization and, 292
Acetoacetyl coenzyme A reductase, spore
  formation and, 83
Actinomycin C, mitosis and, 153
Actinomycin D
  fertilized eggs and, 336, 337
  ribonucleic acid synthesis and, 148–149,
   153
  synchronized cells and, 349–354, 357
Actinosphaerium
  hydrostatic pressure and, 204, 229, 237,
   242
  microtubules,
   colchicine and, 236
   pressure and, 229, 237, 242
Action spectrum, induction of cell division and, 191–192
Active culture technique, synchronous
  sporulation and, 87–88
Adenine nucleotides, levels, hydrostatic
  pressure and, 204
Adenosine deaminase, spore formation
  and, 83
Adenosine diphosphate
  brine shrimp development and, 257
  levels in fertilized egg, 288
  respiratory control and, 288
Adenosine monophosphate
  brine shrimp development and, 257
  incorporation by brine shrimp,
   260–263, 265, 267
  levels in fertilized egg, 288
Adenosine triphosphatase, microtubules
  and, 234
Adenosine triphosphate
  brine shrimp hatching and, 256–259
  coenzyme conversion and, 285
  levels in fertilized egg, 288, 294–295

protein synthesis and, 270
Aeration
  conidial germination and, 126–127
  development of conidiophores and
   conidia, 123–124
Aerobiosis, enzyme repression by, 60
Aging, conidial germination and, 125–126
Alanine racemase, spore formation and,
  83
L-Alanine dehydrase, synthesis of, 56
Alkaline phosphatase
  basal synthesis of, 51–53, 56
  periodic synthesis of, 61, 63
  rate of synthesis, synchronous growth
   and, 41–43
Allosteric controls, steady-state operation and, 12
Allosteric pathways, evolution of, 6
Amino acids
  germination and, 89, 90
  starvation, deoxyribonucleic acid synthesis and, 69
  synthesis, light and, 200
  transport, fertilization and, 283–284,
   293
  growth pattern of, 364, 368
δ-Aminolevulinate dehydrase, synthesis
  of, 56
δ-Aminolevulinate synthetase, synthesis
  of, 56
Ammonia, utilization of, 127
Amoeba
  growth pattern of, 364, 368
  hydrostatic pressure and, 204
Amphiuma, erythrocytes, microtubules
  in, 229–230
Antibiotics, spore formation and, 83, 84
Antiformin, brine shrimp embryos and,
  254
Antigens, spore formation and, 83
Arbacia, cleavage, timing of, 331

C